AQUACULTURE RESEARCH PROGRESS

AQUACULTURE RESEARCH PROGRESS

TAKUMI K. NAKAMURA
EDITOR

Nova Science Publishers, Inc.
New York

For permission to use material from this book please contact us:
Telephone 631-231-7269; Fax 631-231-8175
Web Site: http://www.novapublishers.com

LIBRARY OF CONGRESS CATALOGING-IN-PUBLICATION DATA

Aquaculture research progress / Takumi K. Nakamura (editor).
p. cm.
ISBN 978-1-60456-247-7 (hardcover)
1. Aquaculture--Research. I. Nakamura, Takumi.
SH135.A77 2008
639.8--dc22

2007050780

Published by Nova Science Publishers, Inc. ✢ New York

CONTENTS

PREFACE

This new book presents the latest research on aquaculture which is the cultivation of aquatic organisms. Unlike fishing, aquaculture, also known as aquafarming, implies the cultivation of aquatic populations under controlled conditions. Mariculture refers to aquaculture practiced in marine environments. Particular kinds of aquaculture include algaculture (the production of kelp/seaweed and other algae); fish farming; shrimp farming, shellfish farming, and the growing of cultured pearls.

Expert Commentary A - The present contribution aims to resume the main results obtained within a three years research project, analysing and discussing their potential use for the development of behavioural indicators of "wildness" of the cultured fish. The species the authors used as a study model was the European sea bass (*Dicentrarchus labrax* L.), which is cultured and seeded in the Northern Adriatic lagoons to sustain a traditional, specialised form of Italian aquaculture, named "vallicoltura".The authors focused on the antipredator responses of the juvenile phase of this species (about 2 cm of total length) towards different stimuli associated both to a fish predator and to a potential bird predator (overhead stimuli).

The authors developed a series of experimental apparatuses and protocols to quantitatively assess some components of the antipredator behaviour, which were compared between hatchery-reared and wild fish juveniles. Two components resulted to differ significantly between the wild and the hatchery groups, being good candidates for the development of indices of "wildness". These two components are the shoaling index, that is the level of cohesiveness of the group as a response to a fish predator, and the freezing behaviour, i.e. time spent motionless resting on the bottom as a response to overhead fright stimuli. For both measures it would be necessary to develop a reference system and to further improve and possibly simplify the experimental protocols for a wider and routine use of the behavioural test in the aquaculture practise related to restocking.

Expert Commentary B - The nutritional features of fish are fundamental in the choice of a modern healthy diet. From this point of view, food industries are becoming involved more with quality rather than quantity; Many management factors can influence the qualitative traits of fish. The quality of food is the result of a balance of different parameters, some intrinsically linked to the product, such as taste, colour, smell, and others related to processing food technology as salting, smoking, or canning, which have an effect on the final quality of the product. Last, but not least is the consumer's ability to assess quality.

The environmental sustainability, reducing the impact on fishery products from by-products of agriculture, and the dietetic food quality are among the main topics that the aquaculture productions must consider.

Short Communication A - The induction of gonadal maturation is one of the important aspects of fish aquaculture. Extract of pituitary glands containing two types of gonadotropins (GTH), follicle-stimulating hormone (FSH) and luteinizing hormone (LH) has been administered for this purpose. However, collection of a large amount of pituitaries is not always practically easy, and the pituitary extract is a mixture of FSH and LH whose contents and composition of FSH and LH are usually unknown. It is possible to produce fish GTHs biotechnolgically, and recombinant fish GTHs have been produced using various bioreactors. However, since the amount of hormones produced in these systems is small, they are used only for *in vitro* endocrinological studies of fish and could not be tested *in vivo* as substitutes for the pituitary extract. In order to obtain basic information for aquaculture purpose, studies that examine the *in vivo* effect of recombinant hormones are essential. Recently, the authors successfully produced a large amount of recombinant FSH and LH (rFSH and rLH) of goldfish *Carassius auratus* and Manchurian trout *Brachymystax lenok* by baculovirus in silkworm *Bombyx mori* larvae, and confirmed their biological activities in various *in vivo* bioassays. rFSH and rLH of goldfish and Manchurian trout induced milt production in male goldfish. Goldfish rLH induced ovulation in female bitterling *Rhodeus ocellatus ocellatus*, whereas rFSH showed low potency. rFSH and rLH of goldfish and Manchurian trout induced sex steroid production in goldfish and rainbow trout *Oncorhynchus mykiss*, respectively, in both sexes. Goldfish rFSH and rLH induced spermatogenesis in male Japanese eel *Anguilla japonica*. The yield of rFSH and rLH produced in silkworm larvae is much greater compared to that of fish GTH produced in systems using other bioreactors. These results indicate that a large amount of biologically active recombinant fish GTHs could be produced by a baculovirus-silkworm larvae system. By using these recombinant GTHs, the following possibilities are suggested for the development of aquaculture research: 1) use of rFSH and rLH individually or in combination as substitutes of pituitary extracts, 2) elucidation of *in vivo* biological function of fish FSH and LH, and 3) development of a new efficient method of induction of gonadal maturation by differential use of rFSH and rLH based on physiological profiles of FSH and LH expression.

Short Communication B - Coral reefs, often portrayed as "the rain forest of the sea", provide sustenance to hundreds of million of people worldwide. When severely degraded, reef restoration is expected to uphold the ecosystem biodiversity and retain, at least partially, the original biological and ecological portraits. Such active reef restoration practices are based on aquaculture of coral colonies (mariculture them in underwater nurseries) in as much as silviculture depends on rearing young trees in terrestrial nurseries.

Chapter 1 - Intensification and expansion of aquaculture activities on one hand, emerging problems of diseases on the other pose challenges to the sustained production from aquaculture activities. Increasing awareness on deleterious effects of antibiotics and aquaculture chemicals, leading to the development of bacterial drug resistance and chemical contaminations of the environment, has forced the researchers and planners to look for measures that are host and environment friendly. Immune enhancement of the host is seen as the most effective and long lasting solution for countering the problem of diseases. Delivery of immunogens through various routes has been studied extensively. However, the most natural and non-stressing route of oral delivery has attracted the attention of fish immunologists in recent years. Oral delivery of immunogens such as vaccines, immunostimulants, probiotics and nutritional immune enhancers is being studied with the ultimate aim of enhanced immunity. With a brief background on the digestive interactions that could influence the putative immunogens, this chapter analyses various investigations carried out in the field of oral immunogens in aquaculture and suggest possible future strategies for an improved aquaculture productivity.

Chapter 2 - Control of mollusc movements. Transfer regulations have been developed in order to avoid the introduction of animals from an enzootic area to a pathogen-free area. A such approach needs well suited tools for diagnosing pathogens. Molecular techniques for diagnosing bivalve infectious agents have been developed during the last decade and are now moving from a developmental phase in specialized laboratories for research purposes, to routine application.

I) Understanding complex interactions between animal, environment and pathogen - Marine invertebrates, especially bivalves, are particularly challenged by their environment. Molluscs are typically reared in estuarine environments frequently subjected to fluctuations of environmental factors including temperature, salinity and pollution. Reducing the impact of pathogens is likely to rely on knowledge of their biology.

II) Genetic improvement for disease resistance - Enhancing the mollusc defense ability through their immunity is one of the important approach to prevent and control infectious diseases. Researchs were recenlty developed in order to provide valuable information for further understanding of the defense mechanisms in bivalves and to define genetic markers of interest.

III) Alternative methods of prevention and treatment - The preventive or therapeutical use of chemotherapeutants is highly restricted since mollusc farming is carried out in the natural marine environment. Moreover, efficiency of antibiotics treatemts for ubiquitous pathogens is seriously questioned. Alternative exist, including use of probiotics, although these were not extensively applied until recently.

Chapter 3 - Fatty acids are one of the most important components in the organisms. According to the number of double bonds, fatty acids can be divided into subtypes such as saturated fatty acids, monounsaturated fatty acids (MUFA) and polyunsaturated fatty acids (PUFA). The fatty acid composition in many animals has been identified, and it is found to vary dramatically among tissues. This suggests specific roles for different fatty acids in life systems although their detailed function is far from being revealed. Highly unsaturated fatty acids (HUFAs) are essential for normal physiological actions. The representative fatty acids of HUFAs are DHA, EPA, and AA, which have been intensively investigated in many studies. Conversion processes between fatty acids take place in organisms, which require desaturases and chain elongation and shortening enzymes. Since fatty acids are acquired from food, dietary fatty acids have great effect on the fatty acid composition of tissues and organs, and thus their normal functioning. In this review,the authors summarized the fatty acid pattern of gonads, and the recent progress on their function during reproduction. The authors suppose to emphasize the importance of fatty acids and hope to provide insights into further investigation in-depth.

Chapter 4 - A series of studies have been carried out to establish a light regime giving optimal fish growth for a complete production cycle of red sea bream *Pagrus major*, which is one of the most important fish in Japan because of its multipurpose uses. In different trials, red sea bream were reared under four photoperiods (6L:6D, 12L:12D, 16L:8D and 24L:0D) with constant light intensity 1500 lx on the water surface to investigate their growth performance, apparent digestibility coefficient (ADC) of nutrients and energy, and stress response. Fish were reared for eight weeks in each of three trials. The growth performance and stress response were investigated in trial 1 in fish reared from 1 to 30 g. In trials 2 and 3, growth performance and ADC, and stress response, respectively, were investigated in fish reared from 20 to 100 g.

In trial 1, the results demonstrated that the growth performance, in terms of weight gain, specific growth rate (SGR), feed intake and feed conversion efficiency (FCE) of red sea bream reared from 1 to 30 g can be stimulated significantly by using a continuous (24L:0D) photoperiod without any measurable significant stress response in fish. Similarly, trial 2 demonstrated that the higher growth performance in fish under continuous photoperiod was attributed to improved appetite, greater food intake and higher FCE as well as higher ADC of lipid and energy in red sea bream reared from 20 to 100 g. In addition, trial 3 demonstrated that photoperiod manipulation did not cause significant stress response in red sea bream reared from 20 to 100 g.

These results together help to establish a light regime giving optimal fish growth for a complete production cycle of red sea bream. It is assumed that more attention will be given to on-land and outdoor aquaculture in the near future. Photoperiod manipulation will definitely be the option of choice to yield higher output from on-land and outdoor farms.

Chapter 5 - Omega-3 polyunsaturated fatty acids are essential factors in human health and nutrition that are derived primarily from consuming seafood and marine oil supplements. The commercially important long-chain omega-3 fatty acids, docosahexaenoic acid (DHA; C22:6 n-3) and eicosapentaenoic acid (EPA; C20:5 n-3) are synthesized mainly in phytoplankton and passed onto higher trophic levels in their diets. The proportion of omega-3 fatty acids in bivalves and fishes thus shows strong seasonal fluctuations that correspond with shifts in the quantity and nutritional quality of each dietary component, as well as spawning periodicity, and environmental variability. This chapter reviews the sources of variation in omega-3 content in bivalves and fishes and discusses how farming marine products can overcome some of this variation. Finally, sustainable alternatives for the production of omega-3s are examined to address the growing market for seafood and associated health products.

Chapter 6 - The Industry of Aquaculture has been profiting from fish sperm cryopreservation for many years. Sperm from more than two hundred fish species, most of them with commercial value, has been cryopreserved, allowing year round availability of sperm and the creation of sperm banks from those individuals with special interest. However, to date, fish embryo cryopreservation has not been successfully achieved in any teleost species. The benefits of fish embryo cryopreservation are numerous, not only for aquaculture but also for conservation purposes. This technique will allow for the maintenance of a constant supply of animals, reduce the facilities required on the fish farm, facilitate animal transportation between different farms and enable the preservation of valuable lines. These practical advantages will also report a cost reduction and minimize the impact of epidemics on productivity. This chapter presents an overview of the recent advances in this field and their impact on Aquaculture.

Chapter 7 - The aim of the study presented in this chapter was to develop a new concept in rearing shrimp in cages, by testing different stocking densities and an alternative feed management. Juveniles of the Brazilian pink shrimp *Farfantepenaeus paulensis* were reared in experimental floating cages (1.00m long x 1.00m wide x 1.00m deep), constructed with a polyester/PVC 5mm-mesh and fixed in PVC floating frames, installed in a offshore area of São Paulo's marine coast, in Brazil. The study was carried out according to two different phases. Phase I involved testing three nursery densities of 500, 1,000 and 2,000 postlarvae/m^2, and four grow-out stocking densities of 100, 200, 400 and 800 juveniles/m^2. Commercial pellets of 35 to 40% crude protein was supplied twice a day, and shrimp were harvested after 37 and 54 days of nursery and grow-out stages, respectively. The best results in terms of growth and survival rate were obtained for the densities of 500 postlarvae/m^2 (nursery), and 100 and 200 shrimp/m^2 (grow-out). Phase II was carried out in order to determine growth performance, as well as composition and sensory aspects of shrimp meat from studies of an alternative feed management. In this phase, shrimp were reared at a density of 150 juveniles/m^2, and fed three different diet compositions: diet I = exclusively commercial pellets consisting of 34% crude protein, diet II = exclusively mussel flesh (*Perna perna*), and diet III = mixed diet (50% diet I + 50% diet II). After 122 days of growth, diet II showed the best results in terms of shrimp weight gain, as well as the best chemical and sensory evaluation of shrimp meat. In conclusion, nursery stocking density of 500 postlarvae/m^2 and grow-out stocking densities of 100 to 200 juveniles/ m^2, showed the best results for rearing shrimp in cages, and feed based exclusively on mussel flesh seemed to be the best feed management for producing shrimp within the concepts of organic production and ecological sustainability.

Chapter 8 - A total number of 79 species of fishes, four species of prawn, one species of crabs, one species of snail and five species of reptiles were identified from the Someswari and Nethai river; Gharia and Nidaya beel in Bangladesh. About 11 types of fishing methods were found in operation. Increasing rate of current jal (gill net), Kapuri jal (seine net) and FAD (Fish aggregating device) were identified as detrimental gear killing different species during whole the year. The increasing rate in exploitation of the water bodies a thread to aquatic biodiversity of the open and closed water bodies. The fisheries production of these water bodies declined significantly over the last five (2001-2005) years. The total production of the Someswari and Nethai river; Gharia and Nidaya beel has decreased from 95.79±6.89 to 38.61±2.83; 175.20±11.83 to 76.63±5.59; 104.65±5.50 to 48.03±2.97 and 112.87±5.98 to 50.58±2.98 mt within five years and the percentage of total production had sharply decreased from 19.61% to 59.68%; 22.07% to 56.66%; 21.88% to 59.25% and 19.84% to 56.20% over the same period. As a result, a number of commercially important fish species such as, mohasher (*Tot tor* and *Tor putitora*) nandina (*Labeo nandina*), sharpunti (*Puntius sarana*), Gajar (*Channa marulius*) and reptiles (*Kachuga kachuga* and *Morenia petersi*) were extinct, 52-58 species were at extremely high risk and 7-18 species were in the verge of extinction.

Chapter 9 - At present aquaculture provides over 40% of the world's seafood supply and according to the FAO this figure will approach 54% by the year 2020.

However aquaculture practices cause environmental problems that include the destruction of natural ecosystems and shrimp disease occurrences. Shrimp culture in floating seawater cages is now being considered as an alternative approach for aquaculture.

The biological viability of culturing the Pacific white shrimp *Litopenaeus vannamei* in floating seawater cages has been validated.

The main purpose of our study was to develop a novel model of a culture system for growing *L. vannamei* with environmental sustainability using floating seawater cages of different dimensions and materials; and to evaluate the effect of different stocking densities, nursery, and growout conditions on efficiency of production. We also evaluated if the shrimp developed the most frequent viral diseases.

For this research we used small (9 m^2) and large (200 m^2) floating cages for growing shrimp. The cages were circular (~200 m^2) made of HDPE polyethylene pipes, and square cage (9 m^2) made of PVC pipes. For both type of cages, PVC-covered polyester nets (Sansuy®) were used to hold the shrimp in nursery and growout stages. The project location was Bahia Navachiste, Guasave, Sinaloa, México.

In the 9 m^2 nursery cages, there was no significant difference ($P>0.05$) in the survival rate of *L vannamei* culture at densities of 950, 1300 and 1600 shrimp/m^2 for 38 culture days, nevertheless the final weight (g) was significantly lower at 1600 shrimp/m^2, but the number of shrimp per cage was the highest at the density of 1600 shrimp/m^2.

There were no significant differences ($P>0.05$) in the final weight (g) and survival rate among cultured shrimp in the 9 m^2 growout cages at 200, 250, and 300 shrimp/m^2 densities for 58 days.

The shrimp biomass per square meter obtained at density of 300 shrimp/m^2 was greater (2.6 kg/m^2) when compared with the lowest density assayed.

In 200 m^2 growout cages the growing rate at the lower density (45 shrimp/m^2) is higher (2.1 g/week) and the survival rate is > 80% at both densities assayed (45 and 180 shrimp/m^2) during 62 days. The final weight (g) for each density was 17.7 ± 1.55 g at 45 shrimp/m^2 and 13.3 ± 1.31 g at 180 shrimp/m^2. The FCR as for the 9 m^2 cages was close to 1.

All the monitored shrimp were free of the viral pathogens TSV and WSSV.

In: Aquaculture Research Progress
Editor: Takumi K. Nakamura
ISBN 978-1-60456-247-7

Expert Commentary A

BEHAVIOURAL INDICATORS OF "WILDNESS" AS USEFUL TOOLS FOR RESTOCKING: THE EUROPEAN SEA BASS (*DICENTRACHUS LABRAX* L.) JUVENILE AS A STUDY MODEL

S. Malavasi, V.Georgalas and P.Torricelli
Dipartimento di Scienze Ambientali, Università Ca' Foscari di Venezia, Campo della Celestia, Castello 2737/b, 30122 Venice, Italy

ABSTRACT

The present contribution aims to resume the main results obtained within a three years research project, analysing and discussing their potential use for the development of behavioural indicators of "wildness" of the cultured fish. The species we used as a study model was the European sea bass (*Dicentrarchus labrax* L.), which is cultured and seeded in the Northern Adriatic lagoons to sustain a traditional, specialised form of Italian aquaculture, named "vallicoltura". We focused on the antipredator responses of the juvenile phase of this species (about 2 cm of total length) towards different stimuli associated both to a fish predator and to a potential bird predator (overhead stimuli).

We developed a series of experimental apparatuses and protocols to quantitatively assess some components of the antipredator behaviour, which were compared between hatchery-reared and wild fish juveniles. Two components resulted to differ significantly between the wild and the hatchery groups, being good candidates for the development of indices of "wildness". These two components are the shoaling index, that is the level of cohesiveness of the group as a response to a fish predator, and the freezing behaviour, i.e. time spent motionless resting on the bottom as a response to overhead fright stimuli. For both measures it would be necessary to develop a reference system and to further improve and possibly simplify the experimental protocols for a wider and routine use of the behavioural test in the aquaculture practise related to restocking.

In response to the worldwide decline of fish populations, stock enhancement has been attempted as a way to rejuvenate populations, by rearing a large number of fish under artificial conditions and subsequently releasing them into the natural environment (Kellison et al. 2000).

During the recent years, a growing body of evidence has suggested that the success of stock enhancement programmes is highly related to the behavioural profile of the juvenile fish produced in the hatchery, as the artificial environment of the hatchery may generate a number of behavioural deficits potentially detrimental for the survival of the released fish in the wild (Einung and Fleming, 2001; Brown and Day, 2002; Huntingford, 2004). These behavioural deficits have been documented through a relatively large but constantly increasing number of studies, conducted to compare the behavioural performance between hatchery-reared fish and their wild counterparts. The main and more significant differences in behavioural traits between hatchery and wild fish have been shown especially at the level of feeding, aggressive, and antipredator behaviour (Einung and Fleming, 2001, Huntingford 2004). A large part of these differences seem to have an environmental origin, since hatchery fish experience a psychosensory-deprived environment with respect to the wild counterparts (Olla et al., 1998), and it has been shown that in many cases a relatively short exposure to a relevant and proper experience may adjust the behavioural performance reducing the gap between the wild and the hatchery fish (Huntingford, 2004). On the basis of this body of knowledge, "life skills training" and procedures designed to expose the fish to appropriate stimuli and to increase the complexity of the hatchery environment, have been recently proposed (Brown and Day, 2002). Although these procedures aim to increase both the well-being of captive animals and the related success of the released fish in the wild, it has been noted that they are costly and difficult, in certain cases, to be applied (Huntingford, 2004). On the other hand, if the aim of a release program is to increase the viability of a natural population, the detailed knowledge of the behavioural profile of the reared fish and the eventual differences with respect to the wild counterparts, is an indispensable requisite before fish releasing.

In this framework, it is therefore necessary to develop methods for a rapid and effective evaluation of the behavioural "quality" of the juvenile fish destined to release, as a part of a routine practice of any hatchery/restocking plan.

The behavioural quality of the hatchery fish should be assessed in terms of an array of behavioural indicators of "wildness", enabling to provide a measure of the distance between the hatchery and the wild fish at the different life stages of the fish before releasing. These indicators could constitute the basis to evaluate whether and how to proceed with programs of environmental enrichment, including the life skills training, and to eventually assess, in a second time, the efficacy of these programmes.

Our contribution in this field is based on our participation to a project promoted and funded by the "Italian Ministery of Agriculture and Forestry", aimed at assessing the quality of the juvenile hatchery fish of some brackish species, typical of the Italian coastal and lagoon waters. These species are cultured to sustain traditional forms of lagoon aquaculture, such as the Adriatic vallicultura (Ardizzone et al., 1988). The program was therefore an integral part of the strategies aimed at improving stock enhancement, according to the guide lines for a responsible fisheries (FAO, 1995). On the basis of some results obtained within this research project, we aim, in the present context, to propose some suggestions for a methodological approach to the assessment of the behavioural quality of the hatchery-reared fish. We focused on the juvenile stage of the European sea bass (*Dicentrarchus labrax*), which is routinely seeded in the Adriatic lagoon to sustain "vallicoltura", using both captured wild fish and fish obtained from controlled spawning under hatchery conditions. We compared some behavioural aspects related to the antipredator behaviour between wild and hatchery juveniles of about 2 cm in Total length. We decided to analyse the antipredator behaviour for two main reasons: this is a fundamental component for the individual fitness of the released fish in the wild, and it has been stated, on the basis of previous studies, that the levels of antipredator behaviour generally differ between hatchery and wild fish (Einum and Fleming, 2001). Although details and results of our investigations are partly found in Malavasi et al. (2004), Torricelli et al. (2005) and Georgalas et al. (2007), we aim, in the present context, to point out some aspects of this investigation, especially those related to the methodological approach. These methodological considerations could be potentially useful for a generalised and quantitative evaluation of the behavioural quality of the hatchery fish, using standardised and relatively simple protocols. We found, for example, that a simple behavioural measure, the shoaling index, differed significantly between the groups of hatchery-reared and that of wild fish juveniles, as a response towards a live fish predator behind a transparent partition (Malavasi et al. 2004). Thus, the level of cohesion of the shoal seems to be a reliable measure of the intensity of antipredator behaviour towards piscivourous predation. We measured this behavioural variable, by calculating the dispersion index (ratio between the variance and the mean) of the distribution of the juvenile fish across the rear tank wall, with the aid of a grid, before and after the introduction of the predatory stimuli, that is the fish predator (a common eel in this case). The behaviour of the juvenile shoals was video-recorded and the shoaling index could be assessed through the analysis of subsequent temporal frames. It was enough to analyse a period of 20 seconds both before and after the introduction of the predator to detect an increase of the index from 3 to 5 in the hatchery group and from 3 to 8 in the wild group. The level of shoal cohesion of the wild fish was therefore about 40% higher than in the hatchery fish. This simple behavioural variable could therefore constitute a behavioural index of wildness useful to assess the gap in the intensity of antipredator behaviour between hatchery and wild fish within the juvenile phase. During the same study, we designed other experimental apparatus to test for the differences between the two groups at the level of the response towards overhead fright stimuli, such as a shadow rapidly created on the water surface and a mechanical bill splashing into the water. In this case, we measured the distance from the bottom and the time spent motionless on the bottom, the so called freezing behaviour, as behavioural variables useful to assess the intensity of the antipredator reaction. Even in this case, we video-recorded the behaviour of the juvenile shoals both before and after the exposure to the stimulus. Video-recording is fundamental to analyse in details the behavioural sequences and to assess quantitatively the intensity of the behaviour at the different temporal frames preceding and following the stimulus, by exporting the recorded sequences into a proper software. In this case, the freezing behaviour resulted to differ between hatchery and wild fish, being therefore this measure an other good candidate as behavioural index of wildness (Torricelli et al. 2005). By extending the use of these behavioural variables, it could be possible to build up a sort of reference system, where the optimal value is that reached by the wild counterpart of a given hatchery species at a given life stage, to which the value reached by the hatchery fish could be compared. This reference system could be useful to assess the effects of experience and training on the intensity of the antipredator behaviour, in order to eventually evaluate a "life skills" training program.

Behavioural variables, such as shoaling index and freezing, can be assessed in any fish species, being very basic and "universal" components of the behavioural repertoire of Teleost fishes. If protocols were standardised, these behavioural variables could be used within a wide comparative approach.

Thus, such kind of laboratory test, aimed at assessing behavioural indices of wildness, could be useful for a general use in the hatchery practise related to restocking for 4 main reasons: 1) The experimental apparatuses are simple and easy to set up in the hatchery 2) The experiment can be performed quickly, as it is enough to transfer the experimental fish in the experimental tank just 24 hours before the experiment 3) The intensity of the behavioural response can be assessed quantitatively within a short term scale 4) The use of standardised procedure and the assessment of universal and relatively simple behavioural variables could be used with a comparative approach, taking also into account a large spectrum of factors, such as: hatchery conditions, generations of captivity, quality of parents, species, populations, experience and life stage. This would allow to obtain a reference system and a related scale of "wildness", together with a scale of the gaps between wildness and domestication.

Although we are aware that these behavioural test do not solve the complicated problem of the real efficacy of the restocking programmes and do not make the application of "life skills training" easier and more realistic, we think that they could help at least to generate some results of general interest on the patterns of domestication occurring in the hatchery fish, suggesting more proper solutions to improve the rearing and releasing strategies.

REFERENCES

Ardizzone, G.D.; Cataudella, S. and Rossi, R. (1988). Management of coastal lagoon fisheries and aquaculture in Italy. *FAO technical paper* 293, 1-103.

Brown, C. and Day, R. L. (2002). The future of stock enhancements: lessons from hatchery practice to conservation biology. *Fish and Fisheries* 3, 79-94.

Einum, S. and Fleming, I.A. (2001). Implications of stocking: ecological interactions between wild and released salmonids. *Nordic Journal of Freshwater Research* 75: 56-70.

FAO, (1995). Code for conduct of responsible fisheries. FAO, Rome. 41 pp.

Georgalas, V.; Malavasi, S.; Franzoi, P. and Torricelli, P. (2007). Swimming activity and feeding behaviour of larval European sea bass (*Dicentrarchus labrax* L.): effects of ontogeny and increasing food density. *Aquaculture* 264:418-427.

Huntingford, F.A. (2004). Implications of domestication and rearing conditions for the behaviour of cultivated fishes. *Journal of fish biology* 65(supplement A): 122-142.

Kellison, G.T.; Eggleston, D.B. and Burke J.S. (2000). Comparative behaviour and survival of hatchery-reared versus wild summer flounder (Paralichthys dentatus). *Canadian Journal of Fishery and Aquatic Science* 57:1870-1877.

Malavasi, S.; Georgalas, V.; Lugli, M.; Torricelli, P. and Mainardi, D. (2004). Differences in the pattern of antipredator behaviour between hatchery-reared and wild European sea bass juveniles. *Journal of fish biology* 65(supplement A): 143-155.

Olla, B.L.; Davis, M.W.; and Ryer, C.H. (1998). Understanding how the hatchery environment represses or promotes the development of behavioural survival skills. *Bulletin Marine Science* 62: 531-550.

Torricelli, P.; Malavasi, S.; Georgalas, V.; Lugli, M. and Mainardi, D. (2005). Lo studio del comportamento delle larve dei pesci Teleostei nei grandi volumi di allevamento. *Ministero delle politiche agricole e forestali – Direzione della Pesca e dell'acquacoltura – IV Piano Triennale della Pesca e dell'Acquacoltura.*

In: Aquaculture Research Progress
Editor: Takumi K. Nakamura
ISBN 978-1-60456-247-7

Expert Commentary B

ADVANCES IN PRACTICAL FISH DIETS

Giovanni B. Palmegiano[1] and Ivo Zoccarato[2]
[1] Institute of Sciences of Agricultural Productions
[2] Department of Animal Husbandry Sciences

The nutritional features of fish are fundamental in the choice of a modern healthy diet. From this point of view, food industries are becoming involved more with quality rather than quantity; Many management factors can influence the qualitative traits of fish. The quality of food is the result of a balance of different parameters, some intrinsically linked to the product, such as taste, colour, smell, and others related to processing food technology as salting, smoking, or canning, which have an effect on the final quality of the product. Last, but not least is the consumer's ability to assess quality.

The environmental sustainability, reducing the impact on fishery products from by-products of agriculture, and the dietetic food quality are among the main topics that the aquaculture productions must consider.

At present, high quantity of fishmeal and fish oil used in fish feeds for aquaculture production is not a sustainable way for aquaculture development and it produces a hard pressure on natural fish stocks. Consequently, many studies have been carried out on the substitution of fish meal and fish oil with plant products and it is now possible to make an assessment in order to understand if and how this substitution is possible.

The use of plant source of nutrient is still successful growing; maize, soybean and wheat are the most common plant meal used in fish diet. However, these raw materials are not always utilizable because of technical processing problems: i.e. maize meal tends to transfer a pale yellow colour to the trout flesh, but large scale retail trade demand white fish. An another constraint is represented by many consumers who ask GMO free fish feeding but at present it is not so high the availability of GMO free soybean meal. Microalgae also are potentially the most useful nutrient source of proteins and lipid, but at present the production cost are too high. Often there is the competition of human direct use that makes this production not competitive. Open air Spirulina could be a low cost possibility in developing countries. Table 1 shows some candidates raw materials for substituting fish meal and fish oil in fish diets.

Table 1. Meal and oils, utilized in the experimental trials

Raw materials	Species	Oils	Species
Rice protein concentrate	Sturgeon, Rainbow trout	Maize	Sturgeon, Rainbow trout
Pea protein concentrate	Sturgeon	Echium	Sturgeon
Spiruline	Sturgeon, Rainbow trout	Enotera	Sturgeon
Isochrisys galbana	Seabream	Borage	Sturgeon
Maize gluten	Sturgeon, Rainbow trout	Line seed	Sturgeon, Rainbow trout, Blackspot Seabream
		Tuna stearine	Sturgeon
		Soybean	Rainbow trout
		Peanut	Rainbow trout

Table 2. Main fatty acids of some animal and plant oil. ****Isochrysis galbana* **strain** ***T-iso***

Oils Fatty Acids	Fish	Tuna Stearine	Soya	Lineseed	Peanut	Borrago	T-Iso*
C14:0	10.62	5.31	0.12	3.35	0.06	3.1	15.98
C16:0	19.38	29.61	13.32	5.80	11.95	13.5	11.65
C16:1	9.74	6.03	0.08	0.04	0.08	-	4.13
C18:0	2.20	6.69	4.20	3.25	3.75	4.6	0.4
C18:1n7/n9	1.69	13.86	20.89	17.48	49.07	13.1	17.43
C18:2n6	1.15	1.38	55.49	15.03	27.58	25.3	6.41
C18:3n6	0.37	0.45	-	53.13	-	15.0	T
C18:3n3	1.41	1.09	5.47	0.21	0.10	15.5	4.25
C18:4n3	2.77	0.00	-	0.00	0.06	9.9	2.07
C20:1n9	9.13	0.15	0.38	0.17	0.061.02	-	12.47
C20:5n3	5.54	6.35	-	0.54	-	-	0.73
C22:6n3	11.21	16.38	-	0.18	-	-	10.23
Saturated	34.48	47.41	17.24	12.65	18.24	21.2	28.51
Monounsaturated	23.96	24.03	21.28	17.82	52.90	13.1	34.03
Polyunsaturated	19.98	26.05	60.97	69.33	28.86	65.7	23.69
SFA/UFA	1.73	1.83	0.21	0.18	0.22	0.32	0.49
n3/n6	9.98	13.24	0.10	0.01	0.05	0.63	2.70

Table 2 shows the main fatty acids of some animal and plant oil of experimental fish diets tested by the authors. Each oil has a specific fatty acid, so that, when mixed in the diet in suitable amount, it contributes to characterize the fish fillet composition.

Organic fish demand is growing even if sales are still low (i.e. approx. 4064 tons of salmon and less for mussels and trout). Consumers check fish morphology, i.e. body shape or organoleptic properties, freshness, lack of smells or nutritional value (low lipids and high poly-unsaturated fatty acids), origin and quality, absence of biological or chemical contaminants.

The pollution of the sea, as the North Sea case, makes the fish meal and fish oil often unusable. 750 organizations, fish farmers associations also, in 108 countries are currently united in the International Federation of Organic Agriculture Movements (IFOAM), and its goal is the worldwide adoption of ecologically, socially and economically sound systems that are based on the principles of Organic Agriculture. IFOAM shows the evidence of the economic importance of the organic production.

An emerging topic in fish nutrition is the substitution of artificial antioxidant with natural ones (Dimitrios, 2006; Yanishlieva et al., 2006). Recent studies highlight the presence of natural antioxidants in essential oils (i.e. oregano oil, rosemary oil, thymus oil) and in some agriculture by-products, such as the tomato processing wastes, the dehydrate citrus pulp, the olive oil water mill.

Moreover, the use of these by-product could be meet the need to reduce the environmental impact of these agriculture productions. The antioxidant compositions of rosemary oil and the residue of oil olive oil production are reported in tables 3 and 4.

Table 3. Main phenols and their main effect in rosemary

Polyphenols	Effects	References
Carnosol Carnosic acid Rosmanol Rosmadial	Inhibit lipid peroxidation in microsomal and liposomal systems Inhibit the formation of polar substances and the decomposition of polyunsaturated triacylglycerols.	Aruoma et al., 1992 Yanishlieva et al., 2006

Table 4. Main phenols and their main effect in olive oil mill waste water

Polyphenols	Effects	References
Oleuropein	Vasodilator, Antihypertensive	Ibanez et al., 2000
Hydroxitirosol	Inhibits platelet aggregation	Da-Peng et al., 2007
Tirosol	Scavenges ROS	Sicuro et al., in press
Caffeic acid	Scavenges ROS	Manach et al., 2004
Verbacoside	Anti-inflammatory	Da-Peng et al., 2007

LITERATURE CITED

Aruoma, O.I., Halliwell B., Aeschbach R., Löligers J. (1992). Antioxidant and pro-oxidant properties of active rosemary constituents: Carnosol and carnosic acid. *Xenobiotica* ,22, 257–268.

Da-Peng Yang; De-Xin Kong; Hong-Yu Zhang (2007). Multiple pharmacological effects of olive oil phenols. *Food Chemistry*, 104, 1269-1271.

Dimitrios, B. (2006). Sources of natural phenolic antioxidants. *Food Science and Technology*, 17, 505-512.

Ibanez, E; Palacios, J; Senorans, FJ; Santa-Maria, G.; Tabera, J: Reglero, G. (2000). Isolation and separation of tocopherols from olive by-products with supercritical fluids. *J. Am. Oil Chem. Soc.,* 77,187-190.

Manach, C.; Scalbert, A.; Morand, C.; Jimenez, L. (2004). Polyphenols: food sources and bioavailability. *American Journal of Clinical Nutrition*, 79, 727-747.

Pereira, A.P.; Ferreira, I.C.F.R.; Marcelino, F.; Valentão, P.; Andrade, P.B.; Seabra, R.; Estevinho, L.; Bento, A.; Pereira, J.A. (2007). Phenolic Compounds and Antimicrobial Activity of Olive (*Olea europaea* L. Cv. Cobrançosa) Leaves. *Molecules*, 12, 1153-1162

Sicuro, B.; Daprà, F.; Gai, F.; Palmegiano, G.B.; Schiavone, R.; Zilli, L.; Vitella, S., 2008. In press.

Yanishlieva, N.V.; Marinova, E.; Pokorny´, J. (2006). Natural antioxidants from herbs and spices. *Eur. J. Lipid Sci. Technol.*,108, 776–793.

In: Aquaculture Research Progress
Editor: Takumi K. Nakamura

ISBN 978-1-60456-247-7

Short Communication A

PRODUCTION OF RECOMBINANT FISH GONADOTROPINS BY BACULOVIRUS IN SILKWORM LARVAE AND THEIR *IN VIVO* BIOLOGICAL ACTIVITIES

Makito Kobayashi [1], Young Chang Sohn [2], Tetsuro Morita [3], Goro Yoshizaki [3], Youichi Hayakawa [1], Hidekazu Nagaya [4], and Takeo Suzuki [4]

[1] Department of Biology, Division of Natural Sciences, International Christian University, Mitaka, Tokyo 181-8585, Japan

[2] Faculty of Marine Bioscience and Technology, Kangnung National University, Gangneung 210-702, South Korea

[3] Department of Marine Biosciences, Tokyo University of Marine Science and Technology, Minato, Tokyo 108-8477, Japan

[4] Katakura Industries Co. Ltd., Sayama, Saitama 350-1332, Japan

ABSTRACT

The induction of gonadal maturation is one of the important aspects of fish aquaculture. Extract of pituitary glands containing two types of gonadotropins (GTH), follicle-stimulating hormone (FSH) and luteinizing hormone (LH) has been administered for this purpose. However, collection of a large amount of pituitaries is not always practically easy, and the pituitary extract is a mixture of FSH and LH whose contents and composition of FSH and LH are usually unknown. It is possible to produce fish GTHs biotechnolgically, and recombinant fish GTHs have been produced using various bioreactors. However, since the amount of hormones produced in these systems is small, they are used only for *in vitro* endocrinological studies of fish and could not be tested *in vivo* as substitutes for the pituitary extract. In order to obtain basic information for aquaculture purpose, studies that examine the *in vivo* effect of recombinant hormones are essential. Recently, we successfully produced a large amount of recombinant FSH and LH (rFSH and rLH) of goldfish *Carassius auratus* and Manchurian trout *Brachymystax lenok* by baculovirus in silkworm *Bombyx mori* larvae, and confirmed their biological activities in various *in vivo* bioassays. rFSH and rLH of goldfish and Manchurian trout induced milt production in male goldfish. Goldfish rLH induced ovulation in female bitterling *Rhodeus ocellatus ocellatus*, whereas rFSH showed low potency. rFSH and rLH of goldfish and Manchurian trout induced sex steroid production in goldfish and rainbow trout *Oncorhynchus mykiss*, respectively, in both sexes. Goldfish rFSH and rLH induced spermatogenesis in male Japanese eel *Anguilla japonica.* The yield of rFSH and rLH produced in silkworm larvae is much greater compared to that of fish GTH produced in systems using other bioreactors. These results indicate that a large amount of biologically active recombinant fish GTHs could be produced by a baculovirus-silkworm larvae system. By using these recombinant GTHs, the following possibilities are suggested for the development of aquaculture research: 1) use of rFSH and rLH individually or in combination as substitutes of pituitary extracts, 2) elucidation of *in vivo* biological function of fish FSH and LH, and 3) development of a new efficient method of induction of gonadal maturation by differential use of rFSH and rLH based on physiological profiles of FSH and LH expression.

INTRODUCTION

Artificial Control of Gonadal Maturation of Fish

A stable supply of seedling is critical in fish culture, and control of gonadal maturity of broodstock fish is an important process in aquaculture (Sullivan et al., 2003). However, in some species, gonadal maturation or ovulation does not occur in captivity because of an inappropriate environment for reproduction. For the artificial control of reproductive activity of culture fish, manipulation of environmental factors, such as photoperiod and water temperature (Pankhurst and Porter, 2003) and/or hormonal treatment have been employed depending on the reproductive mode of each species. Various methods for the hormonal control of reproductive activity of culture fish have been developed using different types of hormones based on knowledge of the reproductive endocrinology of fish (Patiño, 1997; Zohar and Mylonas, 2001). As in other vertebrates, the reproductive process in fish is mainly regulated by the brain-pituitary-gonad axis. Gonadotropin-releasing hormone (GnRH) produced in the hypothalamus stimulates secretion of two types of gonadotropins (GTHs), follicle-stimulating hormone (FSH) and luteinizing hormone (LH) in the pituitary gland. FSH and LH are the key regulators of gonadal development, directly acting on the gonads (Kobayashi et al., 1997; Parhar, 2003; Swanson et al., 2003). During the process of gonadal development, FSH is considered to regulate early phases of gametogenesis, such as vitellogenesis in the female and spermatogenesis in the male, and LH the final phases of gametogenesis, such as oocyte maturation and ovulation in the female and spermiation and milt production in the male. In order to mimic the effects of these endogenous hormones, synthetic GnRH or extract of pituitary gland, a mixture of FSH and LH, has been exogenously administered to fish to stimulate gonadal development (Patiño, 1997; Zohar and Mylonas, 2001). Administration of pituitary extract has been successfully conducted for a long time for the induction of gonadal maturation in some species of fishes. However, this method has some disadvantages: collection of a large amount of pituitary glands is quite laborious and it is often impractical to sacrifice so much sexually mature broodstock for the collection of pituitaries. It is also difficult to estimate the contents and composition of FSH and LH. This sometimes causes variability of potency of the pituitary extract. Human chorionic gonadotropin (HCG) has been used as a hormone with LH effect in some aquaculture species, but this mammalian hormone is effective only in a limited number of species because of species specificity (Patiño, 1997; Zohar and Mylonas, 2001). Presently no hormone with FSH effect is commercially available for fish. If a homogeneous pool of FSH and LH for fish could be constantly supplied, FSH and LH could be administered individually to simulate natural profiles of GTH, and a more efficient method of hormonal manipulation of gonadal development could be established.

Recombinant Fish GTH Produced by Various Bioreactors

Since biotechnology for the production of recombinant proteins has been greatly advanced, it is possible to produce fish FSH and LH biotechnologically as substitutes for the pituitary extract. When recombinant GTH is produced for aquaculture purposes, we need to consider following criteria: biological activity on fish *in vivo*, production of a large quantity of hormone for administration to fish, low production cost, and safety to fish and the environment.

To date, recombinant fish GTHs have been biotechnologically produced in many species of fish using various bioreactors, but these hormones are basically for the research of reproductive endocrinology of fish (See Kobayashi et al., 2006, Choi et al., 2005; So et al., 2005; Kamei et al., 2006; Ko et al., 2007; Aizen et al., 2007; Zomora et al., 2007). The bioreactors used to produce recombinant fish GTHs are: bacteria *Escherichia coli*, amoeba *Dictyostelium discoideum*, yeast *Pichia pastoris*, insect cells, and mammalian cells. Since GTH is a heterodimeric glycoprotein, consisting of a common α subunit and FSHβ or LHβ subunit, it is quite difficult to produce such a protein with biological activity in prokaryotic cells (bacteria) that do not carry out post-translational modifications of proteins. Cell culture systems of eukaryotic species (amoeba, yeast, insect cell, and mammalian cell) have been established and biologically active recombinant fish GTHs have been produced in these systems in which association of subunits and glycosylation occur. However, the amount of GTH produced in these systems is small, and the produced GTH is basically for *in vitro* physiological studies. Since a large amount of hormone is required for the administration of aquaculture fishes, cell culture systems are practically and economically inefficient.

We have selected two bioreactors for the production of recombinant fish GTH: rainbow trout *Oncorhynchus mykiss* embryo and silkworm *Bombyx mori* larva. Using such individual organisms rather than culture cells, it is expected that a larger amount of recombinant protein could be produced with post-translational modification. A system of producing recombinant protein using rainbow trout embryo is a novel system, and production of biologically active recombinant FSH and LH of goldfish *Carassius auratus* by this system was reported previously (Morita et al., 2003, 2004). Although this system enabled production of recombinant proteins at low temperature (4-15 °C), the yield of protein was not high enough for *in vivo* study. The baculovirus-silkworm larvae system was established for the production of a large amount of protein at low cost (Maeda, 1994), and using this system, recombinant growth hormone of grass carp *Ctenopharyngodon idellus* was produced in amounts that were enough to administer to juvenile grass carp *in vivo* for growth acceleration (Ho et al., 1998).

For these backgrounds, we attempted to produce recombinant goldfish and Manchurian trout *Brachymystax lenok* GTHs by the baculovirus-sikworm larvae system.

Production of Recombinant Fish GTH by Silkworm Larva

We have produced three types of recombinant fish GTHs by the baculovirus-silkworm larvae system: heterodimeric goldfish GTH (Kobayashi et al., 2006), single-chain goldfish GTH (Hayakawa et al., 2007), and single-chain Manchurian trout GTH with histidine tags (His-tag) (Ko et al., 2007) (Figure 1.) These recombinant GTHs were produced by the KaikoExpress (formerly Superworm System, (Katakura Industries, Saitama, Japan; http://www.katakura.co.jp/kaikoexpress/index-e.htm). cDNA of each GTH subunit or cDNA of β subunit fused to α subunit cDNA were inserted into transfer vectors (pYNG: Katakura Industries, Saitama, Japan). Each transfer vector and baculovirus (*Bombyx mori* nucleopolyhedrovirus; CPd strain: Suzuki et al., 1997) genomic DNAs were co-transfected into *Bombyx mori* culture cells (BmN: Maeda, 1989). Recombinant viruses, into which fish GTH cDNAs were introduced by homologous recombination, were screened by the end-point dilution method (Maeda, 1989).

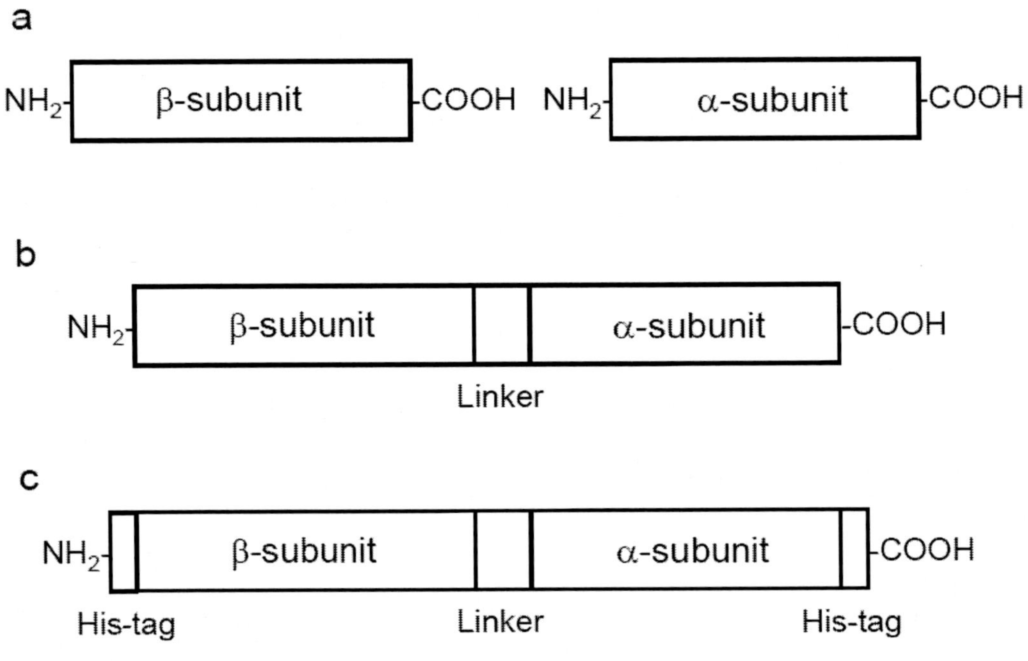

Figure 1. Schematic presentation of recombinant GTHs. a, Heterodimeric GTH. α and β subunits were produced by separate gene constructs, and the association occurred in the silkworm larvae. Goldfish heterodimeric GTHs were produced by this method (Kobayashi et al., 2006). b, Single-chain GTH. cDNA of α and β subunits are connected and recombinant GTHs are produced in the form of single-chain protein. Equine C-terminal peptide (CTP) was used as a linker. Goldfish single-chain GTHs were produced by this method (Hayakawa et al., 2007). c, Single-chain GTH with histidine tags (His-tag). His-tags were added for the purification of the GTH. An artificial *N*-linked glycosylation sequence (NCS) was used as a linker. Manchurian trout single-chain GTHs were produced by this method (Ko et al., 2007).

After propagation of the recombinant baculoviruses in *Bombyx mori* culture cells, the recombinant baculovirus carrying GTH cDNA was infected to silkworm larvae of the 5th instar early stage. To produce heterodimeric recombinant GTH, viruses carrying cDNA of GTHα and either FSHβ or LHβ were co-infected. To produce single-chain GTH, virus carrying fused cDNA was infected to silkworm larvae. Five or six days after inoculation, hemolymph of the larvae containing recombinant FSH or LH was collected. This hemolymph containing recombinant GTHs were used for bioassays, deglycosylation analysis, and western blot analysis. Expression of recombinant GTHs was confirmed by western blot analysis using antisera against goldfish GTH α subunit and goldfish LHβ subunit. Association of subunits was confirmed in the heterodimeric goldfish LH. Since a specific antiserum against goldfish FSH is not available, detection of FSHβ subunit was not conducted. *N*-glycosylation was confirmed by digestion with *N*-glycosydase F.

First we produced heterodimeric goldfish GTH based on the structure of native GTH (Kobayashi et al., 2006). Then, we attempted to produce a single-chain goldfish GTH since some benefits were expected for its simple structure (Hayakawa et al., 2007). In heterodimeric GTH, association of two subunits is one of the rate-limiting steps in hormone expression. It is considered that single-chain GTH is produced more efficiently since a process of assembly of two subunits is bypassed (Narayan et al., 2000; Garcia-Campayo and Boime, 2001). Additionally, the practical labor and cost of producing single-chain GTH is reduced compared to those of producing heterodimeric recombinant GTH. It is also expected that single-chain GTH is considered to be more stable than heterodimeric GTH in the blood circulation (Garcia-Campayo and Boime, 2001).

Although a large amount of heterodimeric or single-chain goldfish GTHs seems to be produced in the hemolymph, we could not establish a good method to measure concentration of these goldfish recombinant GTHs in the hemolymph. One of the reasons is that purification of recombinant protein from hemolymph was difficult because hemolymph of silkworm larva contains various types of own proteins (Ho et al., 1998). Therefore, recombinant GTHs were administered to fish by the volume of hemolymph wihtout calibrating GTH concentration.

Addition of His-tag to recombinant GTH enabled to purify the GTH, and quantification of the hormone became possible by protein measurement. We produced single-chain goldfish GTHs with His-tag in rainbow trout embryo (Morita et al., 2004) and single-chain Manchurian trout GTHs with His-tag in silkworm larvae (Ko et al., 2007). Recombinant Manchurian trout GTHs with His-tag were purified by Ni-NTA agarose beads (Ko et al., 2007) or Ni-Sepharose column (Kobayashi and Sohn, unpublished), and concentration of GTHs could be determined by protein measurement. This enabled to determine specific activity of the GTH preparations and calibrate GTH concentrations in the different batches of the hormones. We are now producing single-chain GTH with His-tag of goldfish and other species of fish in silkworm larvae.

Biological activities of recombinant goldfish and Manchurian trout GTHs were detected by various *in vivo* and *in vitro* bioassays (Table 1). We demonstrated the biological activities of recombinant fish GTH *in vivo* for the first time by injecting recombinant goldfish FSH and LH to male goldfish, and induction of milt production was confirmed both in FSH- and LH-injected fish.

Table 1. Biological activities of recombinant goldfish and Manchurian trout gonadotropins detected by in vivo and in vitro bioassays

	Goldfish				Manchurian Trout	
	Heterodimeric		Single chain		Single chain	
	FSH	LH	FSH	LH	FSH	LH
In vivo						
Sex steroid production						
Testosterone	±	+[1]			+	+[2]
Estradiol	+	+[3]			+	+[2]
Ovarian development[4]	□	+			+	±
Ovulation[5]			±	+		
Spermatogenesis[6]			±	+		
Milt production[7]	+	+	+	+	+	+
In vitro						
Receptor binding[8]						
FSH receptor					+	±
LH receptor					□	+
Sex steroid production[9]					+	+
Testosterone						
Estradiol					+	+

Fish species used for assays: [1] male goldfish, [2] female rainbow trout, [3] female goldfish, [4] rainbow trout, [5] bitterling, [6] Japanese eel, [7] goldfish, [8] amago salmon, [9] rainbow trout follicles. +, strong response, ±, weak response, —, no or trace response.

Although concentrations of hormones in the hemolymph was not measured in the study, the results indicate that they were high enough to induce gonadotropic response in fish under the condition of clearance in the blood that could not be examined by *in vitro* assay. Furthermore, our recombinant GTHs induced both early and final phase of gametogenesis: increase of ovarian weight, spermatogenesis, ovulation, and milt production *in vivo*, and sex steroid production *in vivo* and *in vitro*. Interestingly, recombinant goldfish LH induced complete stages of spermatogenesis in sexually immature Japanese eel *Anguilla japonica*, from spermatogonia to spermatozoa. Recombinant Manchurian trout FSH and LH selectively activate their cognate receptors *in vitro*. These results indicate the recombinant fish GTH in the three different forms, heterodimer, single-chain, and single-chain with His-tag, produced by silkworm larvae possess biological activities that could be used for *in vivo* studies.

The fish species we used for bioassays were relatively small size laboratory species (body weight: bitterling *Rhodeus ocellatus ocellatus*, 1-4 g; goldfish, 5-25 g) to save hormone samples, but we also tried to use aquaculture species (body weight: rainbow trout, 200 g; Japanese eel, 230-280 g). Effectiveness of recombinant GTHs in these aquaculture species indicates that the enough amount of biologically potent recombinant GTHs could be produced in silkworm larvae.

By adding His-tag to recombinant proteins, Manchurian trout GTH could be purified and its amount was measured by protein determination. Our recent study in Manchurian trout GTH revealed that the expression efficiency of GTH is quite high in the baculovirus-silk worm system. Approximately 0.4 ml of hemolymph is collected from one silkworm larva. We obtained 28.6 mg and 20.6 mg of partially purified recombinant Manchurian FSH and LH preparations, respectively from 30 ml of pooled hemolymph (Kobayashi and Sohn, unpublished data). The purity of the preparations is estimated to be 63 and 61 % for FSH and LH preparations, respectively by densitometry of SDS-PAGE and Western blot analysis. Concentrations of FSH and LH in the original hemolymph are estimated as approximately 2,000 μg/ml and 1,500 μg/ml, respectively. This implies that one larva produced approximately 800 μg and 600 μg of FSH and LH, respectively in five or six days. These results indicate that expression levels of baculovirus-silkworm larvae system is quite high compared to those of other systems (Table 2). In addition to high expression rate of recombinant protein in the silkworm larvae, the facilities of Katakura Industries have the capacity of manipulating thousands of larvae per month, which enables large-scale production of recombinant fish GTH.

Cost of production is one of the important matters for future application of recombinant hormones to aquaculture. We have produced GTH of different types of different species several times, and based on the cost of recent batch of Manchurian trout GTH, it cost approximately 600,000 Japanese Yen (5,200 US $) for each of recombinant FSH and LH, for a process of production of 30 ml hemolymph containing GTH and purification of GTH from the hemolymph. Since we obtained 20-30 mg of GTH preparations with approx. 60 % purity, the cost of making 1 mg of recombinant GTH is 30,000 – 50,000 Japanese Yen. Production cost of single-chain GTH is lower than that of hetrodimeric GTH because of its structural simplicity. Since this cost does not include cost of cloning of GTH cDNA, total cost of production of recombinant GTH is much higher when the process starts with cDNA cloning of GTH. Recent development of technology for molecular cloning and large-scale production of recombinant protein would lower the cost.

Table 2. Comparison of yield of recombinant fish GTH produced by various expression systems

Fish	Bioreactor	GTH	Yield (μg/ml)	References
			Culture medium	
Japanese eel	Yeast	FSH	2.0	Kamei et al. 2003
Tilapia	Yeast	FSH	0.1	Aizen et al., 2007
Tilapia	Yeast	LH	0.08	Kasuto and Levavi-Sivan,
Catfish	Amoeba	FSH	0.018-0.061	2005
Catfish	Amoeba	LH	0.031-0.080	Vischer et al., 2003
Catfish	Insect cell	FSH	7.0	Vischer et al., 2003
Catfish	Insect cell	LH	4.0	Zmora et al., 2007
				Zmora et al, 2007
			Hemolymph	
Trout	Silkworm	FSH	2000	Kobayashi and Sohn,
Trout	Silkworm	LH	1500	unpublished data
Carp	Silkworm	GH*	1000	Ho et al., 1998
Goldfish	Trout	LH	0.05**	Morita et al., 2004

* Growth hormone, ** μg per egg.

The safety of recombinant hormones to broodstock and also to the environment should be investigated. Presently, the virus in the hemolymph is removed by ultracentrifugation. By purifying recombinant hormones from the hemolymph, influence of substances of silkworm origin would be further eliminated. Since the use of recombinant GTHs is limited to broodstock, it is unlikely that these hormones would reach human consumption. Apart from recombinant proteins and virus, hemolymph of silkworm larva contains some substances which depress swimming activity of injected fish for a few hours. This side effect was prevented by the removal of substances whose molecular weight is lower than 10 kDa from the hemolymph (Kobayashi, unpublished data).

Potential Advantages and Disadvantages of Use of Recombinant GTH in Aquaculture

Use of recombinant GTHs for aquaculture has some potential advantages. Instead of collecting pituitary glands, a stable supply of hormone could be possible once production systems are established. Obtaining many fish that are sexually mature and large enough to collect pituitaries is not always easy, and collection of a large amount of pituitaries from each fish is quite laborious. It is impractical to sacrifice many broodstock for the collection of the pituitary. It is also difficult to estimate the contents and composition of FSH and LH in the pituitary preparations. This sometimes causes variability of potency of the pituitary extract. Since addition of His-tag to recombinant GTH enabled purification and quantification of the GTH concentration, it is possible to supply calibrated recombinant GTH preparations. Moreover, recombinant GTH preparations are homogeneous pools of FSH and LH unlike pituitary extract, FSH and LH could be administered individually to simulate natural profiles of GTH expression; stimulating early gonadal development by FSH and final maturation by LH. Thus, more efficient method of hormonal manipulation of gonadal development could be established. In fact, it has been reported in Japanese eel that fertilization rate of eggs from fish treated with purified salmon FSH followed by salmon LH was higher than those in fish treated with FSH or LH only (Sato et al., 2003). It is also possible to produce recombinant GTH with higher potency and stability for aquaculture purposes by protein engineering, such as single-chain GTH (Walsh, 2004).

Although recombinant fish GTH produced by baculovirus-silkworm system seems to be a powerful tool for *in vivo* study in aquaculture, there remain several matters to consider in addition to biological activity, yield, cost, safety, and calibration method. Species specificity of GTH is a matter to be examined when GTH is used in fish (Zohar and Mylonas, 2001). Production of recombinant GTH of various species is one of the ways to solve this problem. Presently, goldfish (Kobayashi et al., 2006; Hayakawa et al., 2007), Manchurian trout (Ko et al., 2007), and red seabream *Pagrus major* (Gen et al., 2007) could be produced in the baculovirus-silkworm larvae system, and it would be possible to produce recombinant GTH of other species.

Metabolic clearance rate of recombinant fish GTH should be examined in comparison to native GTH. Since the half-life of administered GTH in circulation highly affects biological potency of GTH, it would be important to examine the profile of administered GTH in the blood as well as biological effects. Glycosylation of GTH is known to be important for the expression of biological activity and metabolic clearance. Especially, addition of sialic acid to

GTH is known to protect GTH from clearance (Boime and Ben-Menahem, 1999). Since mannose is linked to GTH produced in the silkworm (Misaki et al., 2003), recombinant GTH produced in the silkworm might clear more rapidly from the circulation compared to native GTH with sialic acid. One of the solutions to increase the half-life of GTH would be making a single-chain GTH. Since dissociation of subunits inactivates heterodimeric GTH, a single-chain GTH could have a longer half-life *in vivo* (Boime and Ben-Menahem, 1999; Garcia-Campay and Boime, 2001).

CONCLUSION

A large amount of biologically active recombinant GTHs of goldfish and Manchurian trout could be produced by the baculovirus-silkworm larvae system. Expression efficiency of the baculovirus-silkworm larvae system is much higher compared to that of other systems used for producing fish GTH. By using recombinant GTHs produced in silkworm larvae, the following possibilities are suggested for the development of aquaculture research: 1) use of recombinant FSH and LH individually or in combination as substitutes of pituitary extracts, 2) elucidation of *in vivo* biological function of fish FSH and LH, and 3) development of a new efficient method of induction of gonadal maturation by differential use of recombinant FSH and LH based on physiological profiles of FSH and LH expression.

ACKNOWLEDGEMENTS

We thank Professor Shugo Watabe of the Department of Aquatic Bioscience, Graduate School of Agricultural and Life Sciences, the University of Tokyo, for his encouragement throughout this study. This study was supported in part by Grants-in-Aid from the Japan Society for the Promotion of Science for Scientific Research (MK, YH, GY) and for the Japan-Korea Basic Scientific Cooperation Program (MK), and grants from KOSEF for Joint Research Project under the Korea-Japan Basic Scientific Cooperation Program (YCS), and grants from the Ministry of Education, Sports, Science, and Technology for the Academic Frontier Project (2003-2007), International Christian University.

REFERENCES

Aizen, J., Kasuto, H., Golan, M., Zakay, H., and Levavi-Sivan, B. (2007). tilapia follicle-stimulating hormone (FSH): Immunocytochemistry, stimulation by gonadotropin-releasing hormone, and effect of biologically active recombinant FSH on steroid production. *Biol. Reprod.,* 76, 692-700.

Boime, I. and Ben-Menahem, D. (1999). Glycoprotein hormone structure-function and analog design. *Rec. Prog. Hormone Res.* 54, 271-289.

Choi, E.J., Ko, H., Shin, J., Kim, M. A., and Sohn, Y.C. (2005). Expression of gonadotropin genes in Manchurian trout *Brachymystax lenok* and production of recombinant gonadotropins. *Fish. Sci.* 71, 1193-1200.

Garcia-Campayo, V. and Boime, I. (2001). Novel recombinant gonadotropins. *TRENDS in Endocrinol. Metabol.* 12, 72-77.

Gen, K., Soyano, K., Okuzawa, K., and Kagawa, H. (2008). Expression of biologically active red seabream gonadotropins in a baculovirus system: comparison of two different strategies regarding their expression. *Cybium*, in press.

Hayakawa, Y., Morita, T., Kitamura, W., Matsumoto S., Banba, A., Nagaya, H., Komei Hotta, K., Sohn, Y. C., Yoshizaki, G., and Kobayashi, M. (2007). Biological activities of single-chain goldfish follicle–stimulating hormone and luteinizing hormone. *Aquaculture*, in press.

Ho, W.K.K., Meng, Z.Q., Lin, H.R., Poon, C.T., Leung, Y.K., Yan, K.T., Dias, N., Che, A.P.K., Liu, J., Zheng, W.M., Sun, Y., and Wong, A.O.L. (1998). Expression of grass carp growth hormone by baculovirus in silkworm larvae. *Biochim. Biophys. Acta* 1381, 331-339.

Kamei, H., Ohira, T., Yoshiura, Y., Uchida, N., Nagasawa, H., and Aida, K. (2003). Expression of a biologically active recombinant follicle stimulating hormone of Japanese eel, *Anguilla japonica*, using methylotropic yeast, *Pichia pastoris*.

Gen. Comp. Endocrinol. 134, 244-254.

Kamei, H., Kaneko, T., and Aida, K. (2006). *In vivo* gonadotropic effects of recombinant Japanese eel follicle-stimulating hormone. *Aquaculture,* 261, 771-775

Kasuto, H. and Levavi Sivan, B. (2005). Production of biologically active tethered tilapia LHβα by the methylotropic yeast *Pichia pastoris*. *Gen. Comp. Endocrinol.* 140, 222-232.

Ko, H., Park, W.D., Kim, D.J., Kobayashi, M., and Sohn, Y.C. (2007). Biological activities of recombinant Manchurian trout FSH and LH: their receptor specificity, steroidogenic and vitellogenic potencies. *J. Mol. Endocrinol.* 28, 99-111.

Kobayashi, M., Amano, M., Kim, M.H., Yoshiura, Y., Sohn, Y.C., Suetake, H., and Aida, K., (1997). Gonadotropin-releasing hormone and gonadotropin in goldfish and masu salmon. *Fish Physiol. Biochem.* 17, 1-8.

Kobayashi, M., Morita, T., Ikeguchi, K., Yoshizaki, G., Suzuki, T., Watabe, S., 2006. In vivo biological activity of recombinant goldfish gonadotropins produced by baculovirus in silkworm larvae. *Aquaculture*, 256, 433-442.

Maeda, S. (1989). Gene transfer vectors of a baculovirus, *Bombyx mori*, and their use for expression of foreign genes in insect cells. In: Mitsuhashi, J. (Eds), Invertebrate Cell System Applications, CRC Press, Boca Raton, FL, pp. 167-181.

Maeda, S. (1994). Expression of foreign genes in incect cells using baculovirus vectors. In K. Maramorosch, and A. H. McIntosh, A.H. (Eds), *Insect Cell Biotechnology*, pp.1-31, Boca Raton, FL, CRC Press.

Misaki, R., Nagaya, H., Fujiyama, K., Yanagihara, I., Honda, T., and Seki, T. (2003). N-linked glycan structures of mouse interferon-beta produced by *Bombyx mori* larvae. *Biochem. Biophys. Res. Commun.* 311, 979-986.

Morita, T., Yoshizaki, G., Kobayashi, M., and Takeuchi, T. (2003). Production of biologically-active recombinant goldfish gonadotropins in transgenic rainbow trout. *Fish Physiol. Biochem.* 28, 473-474.

Morita, T., Yoshizaki, G., Kobayashi, M., Watabe, S., and Takeuchi, T. (2004). Fish eggs as bioreactors: the production of bioactive luteinizing hormone in transgenic trout embryos. *Transgenic Res.* 13, 551-557.

Narayan, P., Wu, C., and Puett, D. (2000). Genetic engineering of single-chain gonadotropins and hormone-receptor fusion proteins. *Methods* 21, 59-66.

Pankhurst, N.W. and Porter, M.J.R. (2003). Cold and dark or warm and light: variations on the theme of environmental control of reproduction. *Fish Physiol. Biochem.* 28, 385-389.

Parhar, I.S. (2003). Gonadotropin-releasing hormone receptors: neuroendocrine regulators and neuromodulators. *Fish Physiol. Biochem.* 28, 13-18.

Patiño, R. (1997). Manipulations of the reproductive system of fishes by means of exogenous chemicals. *Prog. Fish-Cult.* 59, 118-128.

Sato, N., Kawazoe, I., Suzuki, Y., and Aida, K. (2003). Induction of vitellogenesis. In K. Aida, K. Tsukamoto, and K. Yamauchi (Eds), *Eel Biology*, pp.387-399. Tokyo, Springer-Verlag

Suzuki, T., Kanaya, T., Okazaki, H., Ogawa, K., Usami, A., Watanabe, H., Kadono-Okuda, K., Yamakawa, M., Sato, H., Mori, H., Takahashi, S., and Oda, K. (1997). Efficient protein production using a *Bombyx mori* nuclear polyhedrosis virus lacking the cysteine proteinase gene. *J. Gen. Virol.* 78, 3073-3080.

So, W.K., Kwok, H.F., and Ge, W. (2005). Zebrafish gonadotropins and their receptors: II. cloning and characterization of zebrafish follicle-stimulating hormone and luteinizing hormone subunits spatial-temporal expression patterns and receptor specificity. *Biol. Reprod.* 72, 1382-1396.

Sullivan, C.V., Hiramatsu, N., Kennedy, A.M., Clark, R.W., Weber, G.M., Matsubara, T., and Hara, A. (2003). Induce maturation and spawning: opportunities and applications for research on oogenesis. *Fish Physiol Biochem.* 28, 481-486.

Swanson, P., Dickey, J.T., and Campbell, B. (2003). Biochemistry and physiology of fish gonadotropins. *Fish Physiol. Biochem.* 28, 53-59.

Vischer, H.F., Granneman, J.C.M., Linskens, M.H.K., Schulz, R.W., and Bogerd, J. (2003). Both recombinant African catfish LH and FSH are able to activate the African catfish FSH receptor. *J. Mol. Endocrinol.* 31, 133-140.

Walsh, G., 2004. Second-generation biopharmaceuticals. *Eur. J. Pharm. Biopharm.* 58, 185-196.

Zmora, N., Kazeto, Y., Sampath Kumar, R., Schulzl, R. W., and Trant, J. M. (2007). Production of recombinant channel catfish (*Ictalurus punctatus*) FSH and LH in S2 *Drosophila* cell line and an indication of their different actions. *J. Endocrinol. 194*, 407-416.

Zohar, Y. and Mylonas, C.C. (2001). Endocrine manipulations of spawning in cultured fish: from hormones to genes. *Aquaculture* 197, 99-136.

In: Aquaculture Research Progress
Editor: Takumi K. Nakamura

ISBN 978-1-60456-247-7

Short Communication B

THE UNDERWATER SILVICULTURE APPROACH FOR REEF RESTORATION: AN EMERGENT AQUACULTURAL THEME

Shai Shafir and Baruch Rinkevich

ABSTRACT

Coral reefs, often portrayed as "the rain forest of the sea", provide sustenance to hundreds of million of people worldwide. When severely degraded, reef restoration is expected to uphold the ecosystem biodiversity and retain, at least partially, the original biological and ecological portraits. Such active reef restoration practices are based on aquaculture of coral colonies (mariculture them in underwater nurseries) in as much as silviculture depends on rearing young trees in terrestrial nurseries.

INTRODUCTION- WHAT UNDERWATER SILVICULTURE MEANS?

Coral reefs boast exceptional species diversity but are poorly protected, highly degraded, and exposed to multiple persisting threats (Bruno and Selig 2007). Due to their economic importance and aesthetic attributes, there is increasing academic and public awareness of the worldwide degradation of coral reefs and the associated decline in reef dwelling organisms (Bellwood et al. 2004; Feary et al. 2007). The decline in the abundance and diversity of reef fish through loss of structural reef heterogeneity is well studied. (Jones et al. 2004; Wilson et al. 2006). Not only has approximately one fifth of the world's reefs been significantly destroyed by anthropogenic activities but it also shows ominous prospect of recovery (Wilkinson 2004). A further half of the worldwide reefs is under threats of collapse or, at least, under longer-term degradation. Following the predictions for global change, it has been suggested that some coral reefs will change rather than disappear completely (Bruno and Selig 2007; Hughes et al. 2003), becoming different ecosystems, changing gradually into dissimilar ecological status and new consortia of reef associated vertebrates and invertebrates.

Coral reefs are often portrayed as "the rain forest of the sea". This assertion refers to their being one of the most productive biological ecosystems on earth, to their high level of biodiversity, and above all, to their function as the building blocks of the ecosystem framework (Connell 1978; Goreau et al. 1979; Epstein and Rinkevich 2001; Idjadi and Edmunds 2006). Coral colonies, as trees, share a common ecological role by virtue of their biological properties; both organisms constructed their ecosystems by engineering 3-D structures habitats. As the key organisms in their ecosystems, coral reefs and trees harbor numerous inhabitants and formulate local biological and environmental conditions (i.e., trees stabilize soil, enhance litter production and increase soil organic matter; corals enhance lithification processes and reef rock formation). As damaging anthropogenic activities are growing and intensifying alarmingly, coral reefs and rain forests are losing, their resilience and their recovery ability and self-rehabilitation are impaired. Clearing and logging forests and converting them to agricultural expanses intensify soil erosion and desertification. Similarly, marine recreational activities, pollution, over-fishing and the like, lead to progressive deterioration of coral coverage, collapse of reef communities, phase shift, and development of algal reefs (Lugo et al. 2000; Gunderson 2007; Hughes et al. 2007).

'Underwater silviculture' has a much broader perspective.

Forests cover 30% of the world land area, tending more than 50% of the terrestrial carbon and 60-90% of the animal and plant species. Noteworthy is the deforestation of tropical forests, which causes the lose of 14.2 million hectares a year (www1), sum up to nearly 50% of their area, which in turn, leads to an estimated 19% extinction of forest-associated endemic species (Wright and Muller-Landau 2006). Most forests (90%) in the temperate zone, especially those in the northern hemisphere, are the outcome of silviculture. 'Silviculture is the agriculture of trees; how to grow them, how to maximize growth and return, and how to manipulate tree species compositions to meet landowners' objectives' (Rinkevich 2006). Active rehabilitation of forests through silviculture has become the foremost mode of forest management (Charnley 2006), and planting practiced in 150 countries around the world account for 5% of the forests (www1). In contrast to forests, coral reefs cover less than 1% of the oceanic area (Hughes et al. 2003). However, this productive ecosystem supports almost one third of the world's marine fish species (Newton et al. 2007), yielding around 10% of the total fish consumed by humans (Smith 1978; Pauly et al. 2002), facilitating nursery sites for open sea fish fingerlings (Crépin 2003) and form the ground for tourism industry (Ahmed et al. 2007). More than 100 countries have coral reefs along their coasts, which sustain the livelihood of tens of millions of people (Salvat 1992). Unfortunately, coral reefs are declining much faster than forests, 1% annually before 1997 and 2% annually between 1997-2003 (Bruno and Selig 2007).

Silviculture and reef restoration (Epstein and Rinkevich 2001; Omori and Fujiwara 2004; Rinkevich 2005, 2006; Edwards and Gomez 2007; Skinner 2007) are complex biological and ecological procedures. Essentially, silviculture and reef restoration are expected to maintain the ecosystem biodiversity and retain, at least partially, the original biological and ecological portraits (Rinkevich 1995, 2000, 2006; Parviainen and Frank 2003). However, those disciplines differ in their targeted objectives. Many silvicultural practices favor regeneration of commercial species for the timber industry (Putz et al. 2001) whereas, silviculture endeavors to grow trees not only for timber but also for the environment, restoration of coral reef, at present, is designed solely for the environment. In contrast to silviculture practices and rationale, lack of state-of-the-art remediation protocols for reef restoration impedes the development of this ecological discipline. The following overview wishes to illustrate the practice of 'reef gardening' (Rinkevich 1995, 2000 2005, 2006; Shafir et al. 2006a,b) based on lessons learned from silvics and silviculture, their common features, divergent approaches and ubiquitous rationale as apply in terrestrial (forests) and marine (coral reefs) restoration principles.

The Gardening Concept in Reef Restoration - An Aquaculture Theme

The well-established discipline of silviculture led Rinkevich (1995) to suggest a new marine conservation concept that derives its rationale from silviculture, the two-steps protocol for the "gardening of the coral reef concept". The grounds for this concept is restoration of declined areas by means of active rehabilitation to the stage where conservation acts are sufficient (Rinkevich 1995, 2000, 2005, 2006). The first step in this "gardening concept" entails rearing coral "seedlings", in specially designed nurseries (Rinkevich 1995), to plantable size before applying the second step, rehabilitation (by active transplantation) of damaged area with the maricultured coral colonies (Rinkevich 1995).

Since target coral species need to be cultured in large numbers, the nursery step is pivotal to the success of the rehabilitation process. Careful maintenance ensures faster growth of healthy coral colonies. Issues like nursery structure and shape, temporary substrates for coral colony mariculture, realistic number of newly generated and maintained colonies, duration of nursery phase, growth rates and longevity of farmed colonies, and more (Shafir et al. 2006a,b), are some of the crucial issues associated with nursery care (Rinkevich 1995, 2000, 2005, 2006). The success of coral nursery depends of their initial set-up. The most successful nurseries studied up-to-date are those located in a mid-water, isolated, nutrient-enriched area, away from the natural reef, protected from skin and scuba divers or predation by corallivorous organisms (Shafir et al. 2006a). A protected nursery phase provides cultured corals with a farming period that ensures better survivorship and growth to a size suitable for transplantation. A successful yield of the nursery depends also on management and protocols that produce healthy and plantable coral colonies at profitable cost. The nursery step in the "gardening concept" is the phase drawing most scientific investigation, and which conceptual and technical issues have already been addressed (Shafir et al. 2006a,b; Amar and Rinkevich 2007). The transplantation phase, which is in its preliminary steps, has already revealed interesting results, but its applications are still being developed (unpublished). Both phases are comparable to agricultural concepts and raise new themes in the discipline of farming of marine organisms.

Since the nursery stage in silviculture, in which trees are raised from seeds and seedlings, has been subjected to rigorous scientific testing for more than a century (Dumroese et al. 2005), its concepts and management tools are well established. The theories and practices shared by the "gardening concept" and silviculture led us to approach coral nursery with tree nursery concepts (Rinkevich 2006). Some of the common agricultural aspects are outlined below even though basic protocols have not yet been thoroughly studied.

Economic Considerations- Ecological Goods and Services

The economic value of an ecosystem is a practical tool for consistently evaluating different ecosystems (Naidoo and Ricketts 2006). Goods and services of each ecosystem are classified according to their direct and indirect contribution. For example, timber, fuel and tourism are considered as major direct goods and services derived from the forest ecosystem, whereas indirect benefits are protection of land and carbon storage (Pearce 2001). The total economic value of global forests exceeds US$4.7 trillion year^{-1} and the temperate/boreal forests economic value are evaluated at US$ 1719.2 ha^{-1}year^{-1} (Krieger 2001). There has yet to be developed a single, universal, and uniform model for economic evaluation that will serve as common ground for restoration economy.

However, this is not the case with coral reefs. As specified, coral reefs provide hundreds of million of people with goods and services. These include coastal protection, reef mining, food and proteins, agar and carrageenan, recreational activities, major source material for marine aquarium trade, ornamentals, new drugs for the pharmaceutical industry, raw material (coral skeletons) for medical purposes and cultural benefits (Moberg and Folke 1999; Ahmed et al. 2007). Coral reefs' services are vital for many less developed countries. For example, coral reef fisheries in the Philippines provide livelihood for more than a million small-scale fishermen who contribute almost US$ 1 billion annually to the country's economy (White et al 2000). As reefs come in a great variety of forms, this is not simple to appraise their economic value or to assess the economic outcomes of reef degradation (Moberg and Folke 1999; Subade 2007). The benefits of each square kilometer of coral reef are often valued by parameters like the "willing to pay" factor (which we think is wrong as not based on robust economic considerations), the "total economic value" or the "biodiversity production values" (Subade 2007). For assessing the preferable extent of reef restoration or compensation additional methodologies like 'habitat-equivalent analysis' (Milon et al. 2001, Spurgeon 2001; Precht, et al. 2002) are used. The evaluations of coral reefs therefore vary between reef areas and between estimation methods. The economic value of Hawaii's reef area is estimated at US$ 878 ha^{-1}year^{-1} (Cesar and van Beukering 2004) whereas the economic value of Tubbataha Reefs, the Philippine is estimated at less than US$19 ha^{-1}year^{-1} (Subade 2007). Some methods estimated the economic value of the worldwide reefs to be between US$231 and US$2,700 ha^{-1}year^{-1} (Bruno and Selig 2007). NEED more discussion on the economic values. Examples: values for developed vs. undeveloped countries, future lost

Economic Considerations- Active Restoration Issues

When weighing the economic factors of coral reef restoration, one should favor the priorities of the underdeveloped countries by adopting "low-tech" methodologies that address the type of reef damages and the needs most common in local rural communities (Lindahl 1998; Bowden-Kerby 2001; Rinkevich 2005). Any economic estimation is bound to fluctuate because of the dynamic nature of the coral reef environment (i.e., lose/recruitment of coral colonies, reef invertebrates and fish, natural catastrophes). Anthropogenic loses associated with coral reefs (like fisheries, trade values of coral colonies and other exotic invertebrates or fish.) are not easy to quantify since, in most places, coral reefs are held in public trust by local, or governmental, agencies (Rinkevich 2005), that are not interested in worldwide point of view. Therefore, greater awareness has been put (Spurgeon 2001) for better economic assessments which will improved effectiveness of restoring coral reefs. Innovative economic templates guiding restoration initiatives are urgently needed (Rinkevich 2000).

Some of the economic considerations mirror those of silviculture. Active reef restoration is based on aquaculture of coral colonies (farming them in underwater nurseries) as silviculture depends on rearing young trees in terrestrial nurseries. However, while silviculture has been undergoing rigorous scientific testing for nearly two centuries on a global scale (Graham and Jain 1998), the budding discipline of active reef restoration is less than a decade old. Silviculture is the applied restoration act for rehabilitating terrestrial habitats, where natural self-regeneration of forests is not feasible (Putz et al. 2001). The same applies to the reef arena. Active reef restoration is drawing increasing attention because nearly all other practices for conserving degrading reefs have failed to yield significant results(Rinkevich, 1995; Risk, 1999; Epstein et al., 2001, 2003, 2005; Edwards and Gomez 2007). Such restoration acts carry economic significance. . For example, conserving Southeast Asia reefs is estimated to be worth \$23,100– \$270,000 km^2/year (Burke et al. 2002).

Active restoration deals with scientific questions and with technical challenges. The scientific part focuses on issues like genetics, species combination, landscape manipulation, biological engineering and key species (Palmer et al. 1997; Naveh 2007). The technical part focuses on cost, rates of success in each technical step, maintenance, number of new transplants and their survivorship after transplantation. This overview deals with the first phase of the 'gardening concept', the nursery phase, we outline below some aquaculture perspectives, comparing tree nursery (rearing trees for the timber industry and for reforestation) with the knowledge gained for mid-water coral nursery (rearing coral colonies for rehabilitation of decay reef areas. Efficient nursery maintenance practices (Howell and Harrkngton 2004) are vital to the success of the entire restoration process.

In: Aquaculture Research Progress
Editor: Takumi K. Nakamura
ISBN 978-1-60456-247-7

Chapter 1

ORAL DELIVERY OF IMMUNOGENS IN AQUACULTURE

I.S. Azad, A. Panigrahi, [1], R.A. Dalmo, [2] and K.M. Shankar [3]

Mariculture and Fisheries Department, Kuwait Institute for Scientific Research, PO Box-1638, Salmiya-22017, Kuwait

1. Central Institute of Brackishwater Aquaculture 75-Santhome High Road, Raja Annamalai Puram, Chennai – 600 028.
2. Institute of Marine Biochemistry, Norwegian College of Fishery Science, University of Tromsø, N-9037 Tromsø, Norway
3. College of Fisheries, Mangalore - 575 002. India

ABSTRACT

Intensification and expansion of aquaculture activities on one hand, emerging problems of diseases on the other pose challenges to the sustained production from aquaculture activities. Increasing awareness on deleterious effects of antibiotics and aquaculture chemicals, leading to the development of bacterial drug resistance and chemical contaminations of the environment, has forced the researchers and planners to look for measures that are host and environment friendly. Immune enhancement of the host is seen as the most effective and long lasting solution for countering the problem of diseases. Delivery of immunogens through various routes has been studied extensively. However, the most natural and non-stressing route of oral delivery has attracted the attention of fish immunologists in recent years. Oral delivery of immunogens such as vaccines, immunostimulants, probiotics and nutritional immune enhancers is being studied with the ultimate aim of enhanced immunity. With a brief background on the digestive interactions that could influence the putative immunogens, this chapter analyses various investigations carried out in the field of oral immunogens in aquaculture and suggest possible future strategies for an improved aquaculture productivity.

Keywords: Aquaculture; Oral immunogens; vaccines; immunostimulants; probiotics

INTRODUCTION

Aquaculture contributes substantially to world fish production, the share of which has shown a steep rise due to intensification, in recent years. A major portion of the aquaculture produce comes from the tropics and the sub-tropics, which heavily depend on carp cultivation [1]. One of the gravest of the problems associated with intensive aquaculture is disease. Infection due to a variety of bacteria [2-17], fungi [18-26], viruses [27-31] and parasites [32-36] have been frequently observed. Avoidance, prophylaxis and chemotherapy are in practice for the management of diseases. However, most of the drugs and chemicals used in aquaculture have residual effects on the aquatic environment. Some of the insecticides and biocides used in aquaculture are known to accumulate and concentrate in aquatic organisms while, antibiotics have been known to induce resistance in pathogens [37-41]. Thus, immune enhancement through the application of vaccines, immunostimulants, probionts and immune modulator nutrients have become the most promising and effective methods of preventing diseases in aquaculture (Figure 1). The specific immunity is less competent in the early developmental stages of fish and thus non-specific immune modulation through the use of immunostimulants and probionts is viewed as a suitable method of disease management in the larvae and fry [42-46].

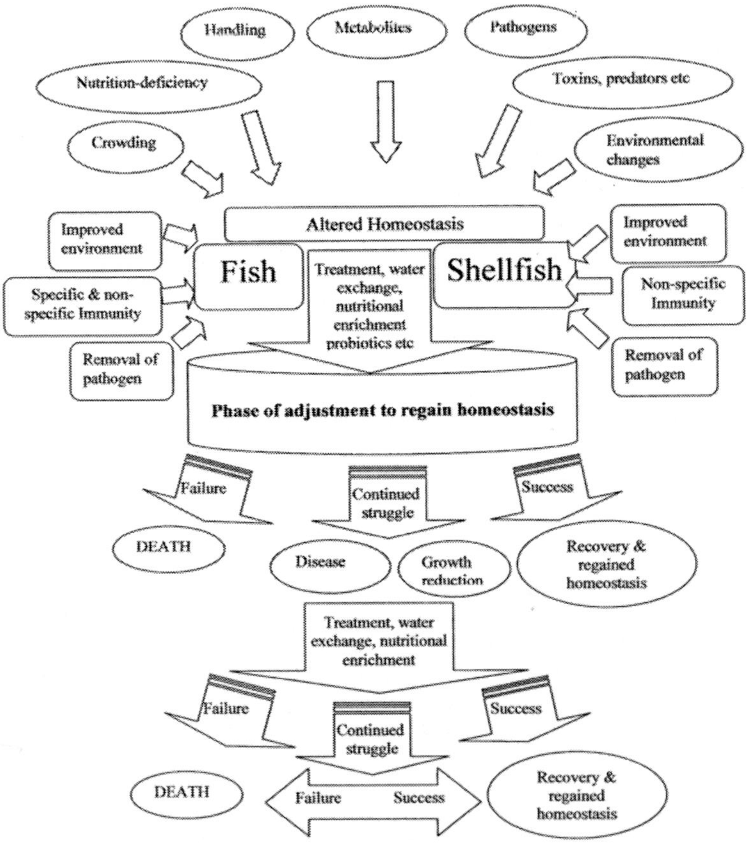

Figure 1. Schematic diagram of possible events under aquaculture and out comes representing complex interactions of host-environment-pathogen.

Immune modulation through external interventions in the form of dietary immunogens (nutritional components in supra-dietary levels have often resulted in enhanced survival and disease resistance), immunostimulants, probiotics and vaccines have been viewed as potential measures of increasing survival and diseases resistance among aquaculture species. In this context, a view at the general physiology of the digestive system in fish and shellfish helps us in understanding the art and science of oral immune modulation in these organisms.

Nutritional intervention aimed at enhanced immunity, and thus reduced infection, is one of the ways of immune enhancement in cultured aquatic organisms. Investigations on the dietary requirements of cultured aquatic animals were concerned primarily with the factors related to growth and reproductive physiology in cultured fish and shellfish. However, it was later realized that the nutritional supplementation, beyond the dietary requirements, helps the cultured aquatic animal in absorbing the stress shock and withstanding the infection by opportunistic pathogens that assume serious proportions in an immune compromised host [47-55]. Several investigations were carried out to elucidate the immunomodulatory effects of unsaturated fatty acids, amino acids, vitamins, micronutrients and other dietary constituents of fish [50, 51, 56-61].

The use of immunostimulants has received increased attention due to the discovery of Toll-like receptors (TLR) or/and pattern recognition receptors (PRR) [62]. These receptors have been found to bind molecules from a range of pathogens including self-molecules. Factors released in the event of damage to the cellular integrity act as so-called "danger" signals possessing pathogen-associated molecular patterns (PAMPs). These danger signals often consist of repeating molecular moieties yielding high molecular weight compounds. Examples are ß-glucans and DNA, but some danger signals possess low molecular weight structures. It has been found that the PPR bind unit structures of PAMPs, and that PAMP-binding involves several other humoral and cell membrane proteins, exemplified by the more or less simultaneous LPS recognition displayed by MD-2, CD-14 and Toll-like receptor 4 that occurs on the cell membrane. Also, the binding of ß-glucans has been shown to include several different cell membrane receptors. Immunostimulants are more often commercially exploited in aquaculture as feed additives. This applies to ß-glucans, alginates and nucleotides. Despite their use as feed additives no targeted investigations have been conducted to include PAMPs as adjuvants in fish vaccines. Interestingly, most of the PAMPs studied activate antigen presenting cells together with naïve T-cells into dendritic cells and Th1 or Th2 cells [63]. In turn, this may activate Th1 and Th2 immune responses with production of Th1 or Th2 signature molecules such as IFN-gamma and IL-4, respectively [64-66]. This review will mainly focus on binding characteristics of ß-glucans to pattern recognition receptors (PPRs), the effects on T helper cell differentiation, out-put gene expression profiles after receptor binding and application of ß-glucan in the aquaculture sector.

With the emergence of shrimp aquaculture in the Asian and Latin American countries, attention of researchers was attracted towards investigations on growth and immune enhancement in shrimp. Use of immunostimulants and probiotics has been more pronounced in shrimp aquaculture than that in finfish. This was necessitated, on one hand, due to the spread and severity of viral diseases in shrimp and on the other due to the non-specificity of immune responses in shrimp. Fungal extracts, bacterial whole cell preparations and several herbal extracts were used for the sole purpose of enhancing disease resistance in shrimp and prevent or reduce losses due to diseases [43, 67-71].

Another important component of health management that has attracted the attention of aquaculturists and researchers is the application of probiotics. Delivery of probiotics has to be done through oral routes as the main aim of this health management strategy is to deliver the live probiotics bacteria to bring about effective colonization of the gut and competent exclusion of the pathogen [72].

Marine and freshwater fish aquaculture is the best benefited sector following the development of vaccines against different bacterial and viral pathogens. There are three widely used routes of immune enhancement in fish *viz.,* injection, immersion and oral. The injection route involves a laborious and time-consuming process of injecting individuals. Size limitation is another bottleneck in the use of this method. Immersion also involves some amount of stress, inconvenience of handling large sized fish and requires larger volumes of vaccine/immunogen [73]. In this context, the oral route has an edge over the other two methods as it is suitable for fish of all sizes, imposes no handling stress and hence, does not interfere with the routine husbandry operations. However, the major problem associated with oral delivery of immunogens is poor and inconsistent immune response and protection, mainly due to the destruction of antigen/immunogen in the stomach and foregut before it reaches the immune responsive lower gut [74]. Hence, there is a need to develop methods that help in delivering immunogens into areas of gut withstanding the acid and enzyme activities with minimal loss of immunogenic epitopes.

For the purpose of discussion, we have divided the chapter into subheads as: dietary immunogens (dietary ingredients beyond the nutritional requirements), immunostimulants (bacterial, fungal and animal extracts), probiotics (live bacterial feed supplements) and vaccines (attenuated pathogens or their extracts, recombinant immunogenic peptides and DNA vaccines). Digestive system of fish and shellfish has been briefly described for a better understanding of the dynamics and role of oral immunogens in aquaculture.

Digestive Systems of Fish and Shellfish – An Overview

Fish

Fish being the first link in the vertebrate evolution; it possesses all the digestive basic structure and functions that are seen in higher vertebrates. Fish can be broadly grouped into herbivores, omnivores and carnivores. Generalized picture of different digestive tracts has been delineated that depicts a comparative analysis of the fish gut [75]. We have briefly described the conceptual representation of the different gut environments seen in fish with different feeding behaviours (Figure 2).

Break down of food starts in the mouth, especially in herbivorous fish. Carnivores may tear or crush harder animal material in the mouth and or pharynx/oesophagus. Though some amount of mucus is secreted in the mouth, enzymatic digestion, at this stage has not been conclusively demonstrated. The process of reducing the digested food into molecules small enough to facilitate their passage into the blood stream across the gut wall is called digestion. Proteins to peptides and or amino acids, carbohydrates to simple sugars and lipids to fatty acids and glycerols can be generally visualized during the process of digestion in fish.

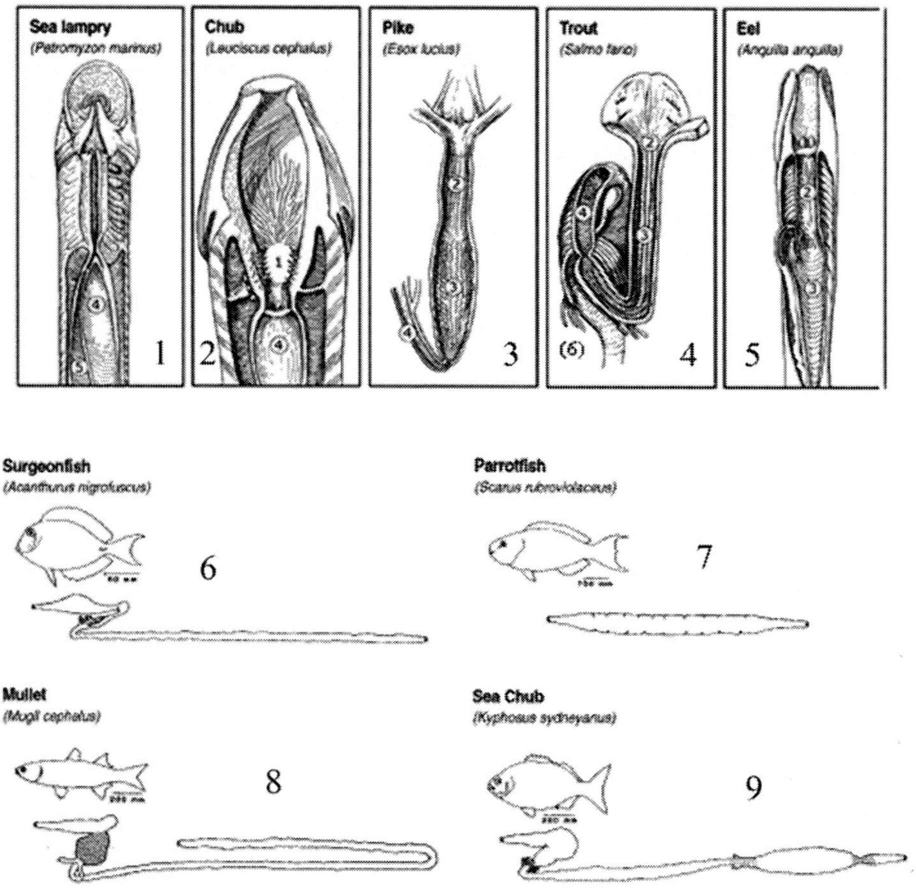

Figure 2. Digestive tract of fish. The digestive articulated jaws are absent in cyclostomes (1), such as the lamprey, and located in the pharynx of some species such as the chub (2). The oesophagus varies in length and the stomach is absent in cyclostomes and some advanced species, such as the chub. Where present, the stomach may be straight as in pike (3), U-shaped as in trout (4), or Y-shaped with a gastric cecum as in the eel (5). The absorptive surface and digestive retention time of the midgut is increased by a spiral valve or pyloric ceca in a number of species. Digestive strategies of herbivorous marine fish; surgeonfish (6) and parrotfish (9) are browsers. Mullet (7) and sea bass (8) are grazers. Gizzard-like stomach of the mullet, pyloric ceca of surgeonfish and sea bass, and two regions of sphincters in distal intestine the seabass have been described representing different feeding habits in fish [76].

Complex sugars such as starch gets digested poorly (5-15%) while, glucose is digested almost completely by the fish. Protein and lipid digestibility in fish varies between 80 and 90 % [77]. Protein digestion in channel catfish is known to occur throughout the gut length with more than 60% digestion occurring in the stomach and more than 25% of the proteins are digested in the hindgut [78]. Pepsin in the stomach and trypsin in the middle and lower gut are the important digestive enzymes in most teleosts. Stomach produces acidic environment (pH ranging between 1.5 and 4) through the secretion of hydrochloric acid facilitating digestion by pepsin. Hence, immunogenic molecules should be protected against digestion in the stomach so that they reach immune-responsive lower gut for mounting optimum immune response. Absorption of soluble food, probably starts from the stomach [77] with predominant absorption taking place in the mid and hind gut portions. Trypsin appears to be the major protein digesting enzyme in the mid and hind gut of fish. Pancreatic collaginase [79], enterokinase of the intestinal wall, chitinase etc., are some of the other digestive enzymes that may play some role in the digestive physiology of fish. Increased numbers of leucocytes in the general circulation following a meal have been demonstrated in sea bream [80].

The age for effective handling of immunogens by the developing fish larvae has been the matter of primary concern for fish immunologists who started working on oral vaccines and immunogens for enhancing larval survival and reducing subsequent losses due to diseases. Several fish were studied revealing valuable information on the gut physiology and probable role of gut-associated lymphoid tissue (GALT) in the handling of oral immunogens. Unlike their terrestrial counterparts, the aquatic vertebrates are constantly exposed to "non-self" molecules and agents forcing the immune system to react. Extensive exposure of fish and shrimp immune system occurs through gills. This is probably the reason why the gills of the young hatchlings of fish have macrophage population that negotiates with the non-self molecules and pathogenic agents through enhanced phagocytic capabilities [81-84]. An interesting and detailed investigation on the role of macrophages in the very early developmental stages of zebra fish embryo was provided by Herbomel *et al* [85]. They showed that the macrophages play a very important role in the clearance of injected bacteria and also produced evidences that the entire macrophage population acquired activated status. Gills, skin and gut have been known to serve as the main sites of active phagocytic activity in many fish after hatch and before the lymphoid organs become functional [86]. The foregut and mid gut of many pelagic fish larvae showed active lipid absorption by the epithelium and pinocytic protein absorption in the hind gut [87-90].

It is clear from the investigations, thus far conducted, that the gut of fish plays an important role in the immunogenicity of oral immunogens in aquaculture providing the most natural mode of administration of immunogens.

Shrimp

A detailed review of digestion in crustaceans was reported by Cecaldi [91]. The digestive system of an adult crustacean (Figure 3) consists of a digestive tube divided into three regions: foregut, midgut and hindgut. Oesophagus and masticating stomach portions occupy the foregut while, the midgut contains the tubuled hepatopancreas (secretes digestive enzymes) and the hindgut is a straight tube ending into rectum and anus. Oesophagus, generally short, connects the mouth (M) to the stomach (S). The stomach with gastric mill and the midgut with digestive gland (hepatopancreas, H) complete the breakdown of food into particles and the enzymatic digestion. Some of the crustaceans contain midgut cecum (MGC) of variable length. Chemical digestion probably starts from the stomach after the physical reduction of particle size and then the enzymatic digestion continues in the midgut followed by nutrient absorption occurring in mid and hindgut portions. A hindgut cecum (HGC) is also seen just before the tail segment.

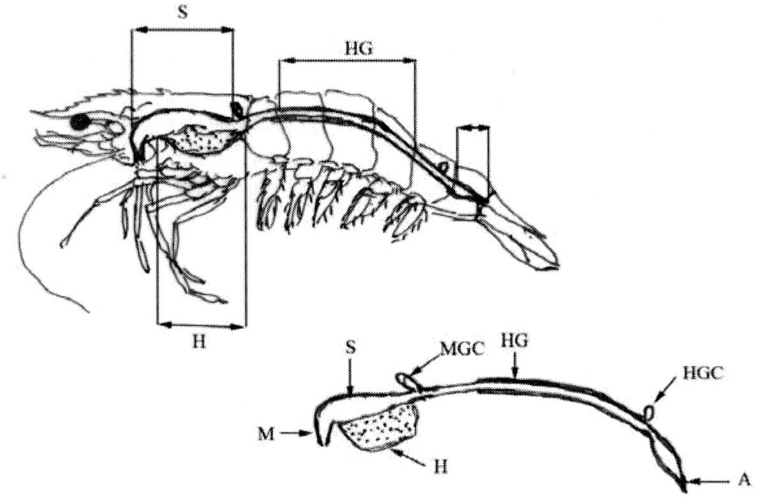

Figure 3. Digestive tract of Shrimp.

Digestive enzymes in different decapods have been studied [92]. Enzymes such as carboxypeptidases A and B and tryptic activity were studied by Gates and Travis [93, 94] and by the team of Galgani [95]. Collaginase activity in the crab, *Uca* sp. [96], principal proteolytic enzyme, trypsin [97] and other aspects of trypsin activity and composition [98, 99] have been reported from different crustacean decapods. Amylase in *Palaeomon serratus* [100], chitinases and chitobiases that are required for the digestion of the exoskeleton of the pray or their own after ecdysis have been reported [101, 102]. Enzyme profile and the gut physiology have to be taken into account while designing immunogens in shrimp aquaculture for effective immune modulation.

DIETARY IMMUNOGENS

In nature, aquatic animals have the liberty of escaping the threats from pathogens, predators, choosing right food and environment. All these liberties are either restricted or absent in captivity and confinement of aquaculture environment. Additionally, the stress of crowding, increased metabolites, right environment for the proliferation of opportunistic pathogens, dietary insufficiencies etc put the aquatic organism into a state of susceptibility. Hence, in addition to fulfilling the nutritional sufficiencies, supplementation of dietary components at supra nutritional levels help the species in aquaculture not only grow normally but withstand surges in stresses of various nature including diseases. Ever since demonstration of the importance of balanced nutrition and nutritional immunogens [50, 52, 103], this facet of immune modulation in different aquaculture species has attracted the attention of fish immunologists and fish nutritionists alike. Different nutritional components have been studied giving detailed information on the immune enhancement in cultured aquatic organism (Table 1). Tacon [104] listed important nutrients and dietary components that are reported to improve the health and/or disease resistance (including immune response) of farmed finfish and crustaceans. These include essential amino acids (lysine, methionine, tryptophan, arginine, histidine, leucine, isoleucine), nucleotides, polysaccharides (peptidoglycans, beta 1,3 and 1,6 glucans, lipopolysaccharides), essential fatty acids (18:2n-6, 18:3n-3; 20:4n-6, 20:5n-3, or 22:6n-3, depending upon species), sterols, phospholipids, essential minerals (P, K, Mg), trace elements (Fe, Zn, Mn, Cu, Se, I), vitamins (B1, B2, B6, B12, pantothenic acid, niacin, biotin, folic acid, inositol, choline, D3, A, K3, E, and C) and carotenoids.

Fatty Acids

Dietary fatty acids are essential for vital cell functions and in the maintenance of cellular integrity. Omega–3 polyunsaturated fatty acids (PUFA) provide strength and fluidity to cellular membranes and thus making them resistant to lysis. Problems associated with the oxidation of dietary poly unsaturated fatty acids have to be countered through the incorporation of anti-oxidants such as vitamins C and E in the diet [105, 106]. Glutathione peroxidase and superoxide dismutase activities depend on the availability of selenium, zinc and copper [107]. It will be difficult to fix fatty acid requirements of a given fish species based on the established fatty acid requirement of another fish species. Cold-water fish have higher n-3 PUFA requirement while, warm-water fish require higher n-6 PUFA content in their diet [108]. Head-kidney macrophage activity was enhanced through higher levels of dietary fatty acids in catfish [109, 110]. Dietary n-3 fatty acids enhanced non-specific cytotoxic cell (NCC) activity in the head-kidney of rainbow trout [111]. Source and composition of essential fatty acids is also known to influence the immune responses in fish. Replacement of fish oil as the source of dietary fat has produced, in many cases, either low survival or higher susceptibility to diseases. A mix of fish and corn oil produced better non-specific immune responses in the juveniles of grouper *Epinephelus malabaricus* than did the individual oils [112].

Role of vitamins, amino acids and PUFA in the diet of cultured shrimp at levels above those required for the normal growth and reproduction for a sustained and alert sate of health has been emphasized by many groups of researchers working in this area. The World Bank-FAO-NACA thematic report submitted by Tacon [104] also viewed the role of vitamins and other nutrients as important dietary components for better health.

Vitamins

Infection, stress and pollution can result in the depletion of anti oxidant molecules such as DNA, proteins, lipids and vitamins. Vitamins and minerals are the most important of all dietary immune modulators due to their complex interactive involvements in both normal physiology and immune modulations. Dietary requirement for tocopherol, thiamine and pyridoxine has been reported to increase with increasing dietary concentrations of polyunsaturated fatty acids, carbohydrate and protein, respectively. However, there are several instances where fish or shrimp under aquaculture, with suitable gut microflora, can synthesize most B vitamins, pantothenic acid, biotin, choline, inositol and vitamin K. Two of the extensively studied vitamins in aquaculture are vitamins C and E.

Table 1. Nutritional immunogens tested in fish and shellfish (A modification of the source: Gallindo-Villegas, J. and Hosakawa, H. 2004)

Immunogen	Fish / Shellfish	Dose (mg/kg, unless stated)	Response	Disease resistance	Reference
PUFAs	Atlantic salmon (*Salmo salar*)	300-5200	No change	Yes	[142]
Vitamin C (or Sodium salt)	Grouper (*E. malabaricus*)	30g/ kg feed	Leucocyte O_2^- ↑ Lysozyme ↑ Complement ↑	-	[112]
	Atlantic salmon (*S. salar*)	5000	Antibody ↑	No	[388]
		2980	Antibody ↑	-	[105]
		4000	Lysozyme ↑	Yes	[52]
	(*Oncorhynchus tshawytscha*)	2000+800 Vitamin E	Macrophage activity ↑	High survival against VHS challenge	[135]
	Coho salmon (*O.kisutch*)	400-1000	Improved wound healing	-	[389]
	Gilthead sea bream (*Sparus aurata*)	2900 + 1200 Vitamin E	Lysozyme ↑ NCCS ↑	No	[400]
		3000	Complement ↑	-	[139]
		250	Tolerant to crowding ↑	-	[391]
	Japanese seabass (*Lateolabrax japonicas*)	489	Growth Lysozyme ↑	Yes	[132]
	Japanese flounder (*Paralychthys olivaceus*)	6100	NBT ↑	Yes	[392]

Immunogen	Fish / Shellfish	Dose (mg/kg, unless stated)	Response	Disease resistance	Reference
Vitamin C (or Sodium salt)	Red sea bream (*Pagrus major*)	10000	Phagocytosis ↑	-	[393]
	Turbot (*Scophthalmus maximus*)	300-2000	Phagocytosis ↑ Lysozyme ↑	-	[390]
	Yellow perch (*Perca flaviscens flaviscens*)	250	Survival ↑	Yes	[118]
	Milkfish (*Chanos chanos*)	1500	Antibody ↑ Memory Factor ↑	Yes	[143]
	Indian major carp (Rohu, *Labeo rohita*)	200	Phagocytosis ↑ Lymphokine ↑	No	[140]
	Freshwater Prawn (*Macrobrachium rosenbergii*)	135	Survival ↑	-	[395]
	Asian catfish (*Clarias batrachus*)	500	Super O_2^- ↑ Haemagglutination ↑ Complement ↑	Yes	[194, 396]
	Channel catfish (*Ictalurus punctatus*)	3000 + Iron supplemented	Haematocrit ↑ Plasma haemoglobin ↑ Macrophage migration ↑	-	[138]

Immunogen	Fish / Shellfish	Dose (mg/kg, unless stated)	Response	Disease resistance	Reference
Other vitamins					
Pyridoxine	Atlantic salmon (*S. salar*)	5	No change	No	[399]
	Chinook salmon (*O. tshawytscha*)	> requirement	-	Yes	[403]
Vitamin A	Gilthead sea bream (*S. aurata*)	50-300	Reactive O_2 species	-	[397]
	Grass shrimp (*P. monodon*)	1500-15000 Retinol Equivalent units/kg	-	Better survival	[405]
Micronutrients and other additives					
Chitin	Gilthead sea bream (*S. aurata*)	25-100	Cytotoxicity Phagocytosis	-	[404]
Nucleotides	Yellow tail (*S. quinqueradiata*)	4	-	-	[235]
	Rainbow trout (*O. mykiss*)	0.03% of diet	-	Better survival to Vibrio, sea lice and ISA virus	[184]
	Turbot (*S. maximus*)	0.4% diet	RAG and immunoglobulin	-	[182]
Levamisole	Rainbow trout (*O. mykiss*)	5 mg/kg (i.p.)	Phagocytosis and NK activity	-	[193]
	Asian catfish (*C. batrachus*)	200mg/kg (i.p.)	Immune response in immune Compromised fish	-	[194]
Eclipta alba (leaf extract)	Tilapia (*Oreochromis mossambicus*)	0.01 to 1% in diet	Reactive O_2 species Myeloperoxidase	RPS of 64-75 against A. hydrophila	[196]

Immunogen	Fish / Shellfish	Dose (mg/kg, unless stated)	Response	Disease resistance	Reference
Vitamin E (α Tocopherol)	Gilthead sea bream (*S. aurata*)	1200	Lysozyme NCCS	-	[400]
	Japanese flounder (*P. olivaceus*)	600	Lysozyme Phagocytic activity	Yes	[392]
	Turbot (*S. maximus*)	500	Phagocytic activity	-	[394]
	Grouper (*E. malabaricus*)	≥100	Leucocyte O_2^- Lysozyme Complement	Yes	[112]
	Japanese seabass (*L. japonicas*)	250	Complement Stress tolerance	-	[391]
	Yellow perch (*P. flaviscens*)	5-160	Survival	Yes	[118]
	Golden shiner (*Notemigonus crysoleucas*)	38 + ≥ 98 Vitamin C	Alternate complement	-	[131]
	Milkfish (*C. chanos*)	50	Antibody Memory Factor	Yes	[143]
	Rohu (*L. rohita*)	100-200	Erythrocyte fragility TBARs value	Better survival	[398]
	Rainbow trout (*O. mykiss*)	≥100 with n-3 HUFA	Complement Phagocytosis Non-specific cytotoxicity	-	[401]
	Grass shrimp (*P. monodon*)	50-200	Superoxide dismutase TBARs	Better survival	[402]

C and E have been known to act in a more or less synergistic manner. Wound healing, haematopoiesis, detoxification of toxins, formation of collagen, role as antioxidant in oxidative metabolism of fats and several other functions in the physiology of fish are both directly and indirectly influenced by vitamin C. Disease resistance in fish due to dietary vitamin C has been well documented [60, 113]. Enhancement of immune function such as stimulated haemolytic complement activity, proliferation of immune cell, Phagocytosis, release of cell signal substances and antibody production [113] have also been reported. Stress, a natural phenomenon in aquaculture, is known to reduce growth [114] adversely affect immune responses in several species such as Atlantic salmon [115], common carp [116] and gilthead sea bream [117]. Growth and survival enhancement of several fish species in aquaculture due to dietary vitamin C has been reported by several investigators. Better growth and feed efficiency in yellow perch *Perca flaviscens* [118], common carp [119], channel catfish [120], juveniles of Korean rockfish [121] and hybrid tilapia [122] has been reported after dietary supplementation of vitamin C. Deficiency of vitamin C resulted in an increased disease susceptibility in several fish species such as rainbow trout [123, 124], channel catfish [125] and Atlantic salmon [48].

Vitamin C is essential in the diet of many fish to counter the adverse effects due to its deficiency and derive beneficial effects from optimized dietary levels. Most teleosts are unable to synthesize ascorbic acid due to the lack of L-gulonolactone oxidase responsible for synthesis of vitamin C *de novo* [126, 127]. Different levels of vitamin C in a variety of cultured fish species have been evaluated by different researchers (Table 2).

Table 2. Recommended vitamin concentrations in commercial shrimp feeds [104]

Vitamin	Quantity/kg feed	Mineral	Quantity/kg feed
Thiamin	50 mg/kg	Calcium	2.3%
Riboflavin	40 mg/kg	Phosphorus (available)	0.8%
Pyridoxine	50 mg/kg	Phosphorus (total)	1.5%
Pantothenic acid	75 mg/kg	Magnesium	0.2%
Niacin	200 mg/kg	Sodium	0.6%
Biotin	1 mg/kg	Potassium	0.9%
Inositol	300 mg/kg	Iron	300 ppm
Choline	400 mg/kg	Copper	35 ppm
Folic acid	10 mg/kg	Zinc	110 ppm
Cyanocobalamine	0.1 mg/kg	Manganese	20 ppm
Ascorbic acid1	1,000 mg/kg (100)1	Selenium	1 ppm
Vitamin A	10,000 IU/kg	Cobalt	10 ppm
Vitamin D	5,000 IU/kg		
Vitamin E	300 mg/kg		
Vitamin K	5 mg/kg		

Fatty acid requirements: 18:2n-6 (0.4% feed), 18:3n-3 (0.3% feed), 20:5n-3 (0.4% feed), 22:6n-3 (0.4% feed) Source: http://govdocs.aquake.org/cgi/reprint/2003/1201/12010440.pdf [164].

Though recommended dietary levels of vitamin C range from 20-50 mg ascorbic acid per kg diet [128], several authors have shown that supra dietary levels of this vitamin produced enhanced immune responses, more importantly non-specific immune responses. Lysozyme and complement activity, phagocytic activity and respiratory burst have been reported in different fish species [49, 125, 129-132]. Metabolic rate is considered as one of the important factors apart from the capabilities of biosynthesis of vitamin C that regulates the requirements of the vitamin in the diet [119]. Role of non-specific immune enhancers in fish and shellfish is of prime concern as larval stages of fish and the entire life cycle in shrimp rely on non-specific defences [133]. The vitamins C and E have an interactive sparing action on each other and hence are considered often together for dietary inclusions. Vitamin E also serves as a biological antioxidant in lipid metabolism and prolongs the life of erythrocytes and plays an important role in cellular respiration [134]. Dietary vitamin C prevented the appearance of vitamin E deficiency symptoms in Atlantic salmon in a dose-dependant manner [61]. Interaction between vitamin C and E in yellow perch (*P. flavescens*) resulted in regeneration and or sparing action of vitamin C on vitamin E [118]. Similar interactions between the dietary vitamins C and E have been reported in many fish species such as rainbow trout [135] and lake sturgeon [136]. Almost all studies [118, 125, 130-132, 137-140] carried out on the dietary vitamin C have suggested a positive immune (non-specific) modulator effect in fish except some of them [141, 142] which reported no beneficial immune modulation with higher levels of vitamin C.

Enhancement of specific immunity has also been reported from different fish fed varying dietary levels of vitamin C. Enhanced antibody production was reported in salmonids following dietary administration of vitamins [49, 52]. Enhancement in the antibody response was recorded in the case of channel catfish against *Edwardsiella ictaluri* [125]. Similarly, dietary vitamin C showed a direct and positive relation with the disease resistance and humoral antibody production in rainbow trout [51]. Antibody production, protective response and immunological memory of juvenile milkfish was enhanced by supra dietary vitamin C levels [143]. The nut-shell beneficial effects of supra dietary levels of vitamins on the immune system of fish can be explained by a buffering action on the reserves of the antioxidant molecules, enhancement of pahgocytosis, super oxide, lysozyme, antibody production and sparing action on other vital micronutrients that become limiting in a stressful condition.

Vitamin E is another vital nutrient essential for immune modulation, more specifically due to its interaction with vitamin C. Hence, most of the investigations were more focussed in dealing with the combined effects of these vitamins in fish and shellfish. However, some of these investigations do not strongly suggest immune modulatory effects of dietary vitamin E. Hardie *et al.* [48] reported that super oxide anion, serum lysozyme, lymphokyne production and humoral immune response of Atlantic salmon were not affected by the dietary vitamin (7 to 880 mg vitamin E/kg diet), but the complement system was compromised in fish fed low levels of vitamin E. Highly positive influence of dietary vitamin E in many of the fish and shellfish diets with high dietary poly-unsaturated fatty acids (PUFA) has been clearly established through many investigations. Recommended levels of dietary vitamin E in fish diet have been 30 mg α-tocopherol/kg diet [128]. Atlantic salmon require 60 mg/kg [144], while channel catfish [145, 146], carp [147], Korean rockfish [148], and sunshine bass [149] have the dietary vitamin E requirement of 30-50, 200-300, 45 and 28 mg/kg diet, respectively.

Recommended levels of vitamins in shrimp diet (Table 2b) are based on several studies conducted in this field. Vitamin C [150-156], vitamin A [157], astaxanthin [154, 156, 157], long-chain polyunsaturated fatty acids and phospholipids [152, 155, 158, 159] and other vitamins and nutrients [160-161] have been studied in shrimp for their immunomodulatory effects apart from growth and survival. Enhanced survival of shrimp fed dietary vitamin E has been reported previously in *Marsupenaeus japonicas* and *Litopenaeus vannamei* [162-163].

Minerals

Iron plays an important role in the oxygen transport and cellular respiration in all higher animals including fish. Iron is an important micronutrient component in the immune system functions and defence against diseases [165, 166]. Though, diffusion and uptake of soluble iron from the water medium surrounding the fish takes place naturally across the gills and intestinal mucosa [167], dietary iron is considered as the major source of the micronutrient due to low concentrations of soluble iron in natural waters [128]. Dietary requirement of iron for optimum growth, feed efficiency, haematological well-being and immune response in juvenile catfish was 30 mg/kg diet [168-170]. Ascorbic acid is also known to be involved in the metabolism of iron in animals including fish [128, 171]. Sub-lethal exposure of copper in fish has been known to suppress specific antibody responses and increased incidence of diseases in fish [172].

In a manner similar to the vitamin C and dietary iron interactions in fish, dietary copper is important in shrimp and other crustaceans. Iron in fish and copper in crustaceans are the major components of the blood pigment. Adequate levels of dietary copper and vitamin C were found to enhance growth response in tiger shrimp, *P. monodon* [173, 174].

Nucleotides

Other components in the dietary mix such as the nucleotides have also been investigated for their immune modulatory roles. Nucleotides, essential components in the DNA replication, have been studied extensively as elements in the nutrition of higher animals [175, 176]. Studies conducted on mice revealed that dietary nucleotides increase resistance to bacterial infection [177, 178], acceleration of T-cell dependant antibody production [179], increased natural killer (NK) cell activity [180] and acceleration of bone marrow cell and peripheral neutrophil proliferation [181]. Hypothesized reason(s) associated with these beneficial effects include dietary provision of physiologically required levels of nucleotides due to limited capacity of certain tissues (e.g. lymphoid), inadequate energetic expenditure for de novo synthesis, immunoendocrine interactions and modulation of gene expression patterns [182, 183]. Role of dietary nucleotides on fish health and immunity has been studied in some cases. Inclusion of extraneous nucleotides to a level of 0.03% in the diet was tested in rainbow trout (*O. mykiss*) that showed a clear immune enhancement [184]. They also showed that the nucleotide fed fish are better protected against infections by *V. anguillarum*, infectious salmon anaemia virus, *Piscirickettsia salmonis* and sea lice (*Lepeoptheirus salmonis*). Dietary oligonucleotides derived from yeast RNA enhanced the disease resistance to *Streptococcus iniae*.

Shrimp nutrition studies were also conducted for evaluating the dietary nucleotides *vis a vis* growth and immune enhancement. Hill and co-workers [185] studied the effects of nucleotide-supplemented diet in different sizes of shrimp. Irrespective of the size, an increase in the total haemocyte count was noticed in shrimp fed nucleotide-supplemented diets compared to those fed control diets. This was more pronounced in juveniles with increases of more than 100% in total haemocyte counts. Larger shrimp registered an increase in total haemocyte count of up to 30% or higher in test subjects compared to controls. The increase in haemocyte number was almost entirely due to the increase in numbers of granular cells in nucleotide-fed shrimp compared to controls. This is particularly relevant as the granular and semi-granular haemocytes are involved in both the oxidative and prophenoloxidase defence cascade. The impact of nucleotides on these processes was further corroborated by replicated, but not statistically significant, increases in prophenoloxidase activity amongst isolated haemocyte populations from nucleotide-supplemented shrimp compared with controls.

Other Additives

Astaxanthin was tested in both fish and shrimp and it was found to have beneficial effects on the disease resistance of the host. Enhanced compelement and phagocytic activity was noticed in rainbow trout fed astaxanthin at levels between 50 and 200 ppm [137]. Wild shrimp were found to contain higher levels (50-80 ppm) of astaxanthin in the whole body composition compared to that in the cultured shrimp (10 ppm). Best growths were reported in *P. japonicus* fed astaxanthin at 200 ppm in the diet [186].

Levamisole is another additive used in the diet of fish for enhancing non-specific immunity. This stimulant has been in use in veterinary and human medicine. Siwiki [187, 188] showed that dietary levamisole produces immunostimulatory activity in carp. Elevated leucocyte and neutrophil counts, phagocytic activity, leucocyte migration, myeloperoxidase activity, lysozyme levels, and natural antibody titers were noticed in carp fed dietary levamisole. Kajita and his team [189] demonstrated, additionally, enhanced natural killer cell activity in rainbow trout. Similar effects were reported in sea bass [190] and gilt-head sea bream [191]. Positive influence of dietary levamisole on the haematocrit and leucocrit values have been reported [192], on the contrary, another study [193] reported that dietary livamisole does not significantly alter the haematocrit and leucocrit values in rainbow trout. Investigations [194] on the Asian catfish with dietary supplementation of ß-1,3-glucan, levamisole, lactoferrin and vitamin C showed that the best immunomodulatory effect was produced by ß-1,3-glucan followed by the other additives in that order.

Microbial levan at 0.5% levels in the diet of common carp was reported [195] to produce enhancement in the total erythrocyte count and haemoglobin content and a very high protective response upon an experimental challenge against *A. hydrophila*. A medicinal plant (*Elipta alba*) leaf extract used in tilapia at levels varying from 0.001 to 1% of the diet produced enhancement in the myeloperoxidase and reactive oxygen species activities. Also the plant extract produced significant resistance to experimental challenge with *A. hydrophila* [196].

There are several other potential immunostimulants that need to be evaluated for applications in aquaculture, thus potentiating efforts of alternate disease control measures to chemicals and antibiotics.

IMMUNOSTMULANTS

Immunostimulants, compounds or components that enhance immune system, have become the integral part of modern aquaculture due to their importance in minimizing the use of chemicals and antibiotics. Vaccines may not be a practical solution in stages of fish with immature specific immunity and in invertebrates with no specific immune functions. Thus, enhancement of non-specific immunity, through "immunostimulants", was the prime method of reducing losses during the aquaculture production stages. Immunostimulants have become more popular with the shrimp aquafarmers world over. This was evidently due to the spread of viral diseases and accompanying severe losses that affected the export-oriented shrimp aquaculture in the Asian countries. With the spread of marine fish farming activities, immunostimulants are spreading their wings into the marine fish larval production and rearing sector as well.

Investigations carried out hitherto have revealed that several proteins are involved in the pro phenol oxidase (PPO) system resulting in the interaction of cellular and humoral factors leading to the successful evasion of pathogenic insults [42, 67, 197, 198]. Microbial infection initiates several cellular and humoral interactions and responses. A variety of immunostimulants, mimicking microbial infections due to fungus, bacteria and parasites, have been tested in aquatic animals.

Blood (haemolymph) clotting is one of the important physiological functions in both invertebrate and vertebrates aimed primarily at preventing blood loss, triggering of cellular signals for migration of lytic factors (immediate bactericidal activity for a possible bacterial attack) and ultimately culminating in cellular co-operation for a long term protection (Immune memory). Most immunostimulants are chemical compounds, similar to the cell wall components of bacteria, fungi and yeasts. Various immunostimulants, their nature and modulation dynamics have been reviewed earlier [53]. Nutritional immunostimulants are one of the important factors affecting immunity in fish and shellfish [50]. These points have been discussed earlier, here, we try to elucidate and discuss the role of non-nutritional or dietary (oral) immunogens in enhancing aquaculture-linked production through improved disease resistance.

Immunostimulants used in aquaculture are derived from either chemical or biological sources [199]. The biological sources include fungal, bacterial and yeast components. One of the most frequently studied components of the biological origin are the glucans (ß-1-3 and ß-1-6) from yeast and the lippopolysaccharides (LPS) from bacteria. Studies have shown that intraperitoneal injection of ß-1, 3-D-glucans, oral administration of peptidoglycan and chitosan [192, 200-202] produced resistance against diseases in fish. A clear emphasis on these compounds has been placed as elicitors of immune response in aquatic organisms [203]. An algal extract, laminaran, was used to activate macrophages [204] and to stimulate the respiratory burst in the leucocytes of the head-kidney of Atlantic salmon [205]. Enhanced survival of yolksac larvae of halibut after a bath exposure to FITC labelled LPS was noticed indicating a positive effect of LPS on the fish larval stages [206]. ß-glucans, LPS and killed Gram-negative bacteria have been used in shrimp and other crustaceans for enhancing survival, non-specific immunity and even growth [68—71, 207-210].

ß-Glucan as Conductor of Immune Symphonies

A majority of the immunostimulants have molecular architecture possessing repeating units of a certain moieties such as glucose residues as in ß-glucans and (deoxy) riboses in DNA/RNA, fatty acid chains in bacterial lipopolysaccharides (LPS) and certain lipoproteins. Such patterns are abundant in microbial communities of prokaryotes, and can be termed pathogen-associated molecular patterns, PAMPS, if they initiate inflammatory responses. During microbial breakdown/degradation, a number of PAMPS may be released that would initiate inflammatory responses upon receptor binding and intracellular activation of signal transducers and transcription factors. The exact composition of released danger signals would be decisive for the out-put gene transcription leading to synthesis and release of pro-inflammatory cytokines. Immunostimulants have been used as feed additives for many years in aquaculture, and yeast ß-glucan may be the one with the longest track record. Homo-polysaccharides, consisting of one sugar, are described by use of the suffix "an". Thus a poly-glucose (poly-glucopyranose) is termed "glucan. Other carbohydrates may be hetero-polysaccharides that may contain several different constituent residues. ß(1,3)-D-glucans (ß-glucans) are the most commonly used term for homo-polysaccharides that has ß(1,3)-D-linkages in the backbone, and may also possess ß-D-glucosidic linkages at position 6 in different, often repeating units (branches). In nature, ß-glucans are widespread and are found in plants, algae, bacteria, yeast and mushrooms. ß-glucans possess differences in molecular weights and degree of branching. Yeast ß-glucan is a particulate carbohydrate that consists of glucose and mannose and is a major constituent of the cell membrane. Of the major fish feed producers, Skretting and Ewos include yeast ß-glucan in their fish feed assortments /sortments, while other fish feed companies include other immunostimulants.

ß-Glucans - Growth Performance

Besides the information presented in comprehensive review articles on immunostimulants [199], several new reports have enlightened the effects of ß-glucans on growth performances in fish. It has been found that ß-glucans such as yeast glucans obtained from *Saccharomyces cerevisiae* incorporated in fish feed increase the growth rate of certain species of fish [211-214]. However, in other scientific experiments no significant growth increase has been found after feeding fish with different ß-glucans [215-217]. It appears that a growth increase is dependent on the amount of ß-glucan incorporated in the diet, duration of feeding, environmental temperature and the species under study. To obtain increased growth performance, feeding strategies should be developed for each fish species with respect to ß-glucan dose and duration. In addition, soluble or particulate ß-glucans may confer different responses since the intestine only absorb soluble ß-glucans. There are no indications that particulate ß-glucans are taken up by the fish intestine or digested by ß-glucan degrading enzymes thereby being nutritious. To date, there are no plausible explanations for the growth enhancing effects displayed by diets containing ß-glucans. One hypothesis may include a local intestinal inflammatory response, resulting from ß-glucan administration that induces resistance against pathogens - that otherwise would result in decreased weight gain and maybe disease. Besides this, any effects may be due to local growth enhancing responses together with certain systemic effects.

Oral Administration of ß-Glucans

A commercial ß-glucan (EcoActiva) was administered as a feed supplement to the snapper, *P. auratus* [211]. The immunostimulant increased macrophage oxygen radical production during the wintertime, but not during the summer time. In contrast to macrophage activity EcoActiva did potentiate either the classical or the alternative complement activity. The results of this study suggest that it may be favourable to include ß-glucan in the feed to snapper during the wintertime both to increase resistance against diseases and to enhance growth rates.

Sea bass were fed a diet supplemented with 2% ß-1,3/ ß-1,6-glucan over a 2-week period every 3 months [218]. Plasma lysozyme and complement activity were measured prior to feeding and after 40 weeks. Both alternative pathway of complement activation and lysozyme activity were significantly enhanced in fish fed the glucan.

Immunomodulation by a yeast ß-1,3/ ß-1,6-glucan in rainbow trout was evaluated with respect to macrophage activities, lysozyme and complement activation. In addition, antibody responeses were analysed after vaccination with formalin-killed *Y. ruckeri* [129]. The non-specific immune parameters were evaluated at the end of a two-week feeding period and 4 weeks later. A significant enhancing effect of dietary ß-1,3/ ß-1,6-glucan was found on concanavalin A-induced proliferation of lymphocytes and the antibody response after vaccination against *Y. Ruckeri*. But glucan showed no effect on oxidative burst, pinocytosis, lysozyme activity and alternative complement pathway activation.

Carp (*C. carpio*) were given ß-glucan by the oral route before vaccination with *A. hydrophila*. The antibody titre was slightly increased. Orally administered glucan did not result in changes of the alternative complement pathway [219]. Short and long-term effects of yeast ß-glucan and alginic acid included in the feed to sea bass (*D. labrax*) were evaluated by analysis of complement activity, percentage of B- and T-lymphocytes in blood and serum lysozyme [216]. Serum complement and lysozyme levels were increased 15 and 30 days after the end of a 15 days feeding period with the glucan, respectively. Similar results were obtained after feeding the fish with alginic acid (Ergosan). ß-Glucan was administered to *L. rohita* fingerlings through different dietary doses for 56 days. Leukocyte counts and both cellular and humoral immune parameters were evaluated at two-week intervals [212]. Administration of the highest dose (500 mg ß-glucan per kg diet resulted in a reduced white blood count compared to fish fed 250 and 100 mg. Maximum superoxide anion production and phagocytic activity in head kidney macrophages were found in fish fed 500 mg and 250 mg ß-glucan per kg diet (for 42 days) respectively. Haemolytic complement activity was significantly higher in fish stimulated with ß-glucan except the 500 mg per kg dose. The lysozyme and bactericidal activity were highest at a dose of 250 mg ß-glucan per kg dry diet fed for 42 days.

The dietary effect of ß-1,3-glucan on immune responses in healthy and aflatoxin-induced immunocompromised rohu (*L. rohita* Hamilton) was investigated. A single injection of aflatoxin reduced the non-specific immunity as measured by neutrophil phagocytosis, serum bactericidal activity and specific immunity against *E. tarda* [220, 221]. Feeding of glucan to aflatoxin-immunocompromised fish for 7 days significantly raised non-specific immunity as measured by serum bactericidal activity, phagocytic activity and antibody titres.

Cyclophosphamide, a multifunctional alkylating agent and cytotoxic drug and a well-known immunosuppressant was used to induce an immunocompromised state in Asian catfish

(*C. batrachus*) [194, 222]. The cyclophosphamide treated fish showed lowered nitroblue tetrazolium, myeloperoxidase and phagocytic activities in blood phagocytes, and a decreased haemagglutination activity. ß-Glucan delivered as feed supplement significantly enhanced the immune parameters

The dietary effect of b-1,3 glucan on innate immune responses of large yellow croaker (*Pseudosciaena crocea*) was studied [213]. The diet was supplemented with 0.09% and 0.18% b-1,3 glucan and the feeding trial lasted for 8 weeks. Low concentration of glucan (0.09%) significantly enhanced the respiratory burst and phagocytic activity in head kidney macrophages while the high concentration (0.18%) did not. The serum lysozyme activity in fish fed diets with both glucan concentrations were significantly higher than the control, and the 0.18% glucan diet significantly stimulated the lysozyme activity compared to 0.09%. No significant differences were seen in alternative complement pathway activity with any of the ß-glucan concentrations.

ß-Glucans - Immune Response

Carp (*C. carpio*) injected intraperitoneally with ß-glucan (100, 500 and 1000 mg) had a significant increase in total leukocyte counts and an increased proportion of neutrophils and monocytes on day 7 [219]. Classical and alternative complement activation were unaffected by glucan injection. Anterior kidney macrophages from both control and glucan injected fish were cultured and assayed for superoxide anion production, bacterial killing of *A. hydrophila* and expression of IL-1ß mRNA. Superoxide anion production was significantly increased in macrophages from glucan-injected fish at all doses. In addition, *A. hydrophila* was more efficiently killed by macrophages from the glucan stimulated fish. Also, expression of IL-1ß mRNA was higher and dose-dependent in glucan treated fish (highest at a dose of 1000 mg per fish).

A combination of ß-glucan and lippopolysaccharides (LPS) injected intraperitoneally to carp (100 mg ß-glucan+10 mg LPS) had a significant increase in total blood leukocyte counts and an increase in the proportion of neutrophils and monocytes [223]. Superoxide anion production by macrophages was also elevated. Classical and alternative complement pathways were unaffected by the glucan. Also, expression of IL-1ß mRNA was not elevated in the stimulated fish.

T Helper Cells and Differentiation

During an immune response, specialized cells known as 'antigen presenting cells' (APC) endocytose antigens, e.g. bacteria or viruses, etc., and process them, then travel from the site of infection to lymphoid tissues/organs where antigen presentation occurs. This allows naïve CD4+ T cells that express specific TcR's against the peptide/MHC complex to be activated. When a Th cell encounters and recognises the antigen on an APC, the TcR-CD3 complex binds strongly to the peptide-MHC complex present on the surface of professional APC. CD4, a co-receptor of the TCR complex, also binds to a different section of the MHC molecule. With the assistance of different kinases associated to TcR, CD3 and CD4 and a phosphatase present on the intracellular section of CD45 (common leukocyte antigen), activation of biochemical pathways in the cytosol of the Th cell occurs. These active pathways are known as signal 1 of T cell activation, as it is the first and primary pro-activation signal in a Th cell. Following a re-stimulation with same antigen memory T-cells use the same TCR pathways to become re-activated. The signal 2 step involves verification step that is a protective measure to ensure non-self response, and involves an interaction between CD28 on the CD4+ T cell and the proteins CD80 (B7.1) or CD86 (B7.2) on the professional APCs. Both CD80 and CD86 activate the CD28 receptor. These proteins are also known as co-stimulatory molecules. Once both signals 1 and 2 are active within the helper T cell, the cells are able to proliferate by releasing a potent T cell growth factor called interleukin-2 (IL-2) that also acts in a autocrine way by binding to up-regulated IL-2 receptors (CD25) on both CD4+ and CD8+ T cells. This binding, when a critical number of receptors are triggered, lead to cell division and cytokine production. After many cell generations, the Th cell progenitors differentiate into effector Th cells, memory Th cells, and suppressor Th cells. Effector Th cells are those that secrete cytokines, proteins or peptides that stimulate or interact with other leukocytes, including Th cells.

Effector T Cell Responses

The T cell response (including cytokine production) may be essential for a successful outcome from vaccination. Understanding exactly how helper T cells respond to infection is of major interest in immunology and such knowledge may be very useful to increase the effectiveness of vaccination. Proliferating helper T cells that develop into effector T cells differentiate into two major subtypes of cells known as Th1 and Th2 cells (also known as Type 1 and Type 2 helper T cells, respectively). While it is established that certain cytokine patterns favour differentiation of Th into Th1 or/and Th2 cells after activation of naïve T cells, it is not fully clear how the various patterns are decided and how the cytokines interplay with each other. In a rather schematic and simple way, Th1 and Th2 responses are facetted by different cytokine profiles. The presence of relatively high amounts of interferon gamma (IFN-g) and tumour necrosis factor beta (TNF-ß) favours a Th1 response characterised by an increased activity of the cellular arm of defence where, killing efficacy of macrophages and proliferation of CD8+ T cells (CTL) are increased (reviewed by O´Garra [224]). IFN-a increases the production of IL-12 by dendritic cells and macrophages in an autocrine manner, to produce more IFN-g in helper T cells promoting a stronger Th1 response. IFN-g inhibits IL-4, an important cytokine within the Th2 response. IL-4, IL-5, IL-6, IL-10 and IL-13 are all involved in a favoured Th2 response that stimulates B cells into proliferation, antibody class switching and to increase antibody production. Of these interleukins, IL-10 inhibits the transcription of a variety of cytokines such as IL-2 and IFN-g by all Th cells, and IL-12 in dendritic cells and APC, but continues to stimulate plasma cells to produce antibodies. While Th1 and Th2 responses are pro-inflammatory, a Th3 response is believed to be immune suppressive by increased production of IL-10 and TGF-ß of Th1/Th2 cells at a later stage in inflammation [224].

Immunostimulant-Vaccine Interactions

One of the key questions for vaccine development against infections in fish and higher vertebrates is to understand the mechanisms underlying the difference between inferior protection attained by inactivated vaccines and the strong immunity often induced by replicating variants. Most inactivated antigens will not induce sufficient immune response when administered in the absence of an adjuvant. Adjuvated vaccines typically work through three principal modes of actions; 1) through enhanced antigen uptake and presentation of antigens (oil emulsion, liposomes etc.), 2) by a "danger effect" (cholera toxin, CpG-ODNs, LPS), and 3) through signal 2 action (co-stimulation; co-stimulatory molecules, cytokines etc.). The use of adjuvants has been largely empirical, but today, we are seeing the empirical approach being superseded by a rational design where one strategy is to deliver an inert antigen in the context of a danger signal [65]. Oil adjuvants mediate their danger signal in forms of varying degrees of tissue damage [63]. In salmon they alarm the immune system likely by causing tissue damage at the injection and depot site, the consequence being scarring and adhesions [225,226].

Some approaches have been made to include immunostimulants as vaccine adjuvants in fish to increase vaccine potency and efficacy. Increased vaccine potency in terms of increased antibody response was observed after injection of ß-glucan Adjuvated inactivated *A. hydrophila* vaccine [227]. Although not addressed in this study, the mushroom ß-glucan they obtained and applied, resulting in the elevated antibody response was probably a water-soluble ß-glucan. Water-soluble ß-glucans, often obtained from fungi, have been thoroughly investigated with respect to biological effects in many mammalian species [228]. Overall, the effects ß-glucans exert include increased mitogenicity, stimulation of hematopoietic stem cells, activation of complement pathways, and activation of lymphocytes, macrophages, dendritic cells, NK cells, Th cells, and cytotoxic T cells as reported in mammalian experimental models. However, some substances may down-regulate certain inflammatory responses thus being novel as anti-inflammatory drugs.

Modern aquaculture would have been impossible without the use of injectable vaccines against common diseases. However, the use of oil emulsions as adjuvants may cause major drawbacks in terms of undesirable effects such as internal organ adhesions, melanin deposits, impaired growth and skeletal deformations. As stated in a state-of-the-art review "Although the majority of the farmed fish seems to possess side- effects within acceptable limits, given the pros of vaccination – it is evident that some fish possess unacceptable levels of side effects" [229]. Adhesions and skeletal deformations are probably the side effects with the most important implications for animal welfare and ethics. This addresses the need to develop immunization strategies using "milder" adjuvants such as ß-glucans and other immunostimulants.

Early findings by Rørstad et al. [230] described protective response of yeast particulate ß-glucan Adjuvated *A. salmonicida* vaccine in Atlantic salmon. Similar protection was also found applying yeast ß-glucan in combination with *A. salmonicida* bacterin in protection against furunculosis [231]. Following on, in another study, Japanese flounder were fed formalin killed *E. tarda* cells together with curdlan, and were assessed for specific antibody response. No increased antibody response was observed although the survival against live *E. tarda* experimental challenge was higher than in fish fed control diet [232]. This indicated that there was no correlation between vaccine potency and efficacy. Other defence mechanisms such as activation of the cellular defence may have been activated. On the contrary, Midtlyng et al. [233] reported no significant protective effects of yeast ß-glucan adjuvated vaccine followed by experimental challenge with *A. salmonicida*. Their explanation was that a high infection pressure in a given challenge experiment may be inferior to find effects of ß-glucan adjuvated vaccines. Following this, absence of adjuvant features of yeast ß-glucan administered with *Streptococcus* bacterin to turbot has been reported [234]. Kawakami and co-workers reported an absence of adjuvant action by using yeast particulate ß-glucan co-injected with *P.piscicida* vaccine [235]. Also oral application of yeast ß-glucan followed by vaccination with *S. iniae* did not confer protection against *S. iniae* challenge [215]. Interestingly, zymosan has been found to increase vaccine potency when co-administered with a HIV-1 DNA vaccine to mouse [236]. Whether this also applies to fish remains unknown, but it opens up for applying non-genetic/non-nucleosidic adjuvants in combination with e.g. VHS and IHN DNA vaccines in aquaculture, especially to DNA vaccines not proven efficacious enough to induce full protection. Apparently, due to the conflicting results obtained by ß-glucans to increase vaccine potency and efficacy it seems that more research is highly warranted in terms of dose, duration, experimental models, and a comprehensive study focusing on what molecular features (molecular weight, conformation, water-solubility and charge) that may be the most potent one for application to aquaculture systems. In addition, there may be species-to-species variation with regard to biological effects, and differential effects against different pathogens.

Non-TLR ß-Glucan Receptors: It has been hypothesised that collaboration among different membrane receptor families, as well as with non-classical opsonins and protein cascades such as complement, is likely to be an important mechanism to enhance affinity of binding and specificity [237]. Several different receptors have been reported to bind ß-glucans, these are the scavenger receptors (SR), complement receptor 3 (CD11b/CD18), dectin-1 and TLR 2/6. All being innate receptors but dectin-1 and TLR2/6 are also highly involved in adaptive responses. There are at least three classes of SR (SR-A, SR-B and SR-C) based on their structural characteristics. Most of the SRs are non-opsonic receptors that display low-affinity and promiscuous binding to many polyanionic and modified substances. Although they may bind anionic ß-glucans (sulphated ß-glucans), they are not considered to be involved in ß-glucan binding and internalization. Dectin-1, belonging to a non-classical C-type lectin-like family that predominantly bind protein ligands, has been considered to be the major ß-glucan receptor [238]. In mouse, dectin-1 is expressed on neutrophils, alveolar macrophages, and inflammatory macrophages, with minor expression on resting macrophages and dendritic cells [239]. Dectin-1 mediates its own signalling and induces specific responses such as IL-10 (anti-inflammatory) production in DC and the respiratory burst in macrophages and neutrophils, but it is speculated that dectin-1 and TLR collaborate to induce other responses to fungal stimuli such as TNF, IL-2 and IL-12p40/p70 production. However, recent findings have suggested that dectin-mediated fungal recognition preferentially induces IL-17 producing Th cells (Th17) in both humans and mice (reviewed by Palm et al. [240]). A Th-17 response is important in epithelial cell activation, neutrophil recruitment and during mucosal homing of leucocytes by modulation of chemokine receptor expression. The resulting immune response following infection with extracellular pathogens and fungi containing dectin-1 ligands may favour Th-17 differentiation and it is hypothesized to be important for eradication of extracellular bacteria and fungi at mucosal sites. A combined stimulation of both TLR and dectin-1 by yeast cell wall extract zymosan results in mainly Th1 cell differentiation. This is probably because of IL-12 – and IFN-g mediated antagonism of Th-17 cell differentiation [240]. The Th-17 differentiation is also dependent of presence of transforming growth factor ß (TGF-ß), IL-6, IL-23. The Th-17 response concept would be highly important to address in fish with use of ß-glucan as a stimulant to increase mucosal defence activity against bacteria and fungi. In line with this, five zebrafish IL-17 genes have been cloned and characterised [241], whereas Atlantic salmon IL-17 is currently being cloned and sequenced (personal communication). In addition, there are available sequence data on putative IL-17 gene sequences from rainbow trout (AJ580842, AJ580843) and fugu (CA588258). Interestingly, relatively high amount of IL-17 transcripts were found in mucosal tissues of the zebrafish [241].

ß-glucan receptors have been found on Atlantic salmon macrophages [242] and channel catfish neutrophils [243], whether these belong to non-TLR or TLR remains to be addressed. Not investigated is the antifungal effect of ß-glucans in fish, even though there exists a reproducible experimental saprolegnia challenge model. Tissue distribution studies on fish have revealed that endothelial cells and macrophages are mainly responsible for uptake of parenterally injected ß-glucan (Figure 4), whereas intestinal epithelial cells (enterocytes) have absorbed ß-glucan after oral and peranal administration (reviewed by Dalmo et al. [82]). Any receptor dependent uptake was not, however, studied. Any uptake of yeast particulate ß-glucan in intestinal cells of fish has not been proved yet.

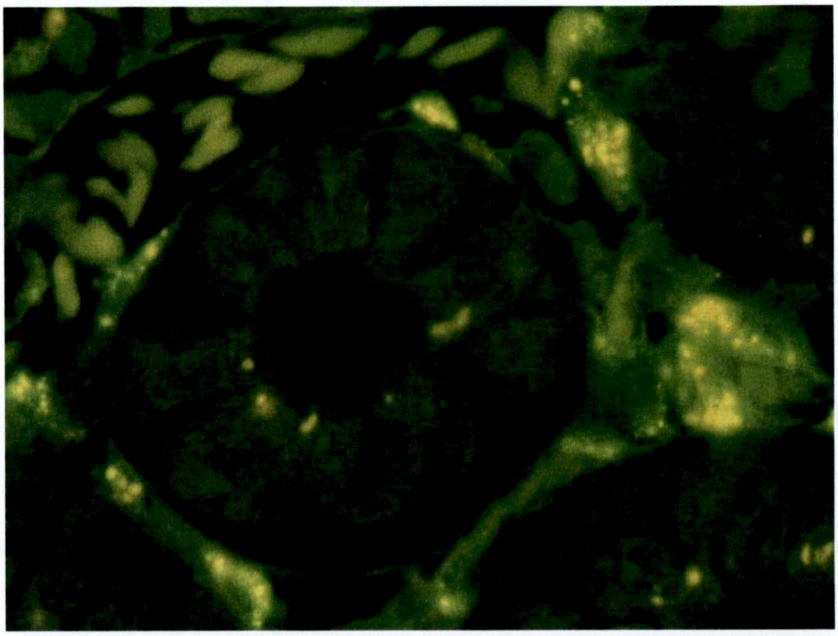

Figure 4. Fluorescence micrograph of head kidney section of Atlantic salmon obtained 2 days after intravenous injection of a soluble ß-glucan (laminaran). Green fluorescence is localization of laminaran. Note localization of ß-glucan in endothelial cells (flattened cells) and macrophage-like cells (larger cells to the right) in the sinusoid. Courtesy: Roy A. Dalmo.

Complement receptors: The complement receptor type 3 (CR3; CD11b/CD18), which belongs to the family of ß2-integrins, is another well-characterized ß-glucan receptor. CR3 has a ß-glucan–binding site within its {alpha}-chain subunit (CD11b) of the heterodimeric protein [244].

Besides increasing the uptake of iC3b and C3b opsonised matter facilitating enhanced antigen presentation, activation of CR3 may control IL-12 synthesis of human monocytes thereby affecting T cell immunity [245]. Functionally, a CR3-dependent rosette formation has been observed in carp [246], while CR3 gene(s) has not been cloned and sequenced yet in any fish species.

Lactosylceramide: Lactosylceramide has been found on leucocytes and endothelial cells, and is a glycolipid containing a hydrophobic ceramide lipid and a hydrophilic sugar component. It may bind specifically ß-glucans with concomitant production of reactive oxygen metabolites [247, 248].

Toll-like ß-glucan receptors: It is widely acknowledged that fungal cells and zymosan, both consisting of many different molecular moieties, may bind TLR2 and TLR4 facilitating myD88 dependent intracellular signalling that induce cytokines favouring Th1 cell differentiation [249]. Zymosan, that is obtained from yeast cell walls is composed of insoluble ß-glucans and mannan, and have been utilized in numerous experiments mainly due to its ability to activate macrophages. Lately, it has been questioned that impurities by e.g. bacterial lipopolysaccharides may be decisive for activation of TLR4 [248].

Gene Expression Following ß-Glucan Administration

Besides gene expression studies of certain acute phase products (TNF-a, IL-1ß, IL-6, complement components 3 and 5 and their subtypes) [214, 223, 250, 251] no targeted approach has been performed to elicit the effects of ß-glucans on adaptive immune responses in fish. In a monocyte cell line (THP-1) a high number of genes were up regulated following ß-glucan stimulation, those of particular interests were genes involved in chemotaxis (chemokine ligands) such as CXCL1, CXCL2, CXCL3 and Th1 associated genes such as IL-12ß, TNF, IL-8 [252]. Taking the Th17 induction capacity by ß-glucans into account together with the induction of chemokine ligands it may be speculated that ß-glucans may have their potential in increasing mucosal disease resistance. The ability of ß-glucans to increase the expression of TLRs would be important to address since it may be hypothesised that a up regulation of e.g. TLR3, TLR7 and TLR9 would in turn, when appropriate stimulus are present, confer increased disease resistance against interferon a/ß susceptible viruses, and an up regulation of most of the other TLRs may increase disease resistance against intracellular bacteria.

Vaccines

Though, immunostimulants and probiotics have produced considerable impact as immune modulators, they produce a very short-lived response and need a repeated stimulation a number of times during an aquaculture production cycle. Vaccines answer these problems by eliciting a long-lasting and effective immune response in vertebrates. Use of vaccines has clearly demonstrated, in salmonid aqua farms, that immune modulation can reduce losses due to diseases and hence the consumption of antibiotics that have deleterious effect on the environment and induce multiple drug resistance among aquatic bacteria (Figure 5).

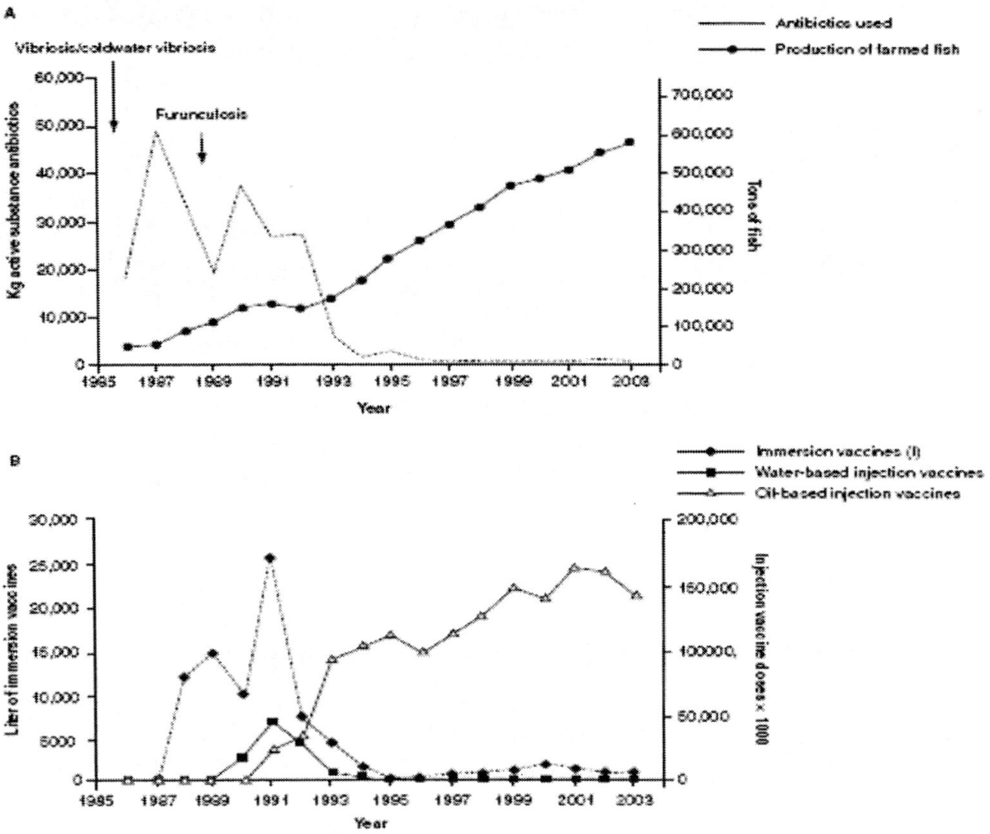

Figure5. Use of antibiotics, production of fish *vis a vis* application of vaccines in Norwegian aquaculture during 1985-2003 clearly indicating positive effect of immunotherapy on the consumption of antibiotic [253]. Reproduced with permission from: Future Drugs Ltd & Future Medicine Ltd.

Duff [254] and Schäperclaus [255] made the earliest of attempts of vaccination in fish. However, only after the eighties concerted efforts were made in the field of fish vaccination. Most of these studies were centred on injection and immersion routes. It is of immense importance to note that Duff [254] made the first attempt of successful vaccination in fish, and that was interestingly, through oral routes. Schaperclaus [255] was the first to attempt vaccination using *A. hydrophila* against bacterial haemorrhagic septicaemia. Schaperclaus [255], Post [256], Song and Kou [257], Thune [258], Thune and Plumb [259] have tried vaccination of different fish species through various immunization routes against infection by the motile aeromonads. Lamers and co-workers have provided different facets of immune response of carps to *A.hydrophila* [260-264]. Contributions of Fryer and his group of researchers, in the field of vaccination against vibriosis, during the seventies are note worthy [265-268]. Among others, studies on pathogenesis and vaccination against important fish viruses such as the Infectious Pancreatic Necrosis virus [269-272], Viral Haemorrhagic Septicaemia virus [273-275] and

Spring Viraemia of carp virus [276-278] are important. Application of biotechnological tools in the production of fish vaccines has also attracted the attention of fish and shellfish immunologists. Sub-unit vaccines, DNA vaccines etc. have been attempted for providing an effective and long-term protection to the vaccinated aquatic species [279]. However, applicability, ease and cost factors weigh heavily in favour of the oral route of vaccination.

Keeping in view the importance for the development of oral vaccines that elicit considerable protective response in fish, the research accomplishments of oral vaccinology (Table 3) and the challenges faced by the fish immunologists are discussed in this paper. Oral vaccination for the prevention of important viral, bacterial, fungal and parasitic diseases has been discussed. Rsearch research results have not been translated into commercial solutions in the form of vaccines as evident from the list (Table 4) of commercially available aquaculture vaccines. Fish vaccination started with the oral route of antigen delivery on trout against *V. salmonicida* by Duff [254]. This was a very successful attempt that reported high protective response in the vaccinated fish. The attempt of Duff encouraged many fish health workers to explore the possibilities of oral vaccination in several species of fish. But all these attempts produced either marginal or poor response in the orally vaccinated fish.

Bacteria

Aeromonas and Vibrio are the two important groups of bacteria that account for over 90% of the freshwater and saltwater (brackish and marine environments) fish and shellfish diseases, respectively. Naturally, these were the pathogens that were addressed during the initial attempts of fish vaccination.

Aeromoniasis

A.hydrophila and *A. salmonicida* are the two most important aeromoads that account for both fresh water and salt water (brackish and marine) aeromoniasis in fish. *A. hydrophila* is a ubiquitous and highly heterogeneous bacterium usually associated with stress induced pathogenicity in freshwater aquaculture. Infection with *A. hydrophila* leads to bacterial haemorrhagic septicemia or motile aeromonad septicemia, a systemic disease producing swelling of the body cavity and haemorrhage of organs. The bacterium may become an opportunistic pathogen following parasitic infestation or injury. Red sore disease of large mouth bass is attributed to a combined assault of *A. hydrophila* and the ciliate, *Epistylis* [280]. The bacterium is frequently associated with the diseases in carp, eel, milkfish, channel catfish, tilapia and ayu [11, 13, 14]. Direct losses due to this motile aeromonad cannot be easily differentiated from the losses to fish crops inflicted due to the secondary infections [13, 39]. It is reported that even under moderate salinity regimes, if injected, it can cause severe lesions in tilapia [281]. The bacterium has been known as *A. liquifaciens, A. punctata, Bacterium punctata* etc., in the past. Heterogeneity of bacteria included in the species *A. hydrophila* has been the major problematic area for the fish immunologists to negotiate. Isolates obtained from the environment and from the diseased fish showed high degree of variability in their virulence [282].

Table 3. Review of Oral Vaccination Investigations In Fish

Pathogen	Reference	Fish species	Dose employed	Challenge route	% Mortality Vaccinated (Control)	Protection (+/-)
1. *A. hydrophila*	[256]	*S. gairdneri*		i.p.	60-90 (80-100)	+
	[284]	*I. punctatus*		Field	34 (32)	-
	[287-290]	Indian major carps	10^7 to 10^9 cells/g feed	Field	55-100 (20-45)	+
				i.p.		+
					RPS of 55 and 63.5	
Liposome entrapped *A.hydrophila*	[371]	*C. carpio*	3 x 10^5 & 10^6 CFU/fish 3 days Oral intubation	Sub-cutaneous injection	RPS 70%	+
A. veroni (extracellular lectin)	[372]	Spotted sandbass, *Paralabrax maculatofasciatus*		i.p challenge		+
3. *A. salmonicida*	[254]	*S. clarkia*		Field	25 (75)	-
	[310]	*S. fontinalis*		i.p.	68 (90)	+
	[373]	Salmonids		Field	-	-
	[306]	*O. kisutch*		Field	7-11 (10)	-
	[374]	*S. gairdneri*		i.m.	82-93 (90)	-
	[375]	*O. kisutch*		Oral	100 (100)	-
	[312]	*Salmo trutta*		Field	3 (0)	+
	[313]	*S. trutta*		Field	35-62 (86)	+
	[311]	*S. fontinalis*		Field	35 (87)	+
	[314]	*S. gairdneri*		Field	0 (58)	+
	[376]			Field	31-53 (60)	-
4. *Flexibacter columnaris*	[339]	*S. gairdneri*		Field	19 (27)	+/-
		O.kisutch		Field	8 (48)	+
5. *Pasteurella piscida*	[340]	*Seriola quinqueradiata*		Artificial	5-12 (100)	+
6. *E. ictaluri*	[341]	*I. punctatus*	0.001%	Artificial	50 (76.3)	+
		5 day + 5 day (with 10 d gap)	0.01%	Artificial	50.3 (76.3)	+
		% killed dehydrated vaccine in	0.1%	Artificial	40 (76.3)	+
		feed.	0.5%	Artificial	30 (76.30	+
			1.0%	Artificial	23.3 (76.3)	+
			10.0 %	Artificial	26.7 (76.3)	+
7. *V. anguillarum*	[265-268]	*O. tshawytscha*		Field	10-45 (98)	+
	[372]	*O. tshawytscha*		Field	80-100 (90-100)	+

Pathogen	Reference	Fish species	Dose employed	Challenge route	% Mortality Vaccinated (Control)	Protection (+/-)
		O. kisutch		Field	0-13 (22-83)	+
	[377]	*S. gairdneri*		Field	(100)	-
	[378]	*A. anguilla* & *S. gairdneri*		Field	7 (100)	+/-
		Plecoglossus altivelis			9 (62)	+
	[379]			Cohabitation	6-53 (20-45)	
		P. altivelis			7 (80)	
					10 (90)	
	[291]		1g wet cell/kg fish/day-	Cohabitation		+
			10 days	ip	23 (58)	+
		O. nerka	-	Field	22 (58)	-
		P. altivelis		Cohabitation	32 (34)	+
	[380]	*S. gairdneri*		Artificial	18 (42)	+
	[381]			ip	2 (100)	+
	[318]	*S. gairdneri*		ip	15-45 (55)d	+
				ip		-
	[382]			Field		+
					10-45 (98)	+
					34 (100)	+/-
		P. altivelis	0.1 to 0.16 g wet			+
			cells/kg fish		30-90 (100)	
	[383]			Experimental at 18 dpv	28	+/-
		Eel	Enteric coated vibrio		30	
				Bath at	0.3-0.4 (4.1-4.7)	
	[320]		Oral through	59 dpv		
		Salmonid	Artemia	Lab	30-32 (56-60)	
	[324]	*C. carpio* and *Sparus aurata*	Oral, anal intubation, i.p.	i.p.		
8. *V. vulnificus*			and immersion		26-32 (52-61)	+
	[384]					
		Flounder	Oral (2mg/fish) for 4			
		Paralichthys olivaceus	weeks, twice a week			++
				-	Higher tieters	
	[333]					

Pathogen	Reference	Fish species	Dose employed	Challenge route	% Mortality Vaccinated (Control)	Protection (+/-)
9. *Y. ruckeri*	[336]	*S. gairdneri*		ip	10 (90)	+
	[337]	*S. gairdneri*		sc	10 (100)a	+
	[338]	*S. gairdneri*			90 (100)b	-
					100 (100)b	-
	[338]	*S. gairdneri*		Field	50 (100)	+
	[318]	*S. gairdneri*		Bath	24 (65)d	+
10. *S. iniae*	[385]	Nile tilapia *(Oreochromis niloticus)*	Up to 2×10^9 CFU/g feed	ip	RPS 63.1	+
12.IPN virus	[269]	*S. gairdneri*				
13.Spring viraemia virus	[277]	*C. carpio*	i.p	-	22 (55)	+
	[277]	*C. carpio* (Field testing)		ip	26 (86)	+
	[386]	*C. carpio*		-	-	+
14.Nervous necrosis virus (DNA vaccine)	[387]	Grouper *E. coides*	Artemia carrying NNV capsid protein gene	oral	RPS of 64 to 69	+

Table 4. Commercial registered inactivated antigens world-wide wide (Source: Sommerset, *et al.* [253])

Disease / Pathogen	Major species affected	Primary region	Commercial vaccine availability
	Bacterial Diseases		
Vibriosis (*L. anguillarum*)	Salmonids, halibut, cod, seabass, bream	Global	Yes
Coldwater vibriosis (*Vibrio salmonicida*)	Salmonids	Northern Europe, Canada, USA	Yes
Furunculosis Salmonicida)	Salmonids	N.Europe, USA Canada	Yes
Enteric red mouth (ERM) (*Yersinia ruckeri*)	Freshwater salmonids	Europe, Chile, Canada, USA	Yes
Bacterial gill disease (*Flavobacterium psychrophilium*)	Salmonids Carp	Canada, Europe USA, Japan	No
Columnaris disease (*Flavobacterium columnaris*)	Catfish Salmonids	USA Chile	Yes
Edwardsiellosis (*Edwardsiella ictaluri*)	Catfish	USA Asia	Yes No
Bacterial kidney disease (BKD) (*Renibacterium salmoninarum*)	Salmonids	Chile, USA Japan	Yes No
Lactococcosis (*Lactococcus graviae*)	Rainbow trout Yellowtail, Amberjack	Italy, France UK, Japan	Yes Yes
Streptococcosis (*Streptococcus* spp.)	Tilapia Barramundi Salmonids	Asia Asia Chile	Yes No No
Pasteurellosis (*Photobacterium damsela*)	Sea bream/bass Yellowtail	Mediterranean Japan	Yes
	Viral Diseases		
Infectious Pancreatic Necrosis Virus (IPNV)	Salmonids	Global	Yes
Pancreatic Disease (PD) Virus	Salmonids	UK, Ireland, Norway	Yes
Infectious Haematopoietic Necrosis Virus (IHNV)	Salmonids	Canada USA	Yes
Viral Haemorrhagic Septicaemia (VHSV)	Trout Turbot, Flounder	Europe Asia	No
Viral Nerve Necrosis (NNV) or related Nodaviruses	Several marine fish, Freshwater aquarium fish	Global	No
Iridovirus	Red sea bream, Amberkjack, yellowtail	Asia	Yes
Channel catfish virus (CCV)	Channel catfish	USA	No
Spring Viremia of Carp Virus (SVCV)	Carps	Europe	No
	Parasites		
Paramoeba	Amoebic gill disease, Australia, Europe, many marine fish, No commercial vaccine, experimental stage.		
Cryptobia, Ichthyobodo	No commercial vaccine		
Freshwater and marine 'itch's (*Ichthyophtherius multiphilis*) (*Cryptocaryon irritans*)	Many freshwater fish Global Marine fish Global No commercial vaccine, experimental stage		
Gyrodactylus, Dactylogyrus, Benedinia, Caligus, fish lice etc.,	Varying severity in many fish both marine and freshwater, No vaccine available		

Some of the earliest attempts of employing simple, inactivated whole cell vaccine of the pathogen in fish by immersion, injection and via oral routes have yielded varied immune responses [39, 256, 260-264, 283, 284, 285]. Post [256] recorded protection levels of 10-40 percent in orally vaccinated *S. gairdneri* against virulent *A. hydrophila vis a vis* 0 to 20 percent protection in control group. However, a negative protection was recorded by Schachte [284] in *I. punctatus,* orally vaccinated against the motile aermonads. Success of oral vaccination obtained by Duff [254] could not be reproduced in several later attempts. Irianto *et al.* [286] have demonstrated that the formalin-inactivated cells of *A. hydrophila* help in controlling infections by atypical *A. salmonicida* in goldfish, *Carrassius auratus*. Commenting on the pros and cons of oral vaccination in fish, Ellis [73], while admitting that oral vaccination is the best suited method for mass field-level immunisation with little or no stress, highlighted the intrinsic limitations of this method such as the heavy requirement for the antigen, differential uptake of bacterin due to variations in feeding pattern among the individuals of the vaccinated stock, variable and low protective response and variable protective response under field conditions. Note worthy contributions in the oral vaccination of fish against the motile aeromonad were made by Azad *et al.* [287-290]. The studies were aimed at addressing the problem of antigen fragility in the foregut of carps as a probable reason for the low response elicited by bacterins of *A. hydrophila* obtained from traditional nutrient-rich broth cultures. These investigations were based on the motile aeromonad isolate grown on fine particles of chitin under sub-optimal nutrient conditions to induce biofilm mode of growth. The biofilm cells were heat-inactivated and incorporated in to the carp pelleted feed and administered orally [287]. The biofilm mode of growth facilitated the production of a polymer coating over the growing microcolonies which served as the capsule. Intact and large quantities of the oral antigen of *A. hydrophila* were demonstrated in various lymphoid organs of the Indian major carps and common carp (Figure 5) using monoclonal antibody-based antigen localization [290]. Also high serum antibody titers were recorded [288] contrary to the belief that low or no serum antibody titers are elicited by the oral antigens [291-294]. These studies have paved way for further research on the approach of using bacterial biofilm as oral antigens in fish vaccines.

The concept of bacterial biofilms as antigens: Microorganisms, in general, get attracted towards inert surfaces in liquid medium and produce an adhesive substance, the glycocalyx, which envelops their growing community. The glycocalyx together with the bacterial inhabitants are referred to as the microbial biofilm [295]. Individual bacterial cells forming constituents of the biofilm differ significantly in their metabolic activity and colony characters from those of the planktonic forms in a broth culture.

Bacteria respond to changes in their environment by way of phenotypic variations in the cell wall composition [296] and surface structure [297], immunologically the most important components of a cell. Bacterial biofilms are highly resistant to antibiotics [298-301], surfactants [302] and antibodies [303]. It is possible that this resistant nature of bacterial biofilms might help in minimizing destruction of vaccine by the gastric acids and the enzymes in the foregut and provide a highly immunogenic material for oral vaccination. The glycocalyx of the biofilm of the pathogen is intended to offer resistance, serving as a microcapsule, to delay destruction in the foregut of fish. Hence the studies conducted on the potentials of biofilm of *A. hydrophila* could yield interesting yet convincing evidences of using these forms of bacteria, wherever possible, as effective oral antigens. The concept of using the biofilms antigens in oral vaccination of fish can be effectively and convincingly answered through trials of oral vaccination in fish with stomach.

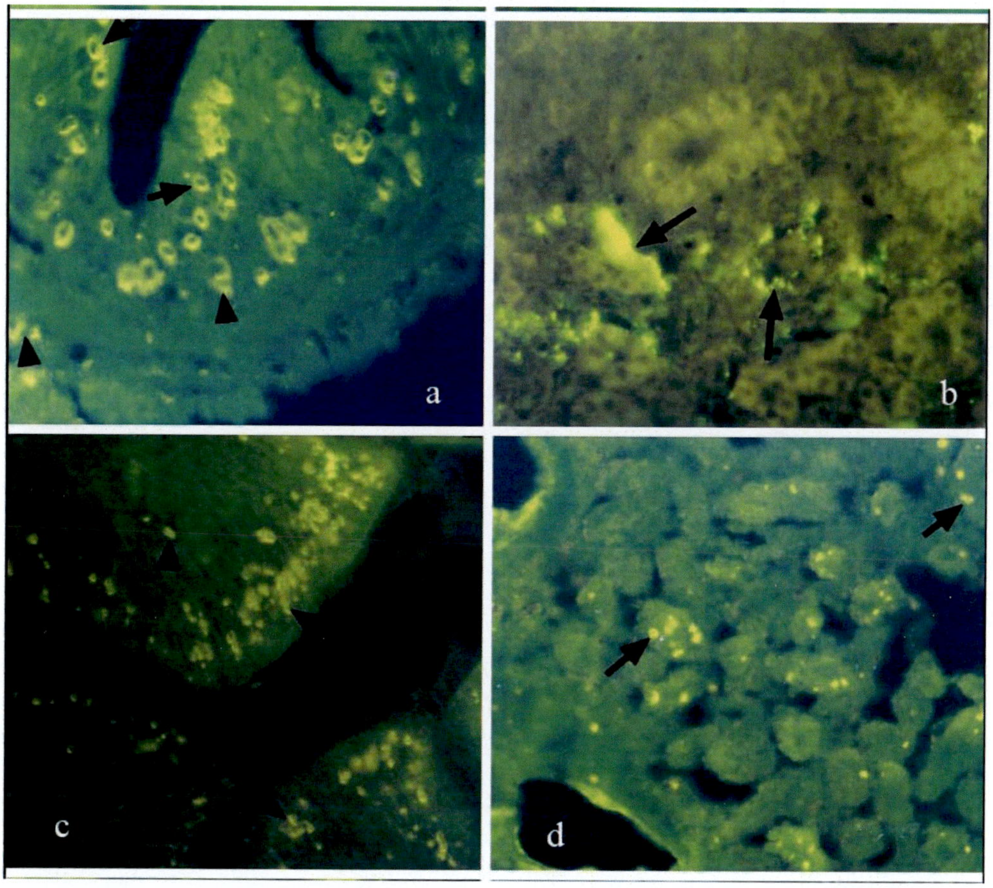

Figure 6. Antigen localization in carps fed with biofilm vaccine of *A. hydrophila* (a and b) compared to that fed free cells (c and d).

Validation of the concept of biofilm antigens was provided by the investigations carried out later using a gastric fish model of freshwater catfish (*C. batrachus*) using biofilm of *A. hydrophila* [304, 305]. The biofilm vaccine produced higher agglutinating antibody titers in catfish. Theses studies indicated on a possible contribution of high molecular weight LPS, expressed in the biofilm mode of growth, in the enhanced serum agglutination antibody titers against *A. hydrophila*. Thus, biofilm antigens, in addition to ensured antigen delivery, helped in enhancing the immune response as well.

Other Aeromonads

Oral vaccination of cutthroat salmon against this bacterial species carried out by Duff, even today, remains a classical example of successful oral vaccination in fish. Orally vaccinated cutthroat trout against furunculosis registered a high level of protection with only 25 percent mortality compared to 75 percent mortality of unvaccinated fish. Several other workers have reported on the oral vaccination using *A. salmonicida* in different fish species. The bacterium attracted the attention of researchers due to the furunculosis disease it caused in salmonids. After the work of Duff, though the out come on vaccination attempts was conflicting and variable, highly useful information on the cell surface and virulence characters of the bacterium was generated. Udey and Fryer [306] were the first to report on the possession of an additional surface layer by the bacterium called A-layer, an important factor associated with the adhesion of bacterial cells to the host cell. This was found to be a 50 kD protein (A-protein) responsible for avoidance and evasion mechanisms against the host defences [307, 308]. The bacteria are not only endowed with the mechanisms for survival in the host due to the above factors, they also possess factors responsible for pathogenesis in fish. Ellis *et al* [309] demonstrated the pathology of Furunculosis and attributed it to the production of extra cellular products (ECPs) within the host.

Attempts of oral vaccination in fish against *A. salmonicida*, after the trials of Duff, did not give encouraging results. Sniezko and Friddle [310] obtained a marginal protection of orally vaccinated salmon, *S. fontinalis* following oral vaccination. Udey and Fryer [306] obtained a negative response in orally vaccinated coho salmon (*O. kisutch*). However, attempts of oral vaccination made in the eighties produced moderate to high levels of protection in different salmonid fish species [311-314]. Infections by atypical *A. salmonicida* was controlled using the formalin-inactivated *A. hydrophila* in goldfish [286] suggesting that cross protection using oral vaccination can also be achieved. Thus, avenues of research using the related bacterial species in vaccination for eliciting protective responses against each other have emerged.

Vibriosis

Vibriosis is the single most important disease caused by the group of Gram negative bacteria belong to the genus *Vibrio*. Vibriosis in various marine aquacultured fish species such as Pacific and Atlantic salmon, rainbow trout, turbot, European seabass, sea bream, striped bass, cod, Japanese and European eel and ayu was extensively studied during the nineties [17, 40, 315-317]. Valuable contributions have been reported by Fryer and his group of workers during the seventies on vaccination against Vibriosis [265-268]. Vibrio vaccines are one of the most extensively used and comparatively more effective antigens in oral vaccination of fish.

A better protective response was observed in salmon and trout administered with antigens by anal intubation than by oral vaccination. Salmon treated with *V. anguillarum* and trout with *Y. ruckeri* showed that the oral route suffers due to antigen destruction by digestive enzymes in the proximal part of the gut [318]. A similar result with oral vaccination in carp against *V. anguillarum* was also recorded [319]. Enteric coated *V. anguillarum* was tested in salmonid fish which produced 80 percent or more survival response [320]. Protection of oral antigens against the digestive processes through microencapsulation of antigens has been demonstrated. Enteric-coated antigen microspheres [321] and chitin-alginate microspheres [322] were very effective in protecting the antigen. However, in one instance of oral intubation of *V. anguillarum* bacterin-layered feed pellet in channel catfish, lower serum antibody titers were recorded [323]. Bioencapsulation of antigenic material using artemia has produced encouraging results [324]. But, the application of these methods becomes limited to the experimental work as more practicable method of antigen delivery is essential for developing a vaccine in aquaculture. Bacterial biofilm is one such option which has shown promising results. Formation of biofilms by *Vibrio* spp and the resistance factor associated with these biofilm forms has been demonstrated in the context of biofilm-associated post harvest contamination of *Vibrio* in marine fish products [325]. Our un-published work showed that *V. alginolyticus* also produces biofilms under sub-optimal nutrient conditions, thus, emphasizing the possibilities of *Vibrio*-associated diseases in re-circulatory and flow-through mariculture and brackishwater aquaculture systems. Hence, the role of biofilm forms of *Vibrio* as oral vaccines in mariculture systems offers scope for further investigations on the immunogenicity of biofilm *Vibrio* in oral vaccination.

Significant pathology and fatal infections due to vibrios other than *V. anguillarum* have also been reported from brackishwater and marine aquaculture systems [326]. *V. alginolyticus* has been reported to cause considerable losses in gilthead sea bream [327, 328], grouper [329] and abdominal swelling-associated mortality in marine fish larvae [330—332]. Hence there is a need for the development of a polyvalent vibrio vaccine for combating the infection due to these vibrio species. An interesting investigation conducted to evaluate the efficacy of acid-resistant diet coating for protecting oral *V. vulnificus* vaccine in flounder [333], showed that the acid-resistant coating produced lower protective efficiency than the oral vaccine delivered without the coating. They suggested that the immunogenic LPS layer of vibrio was more effectively presented in the un-coated antigen. We reported through one of our investigations [334] that a polyvalent Vibrio vaccine was better than respective monovalent vaccines in combating the combined challenge with mixed *Vibrio* spp., indicating the need for developing oral polyvalent *Vibrio* vaccines. A polyvalent vaccine comprising of *V. alginolyticus, V. parahaemolyticus* and *P. damselae* was tested in cobia (*Rachycentron canadum*) for its immunogenicity [335]. The vaccinated fish produced a very high protective response of over 80% when challenged against each of these bacteria both in the laboratory and field conditions. These results provide a proof for the field-level applicability of the polyvalent Vibrio vaccines in mariculture and for designing oral polyvalent vibrio vaccines.

Other Bacterial Diseases

Several other bacterial species have been investigated upon as potential candidate species for the development of the vaccines. *Y. ruckeri* [318, 336—338], *Flexibacter columnaris* [339], *Pasteurella piscida* [340], *E. ictaluri* [341], *P. damsela* [342] have been some of the species of bacteria that attracted the attention of fish immunologists.

Yersiniosis: Yersiniosis is caused by *Y. ruckeri* and the disease is also called as Enteric Redmouth (ERM) disease, mainly occurring in rainbow trout. It gets the name 'redmouth' from the reddening of the mouth. ERM vaccine was the first commercial vaccine to be made available to the aquaculture sector.

BKD: Bacterial Kidney Disease of salmonids is caused by a Gram positive bacterium, *R. salmoninarum*, Piganelli et al [321] produced success in protecting coho salmon orally vaccinated using formalin-killed *R. salmoninarum* after removing the p57 antigen by heat treatment. Live *Arthrobacter* and live Rs TSA1 vaccines were used by Griffiths et al [343] that producing some protection against *R. salmoninarum* in Atlantic salmon. However in their study, the challenge strain of *R. salmoninarum* failed to kill either their nonvaccinated or their vaccinated groups of salmon. Attenuated live bacterial vaccine of *R. salmoninarum* was used in Atlantic salmon to elicit considerable protective response (RPS of 50-76) when challenged against a virulent strain.

Streptococcosis: S. agalactiae and S. iniae are the two important species responsible for streptococcosis in cultured fish. Infections due to *Streptococcus* spp. spread horizontally due to various reasons including those transmitted from infected fish/dead fish. Streptococci released from dead and dying fish into the water are the most important source of infection [344]. Plumb et al [345] opined that the oral route of infection may also significantly contribute for the epizootics involving marine fish. One of the recent fish-kill phenomena occurring in the Arabian Gulf off Kuwait killing over 700 tonnes of fish, principally constituted by mullet, was attributed to a combination of elevated water temperatures and increased load of *S. agalactiae* [346].

Passive (antibodies generated in sheep, rabbit or fish were injected into naïve fish) and active (injection or immersion in formalin killed bacteria) immunization in rainbow trout [347] produced a positive effect of vaccination (>50 % relative percent survival). Vaccination using killed bacterins did not produce protective response against intracellular pathogens [348]. Romalde *et al* [234] could demonstrate long-term protection (2 years) using a toxoid-enriched whole cell enterococcus bacterin administered to turbot. Now efforts are on for the production of DNA vaccines and methods of administering these vaccines more economically and easily.

Lactococcosis: Lactococcosis, one of the serious Gram positive bacterial pathogens, has been reported from the yellowtail (*S. quinqueradiata*) in Japan, rainbow trout in Europe and Australia [349, 350]. Encapsulated and non-encapsulated *Lactococcus graviae* was used in rainbow trout for priming and also for boosting after an intraperitoneal priming with the bacterin [351]. They showed that the boosting using encapsulated bacterin produce a higher protective response (RPS of 87) compared to an all-oral administration (RPS of 50).

Parasites

Parasites are becoming increasingly problematic in aquaculture systems both in the hatchery and grow-out stages. Due to the large number and immune masking of antigens and complexity of life stages, vaccine development against parasites is more challenging. Several studies have been conducted on the protozoan parasites such as the 'itch'. Freshwater 'itch' (*Ichthyophterius multiphilis*) is the most studied parasite. Immobilizing 'I' antigen has been investigated for its immunogenicity and protective response in channel catfish [352-355].

Cryptobia salmositica is a parasite that could cause severe losses in the semi-natural and intensive salmon culture facilities [356]. Attenuation through continuous sub culture and use as vaccine in a number of salmonid species has shown promising results [357, 358]. A single dose of live vaccine in rainbow trout produced protection up to 24 month post vaccination [359]. They attributed the protective response to complement-fixing antibody production and enhancing Phagocytosis and cell-mediated cytotoxicity and the later investigation showed that the live vaccine triggers respiratory burst [360].

Viruses

Like killed bacterial whole cell preparations, attenuated viruses, have been studied in greater details as vaccines. Salmonid aquaculture was the first to reap benefits from these studies. Injection vaccination against many viruses has been studied extensively in different fish species. Infectious pancreatic necrosis (IPN) virus was studied for its immunogenicity in salmonids [269, 270, 361]. Vaccination potential of candidate vaccines against viral haemorrhagic septicaemia (VHS) has been demonstrated, and several a virulent VHS virus vaccines are under development [273, 362, 363]. Almost Formalin-killed irido virus vaccines were field-tested to be effective in red sea bream paving way for the development of a vaccine against RSIV [364, 365]. Formalin-killed virus causing systemic necrosis in carps (VSNC) was also tested to be effective under trials through injection trials [366].

DNA and Recombinant Vaccines

Antigens presented to the immune system are responsible for the immune response and immune memory; this forms the classical basis of vaccination. Now, with the advent of molecular techniques, it is possible to insert the DNA into fish the cells of which take up the DNA and produce the antigen that is encoded by the inserted DNA producing vaccine effect. The exact mechanism with which the fish somatic cells take up the DNA and produce the antigen of interest is not very clear yet [367]. Another way using molecular tools is through the production of recombinant antigens in suitable vectors and bio-encapsulating these antigen carrying bacterial vectors such as *E coli* in natural food of fish (e.g. artemia or rotifers). Zebra fish model was successfully used to demonstrate this possibility [368]. The most important step that leads to the success of developing DNA vaccines in aquaculture is the identification of immunogenic and protective antigenic epitopes and the gene encoding these epitopes/antigens. Surely, the classical approaches have provided this base-line information in many of the bacterial pathogens. Coat proteins of the viral particles have been targeted to provide effective DNA vaccines. Works on viral haemorrhagic septicaemia virus (VHSV) and infectious haematopoietic necrosis virus (IHNV) have shown that viral surface glycoprotein produce protective antibodies in fish [369, 370]. Future of fish vaccinology belongs to the molecular/genetic engineered vaccine development programmes. Investigations on the oral delivery of these next-generation immunogens in aquaculture are still in infancy and are bound to intensify in the years to come.

PROBIOTICS-THE CATCH-WORD IN MODERN AQUACULTURE

Probiotics - What and why in Aquaculture?

"Microbial interference therapy" is the term given to the use of probiotics to prevent antibiotic resistance. The use of antibiotics in aquaculture is to be discouraged as it has led to the appearance of drug-resistant bacteria [406-408], immunosuppression in animals besides harmful effects on the environment and concerns on food safety. The use antibiotics are sometimes banned due to the rejection of export consignments of aquaculture products.

On the other hand microbes, in general and probionts in particular with their unique structure and cell wall components can trigger immunity. Thus microbial intervention through probiotics is fast developing as an alternative way for health management in aquaculture and to improve the aquatic environment quality. "Probiotics" has thus become the 'catch-word in modern aquaculture.

Greek roots of ''probiotic'' (*pro*: positive and *bios*: life) suggest that the substance is beneficial to life. Probiotics are defined as microbial cell preparations or components of microbial cells, which have a beneficial effect on the health and well-being of the host [409]. The Joint FAO/WHO Working Group Report [410] on drafting guidelines for the "Evaluation of Probiotics" in Food defined probiotics as live microorganisms that confer a beneficial physiological effect on the host when administered in adequate amounts.

Taking in to account the mode of action in aquaculture, Verschuere et al. [411] defined probiotics as a live microbial adjunct which has a beneficial effect on the host by (i) modifying the host-associated or ambient microbial community, (ii) ensuring improved use of the feed or enhancing its nutritional value, (iii) enhancing the host's response towards disease, or (iv) improving the quality of its ambient environment.

Probionts as Immunogens

A wide range of microalgae (*Tetraselmis*), yeasts (*Debaryomyces, Phaffia* and *Saccharomyces*) and Gram-positive (*Arthrobacter, Bacillus, Carnobacterium, Enterococcus, Lactobacillus, Lactococcus, Micrococcus, Streptococcus and Weissella*) and Gram-negative bacteria (*Aeromonas, Alteromonas, Photorhodobacterium, Pseudomonas* and *Vibrio*) has been evaluated for probiotic applications. Many studies elucidate the use of Gram positive lactic acid bacteria or Gram negative bacteria as probiotics [44, 45, 72, 411-413]. Probiotic antigens as non-specific immunostimulants [133, 200, 414, 415] has been proposed, in addition to the effort to improve water quality [416, 417] and nutrition [418], as a means to increase larval survival and aquaculture output. Again, the probionts belonging to *Vibrionacea, Pseudomonads* and *Bacillus* are found to give protection to turbot, salmon, cod, prawns and oysters. These bacteria have a wide range of tolerance for salinity, temperature and pH, which usually exists in aquaculture operations. The targeted groups are Pacific oysters [419, 420], Penaeid shrimp [421], Atlantic cod [422], common snook flounder [423], tilapia [424], turbot [425] and salmonids [413].The water probiotics contain multiple strains of bacteria like *Bacillus acidophilus*, *B. subtilis*, *B.licheniformis*, *Nitrobacter* sp., *Aerobacter* sp. and *Saccharomyces cerevisiae* and feed probiotics contain *Lactobacillus* sp., *Bacillus* sp. or *S. cerevisiae*. Both water and soil based probiotics reported to give better survival and growth and improve the protective response especially in the larval stages. Queiroz and Boyd [426] reported that a commercially prepared bacterial mixture of *Bacillus* spp. mixed into the rearing water increased survival and production of channel catfish (*I. punctatus).* According to Austin et al [427] the bacteria *V.alginolyticus* was found to reduce diseases in Atlantic salmon caused by infection of common pathogenic strains (*A. salmonicida*, *V. anguillarum*). Regular use of probiotics in the feed of fish in U. K. and other European countries has been reported to have several health benefits. Atlantic salmon fed with probiotics showed increased survival and reduced mortality caused by Vibriosis, Furunculosis and Enteric Redmouth diseases. Moreover, fish showed enhanced appetite, grew better and had lesser problems with fin and tail rot. *V. alginolyticus* introduced in larval rearing tanks caused a reduction in the incidence and severity of luminous vibriosis caused by *V. harveyi* and improvement in growth of shrimp larvae.

Component of bacterial cells (e.g. lippopolysaccharides, LPS and peptidoglycan, PG) are well known as potent modulator of immune response. In this sense, LPS and PG may be viewed as bacterial derived vitamin, which provide a continuous signal to the defence system to be alert [428]. Nikoskelainen et al [415] demonstrated that the administration of probiotic *L. rhamnosus* ATCC 53101 in diet stimulated the respiratory burst activity, after 2 weeks of feeding. In our study, the LAB supplementation in the diet for longer duration did not sustain the increased superoxide anion production of the leucocytes.

Probionts as Oral Immunogens

The three essential mechanisms proposed for the mode of action of probiotics are:

I) Stimulation of immunity through cellular, humoral and molecular immune response
II) Competitive exclusion by which the probiotic species antagonizes the potential pathogen by the production of inhibitory compounds or by competition for nutrients, space (adhesion sites in the digestive tract) or oxygen.
III) Alteration of microbial metabolism: Enzymes

Other protective mechanisms include antitumorogenic, hypocholesteremic, antimutagenic, anticarcinogenic properties.

Interaction between dietary probitics and immunocompetence has been established with our studies. By increasing the host's specific and non-specific immune mechanisms, lactic acid bacteria (LAB) could increase the immune response [429, 430].

Probiotics – The Mechanisms

The enhancement of innate immune function in healthy subjects given specific probiotic strains is well documented. The innate immune responses represent the first line of host defence against pathogenic organisms. The major cellular components of non-specific immunity include phagocytic cells (monocytes, macrophages and polymorphonuclear cells) and NK cells. These cells possess the ability to recognize molecular patterns shared between groups of pathogens (lipopolysaccharide, lipotechoic acid, peptidoglycans, etc.) with the help of their pattern-recognition receptors (e.g. Toll-like receptors and CD14).

Effect on humoral immunity as observed in increase in lysozyme and complement activity rainbow trout fed with *L. rhamnossus* [430]. The cellular immunity is elicited as seen in marked increase in phagocytic activity of the head kidney leucocytes of rainbow trout with probiotic feeding [414, 415, 429, 430]. Phagocytic cells play a central role in combating microbial infections, whereas NK cells are pivotal for host defence against viral infections and cancers. The intake of specific strains of LAB has also been shown to enhance humoral immune responses to natural infections and systemic or oral immunizations in human subjects 98 – 105.

Probiotic products are usually standardized based on the presumption that the culture viability is a reasonable measure of their activity. Viable live probionts are better than the non-viable heat killed probionts in inducing a higher immune response in rainbow trout especially in the head kidney leucocyte phagocytosis, serum complements, etc [430] and in shrimp with heat killed vibrio bacterin [71].

Probiont and Immune Gene Interaction

Cytokine interaction: Cytokines comprise the largest and the most pleotropic group of immune response mediators. The initiation, maintenance and resolution of both innate and acquired immune responses are regulated by cell-to-cell communication via cytokines. The ability of orally delivered *L. breve* and *B. lactis* to enhance the IFN-α producing capacity in healthy subjects has also been demonstrated. A large variation in the ability of different *Lactobacillus* strains to induce pro- and anti-inflammatory cytokines has also been observed. These cytokines enhances the phagocyte-mediated clearance of micro-organisms (by increasing respiratory burst, nitric oxide production and the expression of Fc receptors on the phagocytic cells), induces MHCI and MHCII expression on a variety of cells, stimulates Th cell function and augments the immunogenicity of vaccines. These cytokines contribute to defence mechanisms of the host in response to bacterial colonization or invasion, through a series of experiments [431], Panigrahi and his team [429-431] made attempts to draw some relation between the different characteristics of probionts and the immune response they elicit in coldwater fish rainbow trout. Three different bacterial species namely; *L. rhamnosus, Bacillus subtilis and Eterococcous faecium* were found to influence the immune response both at local and systemic level. The probiotics bacteria induced non-specific immune responses and the expression of certain cytokine genes, interleukins (IL- β1 and 2), tumour necrosis factor (TNF- α 1 and 2) and transforming growth factor (TGF-β). Temperature is a major environmental factor controlling microbial growth and the ideal conditions differ with the species. A probiotic would be most effective when used in its optimum temperature range. The non-specific cellular immune response increased with probiotic feeding as reflected by the increase of phagocytosis and super oxide anion of head kidney leucocytes. A correlation was found to exist between the probiont optimal temperature requirements and immune gene expression pattern. The immune gene, especially of the cytokine, showed differential expression pattern with the strain of probiotic bacteria.

Probiotic bacteria influence the plasma lipid components, enzyme activities, plasma total cholesterol and triglyceride levels. Alkaline phosphatase activities increased after feeding probiotic *L. rhamnosus* incorporated diet.

Mechanism: Competitive Exclusion and Microbial Metabolism:

Probiotics compete for adhesion sites and nutrients with other microbes in the niches. They antagonise other microbes by producing antimicrobial substances like organic acids, bacteriocins, bacteriocin-like substances, hydrogen peroxide, non-proteinaceous inhibitory metabolites like diacetyl, reuterin and other compounds. They also induce the host defence by inducing the production of antimicrobial proteins like defensins. Manipulation of bacterial species in the water; sediment and animal gut is possible through competitive exclusion. Probiotic bacteria may competitively exclude the pathogenic bacteria or produce substances to inhibit the growth of pathogens. In addition, it provides essential nutrients to the cultured animals, digestive enzyme for better nutrition. Probiotic bacteria directly utilise or decompose the organic matter or toxic material and improve the quality of water. These beneficial microbial species produce a variety of enzymes such as amylase, protease, lipase, xylanase and cellulase along with B group of vitamins in high concentrations than the native bacteria.

This would include the host responses to short chain fatty acid and other bacterial metabolites [432]. The principal probiotic yeast is *S. boulardi (S. cerevisiae),* a non-pathogenic yeast, which has been in use.

Prebiotics: Prebiotics are basically feed for probiotics where they are resistant to attack by indigenous enzymes and hence reach the site for proliferation of gut microflora. Some of the prebiotics which are currently used in animal feed are Mannan-oligosaccharides (MOS), fructo-oligosaccharide (FOS) and mixed oligo-dextran. Bacteria have lectins (glycoprotein) on the cell surface that recognize specific sugars and allow the cell to attach to that sugar. Binding of *Salmonella, E. coli and Vibrio* sp. is shown to be mediated by a mannose specific lectin-like substance present on the bacterial cell surface.

Prospects of Probionts in Aquaculture

In many cases, of late, probiotics are being used in all aquaculture systems, specifically in the shrimp farming sector, with so-called zero/less water exchange probiotic culture systems. Shrimp farmers, it appears, go for various probiotics, immunostimulant products from the market out of desperation than by established or researched technologies. Most of these so-called 'products' lack scientific evaluation. Investigations into microbial interventions were initiated to make the aquatic production and disease management measures more sustainable and environment friendly. Genetically modified microbes can also improve water and soil quality by reducing levels of phosphorus and nitrogen in aquaculture waste and these microbial entities can be evaluated. Acceptance of microbial intervention concept by regulatory bodies will be possible only if the mechanisms of their action are well explained. It is evident that the manner in which probiotics as immunogen can impact as nutritional immunomodulation varies depending on individual strain, its state, its level of consumption and ambient environmental conditions

Probiotic bacteria are increasingly recognized as a component for alternative health management strategy. However the entry of high claiming probiotics in to the market without proper screening is a real concern. The mechanisms of probiotic action in aquaculture, including that at the molecular level as observed through beneficial effects, are not fully or clearly elucidated. Dietary supplementation with defined probiotics may be an effective biotherapeutic or prophylactic means in aquaculture. Recent work on the effects of administering probiotics strains on the GALT of the larval gilthead sea bream suggested a clear beneficial effect on the survival response through enhanced gut leucocytes and immunoglobulin-positive cell proliferation, especially when the probiotics was administered during the gut metamorphosis [433]. Future of this unexplored field will be best justified by taking immediate steps to understand the extent and magnitude of microbial diversity recognizing the aquaculturally important microorganisms, elucidating the metabolic potential of the microbial diversity and their application in aquaculture.

ORAL ROUTES – MORE MEANINGFUL IN AQUATIC ANIMAL HEALTH MANAGEMENT

Its clear from the efforts made thus far that the aquatic environment is more influential than what is realised through extrapolations made based on land-based animal production systems. The medium in which the aquatic animal lives is soup of complex molecules that constantly influence the ability of the aquatic animal to respond to the variations. This happens too sharply and intensely in the aquaculture environment. Crowding, metabolite accumulation, competition for food and availability of choice food make the task of health management a challenging one.

Keeping other things constant, if the ease of application and stress exerted due to the intervention are considered, oral route provides the best option for the delivery of immunogens. Vaccines, immunostimulants, nutritional immune enhancers, prebiotics (food for probionts) and the probionts all can be effectively delivered to elicit immune modulation making the aquacultured organism strong enough to withstand culture-related stress and counter the constant threats by pathogens. Antigen/immunogen presentation to the cellular levels in right quantities and in the valid form becomes more important so that a long lasting effect is achieved. Antigen protection vehicles (APVs) for vaccine delivery, modified antigens (biofilm/bio-encapsulation), engineered antigens (re-combinant immunogenic epitopes) and host-gene interventions (DNA vaccines), all have a role to play in the development of a successful Technology for Oral Immune Modulation (TOIM). As the aquatic species have a varied gut environments and that the gut environment in itself is a hub of biochemical reactions, the efficacy of the TOIM depends on how best the immune responsive cells are primed and sustained to start and continue the task of immune modulation. Role of probiotics seems to be not so well realised in aquaculture health management. These beneficial bacteria have a great role to play in the maintenance of a near-natural condition of the aquatic environment, reducing the proliferation of pathogenic bacteria in the ubiquitous environment and enhancing the non-specific immunity (in some cases even specific immunity) of fish and shellfish.

REFERENCES

[1] FAO (1990). *Aqaculture production* (1985- 1988), FAO Fisheries Circular No.815, Revision 4, FAO, Rome.

[2] Hoshina, T. (1956). An epidemic disease affecting rainbow trout in Japan. *J. Tokyo Univ. Fish 42*, 15-16.

[3] Smith, I.W. (1961). A disease of finnock due to *Vibrio anguillarum*. *J. Gen. Microbiol. 24*, 247 – 252.

[4] Davis, G.H.G. and Park, R.W.A., (1962). A taxonomic study of certain bacteria currently classified as Vibrio species. *J. Gen. Microbiol. 27*, 101 – 119.

[5] Snieszko, S.F. (1964). Remarks on some facets of epizootiology of bacterial fish diseases. *Dev. Ind. Microbiol. 5*, 97 – 100.

[6] Snieszko, S.F., Bullock, G.L., Hollis, E. and Boone, J.G. (1964). *Pasteurella* sp. from an epizootic of white perch (*Roccus americanus*) in Chesapeake Bay tidewater areas. *J. Bacteriol 88*, 1814 – 1815.

[7] Snieszko, S.F., Bullock, G.L., Dunbar, C.E. and Pettijohn, L.L. (1964). Nocardial infection in hatchery-reared fingerling trout (*Salmo gairdneri*). *J. Bacteriol. 88*, 2809 – 2810.

[8] Cisar, J.O. and Fryer, J.L. (1969). An epizootic of vibriosis in Chinook salmon. *Bull Wildl. Dis. Assoc. 5*, 73 –76.

[9] Kaper, J.B., Lockman, H. and Colwell, R.R. (1981). *Aeromonas hydrophila* : Ecology and toxigenicity of isolates from an estuary. *J. Applied Bacteriol. 50,* 359-377.

[10] Horne, M. T. (1982). The pathogenicity of *Vibrio anguillarum* (Bergman). In: *Microbial Diseases of Fish*, R.J.Roberts (ed.), Academic Press, pp. 171 – 187.

[11] Newman, S. (1983). *Aeromonas hydrophila*: A review with emphasis on its role in fish disease. In: *Antigens of Fish Pathogens*, D.P.Anderson, M.Dorson and Ph. Dubourget (eds.), Collection Foundation Marcel Merieux, pp. 87-117.

[12] Colwell, R. R. and Grimes, D. J. 1984. Vibrio diseases of marine fish populations. Helgol.Wiss. Meeresunters, 37: 265 – 287.

[13] Amin, N.E., Abdulla, I.S., Elallawy, T. and Ahmed, S.M. (1985) Motile Aeromonas septicemia among *Tilapia nilotica* (*Sarotherodon nilotica*) in upper Egypt. *Fish Pathol. 20*, 93-97.

[14] Miyazaki, T. and Jo, Y. (1985). A histopathological study of motile aeromonad disease in Ayu, *Plecoglosisus altivelis*, *Fish Pathol. 20*, 55-60.

[15] Egidius, E. (1987). Vibriosis: Pathogenicity and pathology- A review. *Aquaculture 67,* 15 – 28.

[16] Frerichs, G.N. (1989). Bacterial diseases of marine fish. *The Vet. Rec 125*, 315-318.

[17] Toranzo, A.J. and Barja, J.L. (1990). A review of the taxonomy and seroepizootiology of *Vibrio anguillarum*, with special reference to aquaculture in the northwest of Spain. *Dis. Aquat. Org. 9,* 73 – 82.

[18] Tiffney, W.N. (1939). The host range of *Saprolegnia parasitica. Mycologia, 31*, 310 – 321.

[19] Tiffney, W.N. (1939). The identification of certain species of the *Saprolegniaceae* parasitic to fish. *J. Elisha Mitch Scient. Soc. 55*, 134 – 151.

[20] Miyazaki, T. and Kubata, S.S. (1977). Studies on *Ichthyofonus* disease of fishes – III. Life cycle of *Ichthyofonus* affected rainbow trout. *Bull Fac Fish Mie Univ 4*, 67 –80.

[21] Chien, C., Miyazaki, T. and Kubota, S.S. (1979). Studies on Ichthyofonus diseases of Fishes – IV. Comparative study on naturally infected fishes. *Bull. Fac. Mie. Univ .6*, 129 – 146.

[22] Callinan, R.B., Fraser, G.C. and Virgona, J.L. (1989). Pathology of red spot disease in sea mullet, *Mugil cephalus* L., from eastern Australia. *J. Fish Dis .12*, 467 – 479.

[23] Roberts, R.J., Campbell, B. and MacRae, I.H. (eds.).(1994). Proceedings of the ODA regional seminar on Epizootic Ulcerative Syndrome. 25 – 27 Jan. '94. Bangkok, Thailand. 281 p.

[24] Mohan, C.V. and Shankar, K.M. (1995). Role of fungus in epizootic ulcerative syndrome of fresh and brackishwater fishes of Karnataka, India – a histopathological assessment. In : Shariff, M., Arthur,J.R., Subasinghe,R.P. (eds.), *Diseases in Asian Aquaculture* : I. Fish Health Section, Asian Fisheries Soc., Manila. Pp. 299 – 305.

[25] Vishwanath, T.S., Mohan, C.V. and Shankar, K.M. (1997). Mycotic granulomatosis and seasonality are the consistent features of Epizootic Ulcerative Syndrome of fresh and brackishwater fishes of Karnataka, India. *Asian Fish Sci. 10*, 155 – 160.

[26] Vishwanath, T.S., Mohan, C.V. and Shankar, K.M. (1998). Epizootic Ulcerative Syndrome (EUS), associated with a fungal pathogen, in Indian fishes: histopathology – ' a cause for invasiveness'. *Aquaculture 165,* 1-9.

[27] Wolf, K., Sniezko, S.F., Dunbar, C.E. and Pyle, E. (1960). Virus nature of infectious pancreatic necrosis in trout. *Proc .Soc. Exper .Biol. Med. 104*, 105-108.

[28] Wolf, K. and Darlington, R.W. (1971). Channel catfish virus: a new herpes-virus of ictalurid fish. *J.. Virol. 8(4*), 525-533.

[29] Wolf, K. (1988). Fish Viruses and Fish Viral Diseases. Cornell Univ. Press, Ithaca, NY.

[30] Munday, B. and Nakai, T. (1997). Special topic review: nodaviruses as pathogens in larval and juvenile marine fin fishes. *Worl. J. Micrbiol. Biotech. 13*, 375-381.

[31] Azad, I.S., Shekhar, M.S., Thirunavukkarasu, A.R., Poornima, M., Kailasam, M., Rajan, J.J.S., Ali, S.A., Mathew Abraham and Ravichandran, P. (2005). Nodavirus infection causes mass mortalities in hatchery produced larvae of Asian seabass, *Lates calcarifer*, Bloch: first report from India. *Dis. Aquat. Org 63*, 113-118.

[32] Woo, P.T.K. (1979). *Trypanoplasma salmositica* : experimental infection in rainbow trout , *Salmo gairdneri*. *Exper. Parsitol. 47*, 36 – 48.

[33] Nagasawa, K. (1987). Prevalence and abundance of *Lepeophtherius salmonis* (Copepod: Caligidae) on high-sea salmon and trout in the North Pacific Ocean. *Nippon Suisan Gakkaishi 53*, 2151 – 2156.

[34] Dezfuli, B.S., Manera, M., Onestini, S. and Ross, R. (1997). Histopathology of the alimentary canal of *Anguilla anguilla* L. associated with digenetic trematodes: a light and electron microscopic study. *J. Fish Dis. 20*, 317 – 322.

[35] Company, R., Sitja-Bobadil, A., Pujalte, M.J., Garay, A., Alvarez-Pellitero, P. and Perez-Sanchez, J. (1999). Bacterial and parasitic pathogens in cultured common dentex, *Dentex dentex* L. *J. Fish Dis. 22(4)*, 299 – 309.

[36] Azad, I.S., Al-Marzouk, A., James, C.M., Almatar, S. and Al-Gharabally, H. (2007). Scuticociliatosis-associated mortalities and histopathology of natural infection in cultured silver pomfret (*Pampus argenteus* Euphrasen) in Kuwait. *Aquaculture 262*, 202-210.

[37] Aoki, T. Kitao, T and Arai, T. (1977). R plasmids in fish pathogens, In S. Mitsuhashi, L. Rosival, and W. Krcmery (ed.), Plasmids: medical and theoretical aspects. Avicenum Czechoslovak Medical Press, Prague. p. 39-45.

[38] Mitchell, A.J. and Plumb, J.E. (1980). Toxicity and efficacy of furnace on channel catfish infected experimentally with *Aeromonas hydrophila*. *J. Fish. Dis. 3*, 93-110.

[39] Ruangapan, L., Kitao, T. and Yoshida, T. (1986). Protective efficacy of *Aeromonas hydrophila* vaccine in Nile tilapia. *Vet. Immunol. Immunopathol. 12,* 345-350.

[40] Toranzo, A.E. and Barja, J.L. (1993). Virulence factors of bacteria pathogenic for cold water fish. *Annu. Rev. Fish. Dis. 3*, 5– 36.

[41] Alcaide, E., Blasco, M-D. and Esteve, C. (2005). Occurrence of Drug-Resistant Bacteria in Two European Eel Farms. *Appl. Environ. Microbiol. 71*, 3348–3350.

[42] Söderhäll, K. and Häll, L. (1984). Lipopolysaccharide induced activation of prophenoloxidase activating system in crayfish haemocyte lysate. *Biochem. Biophys. Acta. 797*, 99-104.

[43] Sung, H.H., Yang, Y.L. and Song, Y.L. (1996). Enhancement of microbicidal activity in the tiger shrimp *Penaeus monodon* via immunostimulation. *J. .Crust. Biol. 16(2)*, 278-284.

[44] Ringø, E. and Gatesoupe, F.J. (1998). Lactic acid bacteria in fish: a review. *Aquaculture 160*, 177-203.

[45] Gatesoupe, F.J. (1999). The use of probiotics in aquaculture. *Aquaculture 180*, 147-165.

[46] Vine, N.G. (2004). Towards the development of a protocol for the selection of probiotics in marine fish larviculture. PhD thesis, Rhodes University, Grahamstown, South Africa.

[47] Hardie, L.J, Marsden, M.J, Fletcher, T.C. and Secombes, C.J. (1983). *In vitro* addition of vitamin C affects rainbow trout lymphocyte responses. *Fish Shellfish Immunol. 3*, 207-219.

[48] Hardie, L.J, Fletcher, T.C. and Secombes, C.J. (1991). The effect of vitamin E on the immune response of the Atlantic salmon (*Salmo salar* L.). *Aquaculture 87*, 1-13.

[49] Navarre, O. and Halver, J.E. (1989). Disease resistance and humoral antibody production in rainbow trout fed high levels of vitamin C. *Aquaculture 79*, 207-221.

[50] Blazer, V.S. (1992). Nutrition and disease resistance in fish. *Ann. Rev. Fish Dis. 1*, 309-323.

[51] Lall, S.P. and Olivier, G. (1993). Role of micronutrients in immune response and disease resistance in fish. In: Kaushik, S.J., Luquet, P., Eds.) *Fish Nutrition in Practice*, INRA, Paris. pp. 101-118.

[52] Waagbø, R., Glette, J., Nilsen, E.R. and Sandnes, K. (1993). Dietary vitamin C, immunity and disease resistance in Atlantic salmon (*Salmo salar*). *J. Fish. Physiol. Biochem. 12*, 61-73.

[53] Raa, J. (1996). The use of immunostimulatory substances in fish and shellfish farming. *Rev Fish Sci 4(3)*, 229-288.

[54] Gannam, A.L. and Schrock, R.M. (1999). Immunostimulants in fish diets. *J. Appl. Aquaculture 9(4)*, 53-89.

[55] Gallindo-Villegas, J. and Hosokawa, H. (2004). Immunostimulants: Towards temporary prevention of diseases in marine fish. In: Cruz Suárez, L.E., Ricque Marie, D., Nieto López, M.G., Villarreal, D., Scholoz, U. González, M. 2004. *Advances en Nutición Acuicola* VII. Memorias del VII Simposium Internacional de Nutición Acuicola. 16-19 Noviembre, 2004. Hermosillo, Sonora, México.

[56] Bell, J.G., Raynard, R.S. and Sargent, J.R. (1991). The effect of dietary linoleic acid on the fatty acid composition of individual phospholipids and lipoxygenase products from gills and leucocytes of Atlantic salmon (*Salmo salar*). *Lipids* 26, 445-450.

[57] Woodward, W. (1994). Dietary vitamin requirements of cultured young fish, with special emphasis on quantitative estimates for salmonids. *Aquaculture 124*, 133-168.

[58] Kiron, V., Fukuda, H., Takeuchi, T. and Watanabe, T. (1995). Essential fatty acid nutrition and defence mechanisms in rainbow trout *Oncorhynchus mykiss*. *Com. Biochem. Physiol. 111A,* 361-367.

[59] Kiron, V., Fukoda, H., Okamoto, N. and Takeuchi, T. (1995). Protein nutrition and defence mechanisms in rainbow trout *Oncorhynchus mykiss*. *Comp. Biochem. Physiol. 111*, 351-359.

[60] Waagbø, R. (1997). The impact of nutritional factors on the immune system in Atlantic salmon, *Salmo salar* L.: A review. *Aqauatic Fish Manag 25*, 175-197.

[61] Hamre, K., Waagbø, R., Berg, R. and Lie, Ø. (1997). Vitamin C and E interact in juvenile Atlantic salmon (*Salmo salar* L.). *Free Radic. Biol. Med., 22,*137-140.

[62] Janeway Jr Ch, A. (1992). The immune system evolved to discriminate infectious self from noninfectious self. *Immunol. Today 13*, 11-16.

[63] Aderem, A. and Ulevitch, R.J. (2000). Toll-like receptors in the induction of the innate immune response. *Nature 406*, 782-787.

[64] O'Hagan, D.T., MacKichan, M.L. and Singh, M. (2001). Recent developments in adjuvants for vaccines against infectious diseases. *Biomol Engineering 18*, 69-85.

[65] Schinjs V. E.J.C (2003). Mechanisms of vaccine adjuvant activity: initiation and regulation of immune responses by vaccine adjuvants. *Vaccine 21*: 829-831.

[66] Bricknell, I.R. and Dalmo, R.A. (2005). The use of immunostimulants in fish larval aquaculture. *Fish Shellfish Immunol. 19*: 457-472.

[67] Johansson, M.W. and Söderhäll, K. (1989). Cellular immunity in crustaceans and the proPO system. *Parasitol.Today 5*, 171-176.

[68] Johnson, M.W. and Söderhäll, K. (1989). A cell adhesion factor from cray fish haemocyte has degranulating activity towards cray fish granular cells. *Insect Biochem. 19*, 183– 190.

[69] Ashida, M. and Yamazaki, H.I. (1990). Biochemistry of phenoloxidase system in insects with special reference to its activation. In: Editors Omishi, E., Ishizaki, H., Molting and Metamorphosis. Japan Sci Soc Press Tokyo, pp. 239– 265.

[70] Azad, I.S., Panigrahi, A., Natarajan, M., Kulasekarapandian, S., Shivagnanam, S., Ravichandran, P. and Rao, L.H. (2002). Booster dose of immunostimulation in larvae of white shrimp and tiger shrimp: protective response and growth. In: Editors. Ayyappan, S., Jena, J.K., Mohan Joseph, M., Proceedings of the Fifth Indian Fisheries Forum. Asian Fisheries Society, Indian Branch, Mangalore, pp. 147–150.

[71] Azad I. S., Panigrahi A., Gopal C., Paulpandi S., Mahima C. and Ravichandran P. (2005). Routes of immunostimulation *vis-a vis* survival and growth of *Penaeus monodon* (Fabricius) postlarvae. *Aquaculture 248*, 227-234.

[72] Olsson, J.C., Joborn, A., Westerdahi, A., Blomberg, L., Kjelleberg, S. and Conway, P.L. (1998) Survival, persistence and proliferation of *Vibrio anguillarum* in juvenile turbot, *Scophthalmus maximus*, intestine and faeces. *J. Fish Disease 21*,1-9.

[73] Ellis A.E. (1988). Ontogeny of the immune system in teleost fish. In: General principles of fish vaccination. Academic Press, New York, pp 20-31.

[74] Rombout, J.H.W.M., Lamers, C.H.J., Helfrich, M.H., Dekker, A. and Taverne-Thiele, J.J. (1985). Uptake and transport of intact macromolecules in the intestinal epithelium of carp (*Cyprinus carpio* L.) and the possible immunological implications. *Cell Tissue Res 239*, 519-530.

[75] *http://www.cnsweb.org/digestvertebrates/WWWEdStevensTopicsAnatomyGIT.html*

[76] Horn, M. H. (1989). Biology of marine herbivorous fishes. Oceanogr. *Mar. Biol. Annu. Rev. 27*, 167-272.

[77] Smith, L.S. (2007). *http://www.fao.org/docrep/X5738E/x5738e02.htm*

[78] Smith, B.W. and Lovell, R.T. (1973). Determination of apparent protein digestibility in feeds for channel catfish. *Trans Am. Fish Soc. 102*, 831-835.

[79] Yoshinaka, R., Sato, M and Ideka, S. (1973). Studies on collaginase of fish. I. Existence of colla-genolytic enzyme in pyloric caecae of *Seriola quinqueradiata. Bull. Japan Soc. Sci Fish 39(3)*, 275-281.

[80] Smirnova, L. I. (1966). Digestive leucocytosis of bream (*Abramis brama*). In: Biology of fishes of the Volga reservoirs. Tr.Inst.Biol.Vnutr.Vod./Trans. Inst. Biol. *Inland Waters. 10 (13):*143-147.

[81] Rombout, J.H.W.M., Tavernethiele, A.J. and Villena, M.I. (1993). The gut-associated lymphoid tissue (GALT) of carp *(Cyprinuscarpio* L.) - an immunocytochemical analysis. *Dev. Comp. Immunol. 17*, 55–66.

[82] Dalmo, R.A, Ingebrigtsen, K, and Bøgwald. J. (1997) Non-specific defence mechanisms in fish, with particular reference to the reticuloendothelial system (RES). *J Fish Dis 20*, 241–273.

[83] Lin, S.H, Davidson, G.A., Secombes, C.J. and Ellis, A.E. (1998). A morphological study of cells isoltated from the perfused gill of dab and Atlantic salmon. *J. Fish Biol. 53,* 560–568.

[84] Press, C. and McL Evensen Ø (1999) The morphology of the immune system in teleost fish. *Fish Shellfish Immunol. 9*, 309–318.

[85] Herbomel, P., Thisse, B. and Thisse, C. (1999). Ontogeny and behavior of early macrophages in the zebrafish embryo. *Development 126*, 3735-3745.

[86] Vadstein, O., Mo, T.A. and Bergh, Ø. (2004). Microbial interactions, prophylaxis and diseases. In: Moksness, Kjørsvik, Olsen, editors. *Culture of cold-water marine fish.* Oxford Blackwell Publishing Ltd.; p 28-64.

[87] Kjørsvik, E. van der Meeren, T., Kryvi, H., Arnfinnsson, J. and Kvenseth, P. (1991). Early development of digestive tract of cod larvae (*Gadus marhua* L.) during start-feeding and starvation. *J. Fish Biol. 39*: 1-51.

[88] Olafsen, J.A. and Hansen, G.H. (1992). Intact antigen uptake in intestinal epithelial cells of marine fish larvae. *J. Fish Biol. 40,* 141-156.

[89] Sarasquete, M.C., Polo, A. and Gonzales de Canales, M.L. (1993). A histochemical and ultrastructural study of digestive enzymes and hormones during larval development of the sea bream, *Sparus aurata* L. *Histochemical J. 25*, 430-437.

[90] Elbal, M.T., Garcia Hernandez, M.P., Lozano, M.T. and Angulleiro, B. (2004). Development of the digestive tract of gilthead sea bream (*Sparus aurata* L.). Light and electron microscopic studies. *Aquaculture 234,* 215-238.

[91] Ceccaldi, H.J. (1989). Anatomy and physiology of digestive tract of Crustacean Decapods reared in aquaculture. *Adavances in Tropical Aquaculture*, Tahiti, Feb 20-March 4, 1989, AQUACOP, IFREMER, *Actes de Colloque* 9 pp. 243-259.

[92] Hoppe-Seyler, F. (1877). Ueber Unterschiede in chemischen Bau und Verdauung höherrer und niederrer Tiere. Pflüger's *Arch. Physio.l 14*, 395-400.

[93] Gates, B. and Travis, J. (1969). Isolation and comparative properties of shrimptrypsin. *Biochemistry 8,* 4483-4489.

[94] Gates, B. and Travis, J. (1973). Purification and characterization of carboxypeptidases A and B from the white shrimp *Penaeus setiferus*. *Biochemistry 12*, 1867-1874.

[95] Galgani, F., Benyamin, Y and Van Wormhoudt, A. (1984). Identification of digestive proteases of *Penaeus kerathurus* (Forskal): A comparison with *Penaeus japonicas* Bate. *Comp. Biochem. Physiol. 78B*,335-361.

[96] Eisen, A.Z., Henderson, K.O., Jeffery, J.J. and Bradshaw, R.A. (1973). Acollaginolytic protease from the hepatopancreas of the fiddler crab *Uca pugilator*, purification and properties. *Biochemistry 12*,1874-1822.

[97] Galgani, F., Benyamin, Y., Van Wormhoudt, A. and Ceccaldi, H.J. (1985). Purification, propertics and immunoassayof trypsin from the shrimp *Penaeus japonicas*. *Comp. Biochem. Physiol. 81B*, 447-452.

[98] Zwilling, R. and Tomasek, V. (1970). Amino acid composition of crayfish trypsin. *Nature (London). 228,* 57-58.

[99] Titani, K., Sasagawa, T., Woodbury, R., Erisson, L. Dorsam, H., Kraemer, M. Neurath, H. and Zwilling, R. (1983). Amino acid sequence of crayfish (*Astacus fluviatilis*) trypsin I.F. *Biochemistry 22*, 1459-1463.

[100] Van Wormhoundt, A. (1980). Adaptation des activities digestives, de leurs cycles et de leur controle aux facteurs du milieu chez *Palaeomon serratus* (Crustacea, Natantia). Thése Doct. D'Etat, Univ. Aix Marseille II, 351 p.

[101] Jeuniaux, C. (1963). Chitine et chitinolyse. Masson Ed., Paris.

[102] Muzzarelli, R.A.A. (1977). *Chitin*. Pergamon Press Ed., Oxford.

[103] Landolt, M.L. (1989). The relationship between diet and the immune response of fish. *Aquaculture 79*, 193-206.

[104] Tacon, A.G.J. (2002). Thematic Review of Feeds and Feed Management Practices in Shrimp Aquaculture. Report prepared under the World Bank, NACA, WWF and FAO Consortium Program on Shrimp Farming and the Environment. Work in Progress for Public Discussion. Published by the Consortium. 69 pages.

[105] Erdal, J.I., Evensen, O., Kaurstad, O.K., Lillehaug, A. Solbakken, R. and Thorud, K. (1991). Relationship between diet and immune response in Atlantic salmon (*Salmo salar L.*) feeding various levels of ascorbic acid and omega-3 fatty acids. *Aquaculture 98*, 363-379.

[106] Calder, P.C. (1995). Fatty acids, dietary lipids and lymphocyte functions. *Biochem Soc trans. 2*, 302-309.

[107] Shils, M.E. and Young, V.R. Editors. (1988). Modern Nutrition in Health and Disease. 7th edition. Lea Febiger, Philedelphia, Pennsylvania.

[108] Chaou, B.S. and Shiau, S.Y. (1999). Both n-6 and n-3 fatty acids are required for maximal growth of juvenile hybrid tilapia. *N. Am. J. Aquacul. 61*,13-20.

[109] Blazer, V.S., (1991). Piscine macrophages function and nutritional influences: a review. *J. Aquat. Animal Health. 3*, 77-86.

[110] Sheldon, W.M. Jr. and Blazer, V.S. (1991). Influence of dietary lipid and temperature on bactericidal activity of channel catfish macrophages. *J. Aquat. Animal. Health. 3*, 87-93.

[111] Kiron, V., Gunji, A., Okamoto, N., Satoh, S., Ikeda, Y. and Watanabe, T. (1993). Dietary nutrient dependent variations on natural-killer activity of the leucocytes of rainbow trout. *Fish Pathol. 3*,28: 71–76.

[112] Lin, Y.H. and Shiau, S.Y. (2007). The effects of dietary selenium on the oxidative stress of grouper, *Epinephelus malabaricus*, fed high copper. *Aquaculture 267*: 38-43.

[113] Verlhac, V. and Gabaudan, J. (1997). The effect of vitamin C on fish health. Roche Technical Bulletin, Hoffman-La Roche Ltd, Basel, Switzerland, 30p.

[114] Pickering, A.D. and Stewart, A. (1984). Acclimation of interregnal tissue of the brown trout, *Salmo trutta* L. to chronic crowding stress. *J. Fish Biol. 24*,731-740.

[115] Mazur, C.F. and Iwama, G.K. (1993). Handling and crowding stress reduces number of plaque-forming cells in Atlantic salmon. *J. Aquat. Anim. Health 5*, 98-101.

[116] Yin, Z., Lam, T.J. and Sin, Y.M. (1995). The effect of crowding stress on the non-specific immune response in fancy carp (*Cyprinus carpio* L.). *Fish Shellfish Immunol., 5*, 519-529.

[117] Tort, L., Sunyer, J.O., Gomez, E. and Malinero, A. (1996). Crowding stress induces changes in serum haemolytic and agglutinating activity in gilthead seabream *Sparus aurata*. *Vet. Immunol. Immunopathol. 51*, 179-188.

[118] Lee, K.J. and Dabrowski, K. (2003). Interaction between vitamin C and E affects their tissue concentrations, growth, lipid oxidation, and deficiency symptoms in yellow perch (*Perca flaviscens*). *Br. J. Nut. 89*, 589-596.

[119] Dabrowski, K. (1990). Absorption of ascorbic acid and ascorbic sulfate and ascorbate metabolism in stomachless fish, common carp. *J. Comp. Physiol. B 160*, 549-561.

[120] Dabrowski, K., Moreau, R. and El-Saidy, R. (1996). Ontogenic sensitivity of channel catfish to ascorbic acid deficiency. *J. Aquat. Anim. Health 8*, 22-27.

[121] Lee, K.J., Kim, K.W. and Bai, S.C. (1998). Effect of different dietary levels of L-ascobic acid on growth and tissue vitamin C concentration in juvenile Korean rockfish, *Sebestes schlegeli* (Hilgendorf). *Aquacult Res 29,*137-244.

[122] Shiau, S.Y. and Hsu, T.S. (1999). Quantification of vitamin C requirement for hybrid juvenile tilapia. *Oreochromis niloticus* x *Oreochromis aureus*, with L-ascorbyl-2-monophosphate-Mg. *Aquaculture 175*, 317-326.

[123] Blazer, V.S. and Wolke, R.E. (1984). The effect of α-tocopherolon the immune response and non-specific resistance factors of rainbow trout (*Salmo gairdneri*). *Aquaculture 37*, 1-9.

[124] Blazer, V.S. (1982). The effect of marginal deficiencies of ascorbic acid and α-tocopherol on the natural resistance and immune response of rainbow trout (*Salmo gairdneri*). PhD Dessertation. Univ. Rhode Island, USA.

[125] Li, Y. and Lovell, R.T. (1985). Elevated levels of dietary ascorbic acid increase immune responses in channel catfish. *J. Nutr. 115*,123-131.

[126] Wilson, R.P. (1973). Absence of ascorbic acid synthesis in channel catfish (*Ictalurus punctatus*) and blue catfish (*Ictalurus furcatus*). Comp Biochem Physiol B46, 635-638.

[127] Fracalossi, D.M., Allen, M.E., Yuyama, L.K. and Oftedal, L.T. (2001). Ascorbic acid biosynthesis in Amazonian fishes. *Aquaculture 192,* 321-332.

[128] National Research Council (NRC) (1993). *Nutritional Requirements of Fish.* National Academy Press, Washington DC, pp. 114.

[129] Verlhac, V., Obach, A., Gabaudan, J., Schüep, W. and Hole, R. (1998). Immunomodulation by dietary vitamin C and glucan in rainbow trout *Oncorhynchus mykiss*). *Fish Shellfish Immunol. 8*, 409-424.

[130] Anbarasu, K, and Chandran, M.R. (2001). Effect of ascorbic acid on the immune response of the catfish, *Mystus gulio* (Hamilton), to different bacterins of *Aeromonas hydrophila*. *Fish Shellfish Immunol. 11*, 347-355.

[131] Chen, R., Lochmann, R., Goodwin, A., Praveen, K., Dabrowski, K. and Lee, K.J. (2004). Effects of dietary vitamins C and E on alternative complement activity, hematology, tissue composition, vitamin concentrations and response to heat stress in juvenile golden shiner (*Notemigonus crysoleucas*). *Aquaculture 242,*553-569.

[132] Ai, Q., Mai, K., Zhang, C., Xu, W., Duan, Q., Tan, B. and Liufu, Z. (2004). Effects of dietary vitamin C on growth and immune response of Japanese seabass, *Lateolabrax japonicas*. *Aquaculture 242*, 489-500.

[133] Anderson, D.P. (1992). Immunostimulants, adjuvants, and vaccine carriers in fish: application to aquaculture. *Ann. Rev. Fish. Dis. 2*, 281-307.

[134] Hung, S.S.O., Cho, S.Y. and Slinger, S.J. (1981). Effect of oxidized oil , DL-α tocopherol acetate and ethoxyquin supplementation on the vitamin E nutrition of rainbow trout (*Salmo gairdneri*) fed practical diets. *J. Nutr. 11*, 648-657.

[135] Wahli, T., Verlhac, V., Gabaudan, J., Meier, W. and Schüep, W. (1998). Influence of combined vitamins C and E on immunity and disease resistance of rainbow trout (*Oncorhynchus mykiss*). *J. Fish Dis. 21*, 127-137.

[136] Moreau, R., Dabrowski, K., Czesny, S. and Cihla, F. 1999. Vitamin C and E interaction in juvenile lake sturgeon (*Acipenser fulvescens* R.), a fish able to synthesize ascorbic acid. J. Appl. Ichthyol. 15:250-257.

[137] Verlhac, V., Gabaudan, J. and Schielre, J. (1996). Influence of astaxantin on nonspecific immune response of rainbow trout. In Book of Abstracts of the 1996 Annual Meeting of the World Aquaculture Society, Bangkok, Thailand. LeRoy Creswell, R. (ed), Ft. Pierce, Florida, pp. 426.

[138] Lim, C., Klesius, P.H. and Li b, M.H. (2000). Edwin H. Robinson Interaction between dietary levels of iron andvitamin C on growth, hematology, immune response and resistance of channel catfish (*Ictalurus punctatus*) to *Edwardsiella ictaluri* challenge. *Aquaculture 185*, 313-327.

[139] Oruñto, J., Esteban, M.A. and Meseguer, J. (2002). High dietary intake of α-tocopherol acetate enhances the non-specific immune response of gilthead sea bream (*Sparus aurata* L.). *Fish Shellfish Immunol. 10*, 293-307.

[140] Misra, K.C., Das, B.K., Mukherjee, S.C. and Pradhan, J. (2007). Effect of dietary vitamin C on immunity, growth and survival of Indian major carp *Labeo rohita* fingerlings. *Aquacul. Nut. 13*, 35-44.

[141] Li, M.H., Johnson, M.R. and Robinson, E.H. (1993). Elevated dietary vitamin C concentrations did not improve resistance of channel catfish, *Ictalurus punctatus* against *Edwardsiella ictaluri* infection. *J. World Aqua Soc. 29,* 1-8.

[142] Thompson, A., White, A., Fletcher, T. C., Houlihan, D. F. and Secombes, C. J. (1993). The effect of stress on the immune response of Atlantic salmon (*Salmo salar* L.) fed diets containing different amounts of vitamin C. *Aquaculture 114*, 1-18.

[143] Azad, I. S., Syama Dayal, J., Poornima, M. and Ali, S.A. (2007). Supra dietary levels of vitamins C and E enhance antibody production and immune memory in juvenile milkfish, *Chanos chanos* (Forsskal) to formalin-killed *Vibrio vulnificus*. *Fish Shellfish Immunol 23*, 154-163.

[144] Hamre, K. and Lie, Ø. (1995). Minimum requirement of vitamin E for Atlantic salmon *Salmo salar* L., at first feeding. *Aqua Res. 52*, 119-123.

[145] Murai, T. and Andrews, J.W. (1974). Interaction of dietary α-tocopherol, oxidizing menhaden oil and ethoxyquin on channel catfish (*Ictalurus punctatus*). *J. Nutr. 104*, 1416-1431.

[146] Wilson, R.P., Bowser, P.R. and Poe, W.E. (1984). Dietary vitamin E requirement of fingerling channel catfish. *J. Nutr. 114*, 2053-2058.

[147] Watanabe, T., Takeuchi, T. and Wada, M. (1977). Effect of α-tocopherol deficiency on carp: VII. The relationship between dietary levels of linoleate and α-tocopherol requirement. *Bull. Jpn. Soc. Sci. Fish .43*, 935-946.

[148] Bai, S.C. and Lee, K.J. (1998). Different levels of dietary DL-α-tocopheryl acetate affect the vitamin E status of juvenile Korean rockfish, *Sebestes schlegeli*. *Aquaculture* 161, 405-414.

[149] Kocabas, A.M. and Galtin III, D.M. (1999). Dietary vitamin E requirement of hybrid striped bass (*Morone chrysops* female x *M. saxatilis* Male). Aquac. Nutr., 5: 3-7.

[150] Boonyaratpalin, M. 1996. Nutritional requirements of commercially important shrimps in the tropics, In: Santiago, C.B., Coloso, R.M., Millamena and Borlongan, I.G. (Eds) *Feeds for Small-Scale Aquaculture*. Proceedings of the National Seminar-Workshop on Fish Nutrition and Feeds. Tigbauan, Iloilo, Philippines, 1-2 June 1994. Southeast Asian Fisheries Development Center, Iloilo, Philippines. pp. 10-28.

[151] Chou, R. (1996). Recent developments in marine shrimp and fish nutrition. *Victam Asia Conference,* 14-15November 1996, Bangkok, Thailand, pp.9-42.

[152] Kanazawa, A. (1996). Shrimp nutrition. *The 1996 Annual Meeting of the World Aquaculture Society Book of Abstracts*, 29 January to 2 February 1996, Bangkok, Thailand pp.191-192.

[153] Kontara, P., Merchie, G. Lavens, P. Robles, R. Nelis, H. De Leenheer, A. and Sorgeloos, P. (1997). Improved larviculture outputs of postlarval shrimp *Penaeus vannamei* through supplementation of l-ascorbyl-2-polyphosphate in the diet. *Aquaculture International, 5*:127-136.

[154] Kurmaly, K. and F.C. Guo. (1996). Effect of environmental stressors; high ammonia, low dissolved oxygen and low temperature shock, on vitamin C and astaxanthin contetn of shrimp tissues. *1996 Annual Meeting of the World Aquaculture Society, Book of Abstracts*, 29 January to 2 February 1996, Bangkok, Thailand, pp.207-208.

[155] Lavens, P.L. and P. Sorgeloos. (2000). Advances in shrimp postlarval nutrition. *The Global Aquaculture Advocate, 3(6)*, 37-39.

[156] Merchie, G., E. Kontara, P. Lavens, R. Robles, K. Kurmaly and P. Sorgeloos. (1998). Effect of vitamin C and astaxanthin on stress and disease resistance of postlarval tiger shrimp, *Penaeus monodon* (Fabricus). *Aquaculture Research, 29*, 579-585.

[157] Pangantohon, M.P., Chiu-Chern, Y. and Millamena, O.M. (1996). Effect of dietary astaxanthin and vitamin A on the reproductive performance of *Penaeus monodon* broodstock, (Abstract). In Santiago, C.B., Coloso, R.M., Millamena and Borlongan, I.G. (Eds) *Feeds for Small-Scale Aquaculture*. Proceedings of the National Seminar-Workshop on Fish Nutrition and Feeds. Tigbauan, Iloilo, Philippines, 1-2 June 1994. Southeast Asian Fisheries Development Center, Iloilo, Philippines, pp.109.

[158] Chim, L., Galois, R. Lemaire, P. Delaporte, M. and Martin, J.L.M. (2000). Effects of dietary fatty acids on the adaptability of the farmed shrimp *Penaeus stylirostris* to environmental rearing variations. (T°C and S%). *AQUA 2000, Nice, France, May 2-6,2000. Book of Abstracts.* European Aquaculture Society special.pub.28. Oostend, Belgium, p.129.

[159] Coutteau, P. E., Kontara, K.M., and P. Sorgeloos. (2000). Comparison of phosphotidylcholine purified from soybean and marine fish roe in the diet of postlarval *Penaeus vannamei* Boone. *Aquaculture 181(3-4)*:331-345.

[160] Bird, J.N. (1997). Vitamin supplements revised. *Feed Milling International, March*, pp.10-13.

[161] Devresse, B. 1998. Production of water stable shrimp feeds. *Manuscritos de Conferencias y Resumenes de Carteles, IV Simposium Internacional de Nutricion Acuicola*, Noviembre 15-18, 1998. La Paz, Mexico, p.11.
[162] Kanazawa, A. (1985). Nutrition of penaeid prawns and shrimps. In: Taki, Y., Primavera, J.H., Llobrera, J.A. (eds.) *Proceedings of the First International Conference on the Culture of Penaeid Prawns/Shrimps*. Iloilo, Philippines: Aquaculture Development, Southeast Asian Fisheries Development Centre; p.123-130.
[163] He, H., Lawrence, A. L. and Liu, R. (1992). Evaluation of dietary essentiality of fat-soluble vitamins, A, D, E and K for penaeid shrimp (*Penaeus vannamei*). *Aquaculture 103*, 177-185.
[164] Akiyama, D.M., Dominy, W.G. and Lawrence, A.L. (1991). Penaeid shrimp nutrition for the commercial feed industry revised. In: Editors D.M. Akiyama and R.K.H. Tan, *Proceedings of the Aquaculture Feed Processing and Nutrition Workshop*. Thailand and Indonesia, Sept. 19–25, American Soybean Association, Singapore (1991), pp. 80–90.
[165] Beisel, W.R. (1982). Single nutrient and immunity. *Am. J. Clin. Nutr .35*, 417-468.
[166] Bhaskaram P. (1988). Immunology of iron-deficient subjects. In: Chandra, R.K. (ed.), *Nutrition and Immunology*, Alan R. Liss, New York, pp. 149-168.
[167] Roedar, M. and Roedar, R.H. (1968). Effect of iron on the growth rate of fish. *J. Nutr. 90*, 86-90.
[168] Galtin, D.M. III and Wilson, R.P. (1986). Characterization of iron deficiency and dietary iron requirement of fingerling channel catfish. *Aquaculture 52*, 191-198.
[169] Lim, C., Sealey, W.M. and Klesius, P.H. (1996). Iron methionine and iron sulfate as source of dietary iron for channel catfish *Ictalurus punctatus*. *J. World Aquacult. Soc. 27*, 290-296.
[170] Sealey, W.M., Lim, C. and Klesius, P.H. 91997). Influence of dietary iron from iron methionine and iron supfate on immune response and resistance of channel catfish to *Edwardsiella ictaluri*. *J. World Aquacult. Soc. 28* 142-149.
[171] Hilton, J.W. (1989). The interaction of vitamins minerals and diet composition in the diet of fish. *Aquaculture 79*, 223-244.
[172] Anderson, D. P., Dixon, O.W., Bodammer, J.E. and Lizzio, C. (1989). Suppression of antibody-producing cells in rainbow trout spleen sections exposed to copper in vitro. *J. Aquat. Anim. Health 1*, 57–61.
[173] Lee, M.H. and Shiau, S.Y. (2002). Dietary copper requirement of juvenile grass shrimp, *Penaeus monodon*, and its effects on non-specific immune responses. *Fish Shellfish Immunol. 13,* 259-270.
[174] Lee, M.H. and Shiau, S.Y. (2003). Increase of dietary vitamin C improves haemocyte respiratory burst response and growth of juvenile grass shrimp, *Penaeus monodon*, fed with high dietary copper. *Fish Shellfish Immunol. 14,* 305-315.
[175] Uauy, R. (1989). Dietary nucleotides and requirements in early life. In: Lebenthal, E. (Ed.), Textbook of Gastroenterology and Nutrition in Infancy. 2nd edn. Raven Press, New York, pp. 265-280.
[176] Barnes, L.A. (1994). Dietary sources of nucleotides-from breast milk to weaning. *J. Nutr. 124* (Suppl.IS), 128S-130S.

[177] Kulakarni, A.D., Fanslo, W.C., Drath, D.B., Rudolph, F.B. and Van Buren, C.T. (1986). Influence of dietary nucleotide restriction on bacterial sepsis and phagocytic cell function in mice. *Arch. Surg. 121*, 169-172.

[178] Kulakarni, A.D., Fanslo, W.C., Rudolph, F.B. and Van Buren, C.T. (1986). Effect of dietary nucleotides on response to bacterial infections. *J. Parenter. Enteral. Nutr. 10*, 169-171.

[179] Jyonouchi, H. (1994). Nucleotide actions on humoral immune response. J. Nutr., 124 (Suppl. IS): 138S-143S.Kamilya, D., Maiti, T.K., Joardar, S.N. and Mal, B.C. (2006). Adjuvant effect of mushroom glucans and bovine lactoferrin upon *Aeromonas hydrophila* vaccination in catla, *Catla catla* (Hamilton). *J. Fish Dis. 29*, 331-337.

[180] Carver, J.D. (1994). Dietary nucleotides: cellular immune, intestinal and hepatic system effects. *J. Nutr. 124* (Suppl. IS), 144S-148S.

[181] Matsumoto, Y., Adje, A.A., Yamauchi, K., Kise, M., Nakasone, Y., Shinegawa, Y., Yokoyama, H. and Yamamoto, S. (1995). Mixture of nucleosides and nucleotides increases bone marrow cell and peripheral neutrophil number in mice infected with methicillin-resistant *Staphylococcus aureus*. Biochemical and molecular roles of nutrients. *J. Nutr. 125*, 815-822.

[182] Low, C., Wadsworth, S., Burrells, C. and Secombes, C.J. (2003). Expression of immune genes in turbot (*Scophthalmus maximus*) fed a nucleotide-supplemented diet. *Aquaculture 221*, 23-40.

[183] Li, P. and Gatlin III, G.M. (2006). Nucleotide nutrition in fish: Current knowledge and future applications. *Aquaculture 251,* 141-152.

[184] Burrells, C, Williams, P.D., Southgate, P.J. and Worsworth, S.L. (2001). Dietary nucleotide: a novel supplement in fish feeds. Effects on vaccination, salt water transfer, growth rate and physiology of Atlantic salmon (*Salmo salar* L.). *Aquaculture 199*, 177-184.

[185] Hill, J., Smullen, R., Daniel R. and Barnes, A.C. (2006). Highly purified dietary nucleotide supplements improve growth performance and health status of penaeid shrimp. *https://www.was.org/meetings/AbstractData.asp?AbstractId=11045*

[186] Tacon, A.G.J. and Kurmaly, K. (1996). Nutrition and health management in shrimp culture. In *Book of Abstracts of the 1996 Annual Meeting of the World Aquaculture Society,* Bangkok, Thailand. LeRoy Creswell, R. (ed), Ft. Pierce, Florida, pp. 397.

[187] Siwiki, A.K. (1987). Immunomodulating activity of levamisole in carp spawners, *Cyprinus carpio* L. *J. Fish Biol. 31(Supp. A)*, 245-246.

[188] Siwiki, A.K. (1989). Immunomodulating activity of levamisole on non-specific immunity in carp, *Cyprinus carpio* L. *Dev. Comp. Immunol 13*: 87-91.

[189] Kajita, Y., Sakai, M., Atsuta, S. and Kobayashi, M. (1990). The immunomodulatory effects of levamisole on rainbow trout, *Oncorhynchus mykiss*. *Fish Pathol. 25*, 93-98.

[190] Jeney, G. and Anderson, D.P. (1993). Glucan injection or bath exposure given alone or in combination with a bacterin enhance the non-specific defence mechanisms in rainbow trout (*Oncorhynchus mykiss*). *Aquaculture 116,* 315-329.

[191] Mulero, V., Esteban, M.A., Munoz, J. and Meseguer, J. (1998). Dietary intake of levamisole enhances the immune response and disease resistance of the marine teleost gilt-head seabream (*Sparus aurata* L.). *Fish Shellfish Immuno., 8*, 49-62.

[192] Siwicki, A.K., Anderson, D.P. and Rumsey, G.L. (1994). Dietary intake of immunostimulants by rainbow trout affect nonspecific immunity and protection against furunculosis. *Vet. Immunol. Immunopathol. 41*, 125– 139.

[193] Unal, I. and Mustafa, D. (2005). A study on the effect of levamisole on the immune system of rainbow trout (*Oncorhynchus mykiss*, Walbaum). *Turk J. Vet. Anim. Sci., 29*, 1169-1176.

[194] Kumari, J. and Sahoo, P.K. (2006). Dietary immunostimulants influence specific immune response and resistance of healthy and immunocompromised Asian catfish *Clarias batrachus* to *Aeromonas hydrophila* infection. *Dis. Aqua Org. 70*, 63-70.

[195] Rairakhwada, D., Pal., A.K., Bhathena, Z.P., Sahu, N.P., Jha, A. and Mukherjee, S.C. (2007). Dietary microbial levan enhances cellular non-specific immunity and survival of common carp (*Cyprinus carpio*) juveniles. *Fish Shellfish Immunol. 22*, 477- 486.

[196] Christybapita, D., Divyagnaneswari, M. and Michael, R.D. (2007). Oral administration of *Eclipta alba* leaf aqueous extract enhances the non-specific immune responses and disease resistance of *Oreochromis mossambicus*. *Fish Shellfish Immunol. 23*, 840-852.

[197] Ashida, M.A. (1981). A cane sugar factor suppressing activation of phenoloxidase in hemolymph of the silkworm, *Bombyx mori*. *Insect Biochem 11*,57-65.

[198] Söderhäll, K. (1982). Phenoloxidase activiating system of arthropods? A review. *Dev Comp Immunol 6*, 601-611.

[199] Sakai, M. (1999). Current research status of fish immunostimulants. *Aquaculture. 172*, 63-92.

[200] Robertsen, B., Rorstad, G., Engstad, R. and Raa, J. (1990). Enhancement of non-specific disease resistance in Atlantic salmon (*Salmo salar* L.) by a glucan from *Saccharomyces cerevisae* cell wall. *J. Fish Dis. 13*, 391-400.

[201] Itami, T., Kondo, M., Vozu, M., Suganuma, A., Abe, T., Nakagawa, A., Suzuki, N. and Takahashi, Y. (1996). Enhancement of resistance against *Enterococcus seriolicida* infection in yellowtail, *Seriola quinqueradiata* (Temminck and Schlegel) by oral administration of peptidoglycan derived from *Bifidobacterium thermophilum*. *J. Fish Dis. 19*, 185–187.

[202] Dalmo, R.A., (2000). Immunostimulatory ß(1,3)-D-glucans: prophylactic drugs against threatening infectious diseases in fish. In: Paulsen, B.S. (Ed.), *Bioactive Carbohydrate Polymers*. Proceedings of the Phytochemical Society of Europe Kluwer. Academic Publishing, Dordrecht, pp. 95– 106.

[203] Robertsen, B., Engstad, R.E. and Jørgensen, J.B. (1994). ß-glucans as immunostimulants in fish. Pages 83-99 in J.S. Stolen and T.C. Fletcher, eds, *Modulators of Fish Immune Responses*. Volume 1. Fair Haven, NJ, SOS Publications.

[204] Dalmo, R. A. and Seljelid, R. (1995). The immunomodulatory effect of LPS, laminaran and sulphated laminaran [β(l,3)-D-glucan] on Atlantic salmon, *Salmo salar* L., macrophages in vitro. *J. Fish Dis.* ***18***, 175–185.

[205] Dalmo, R. A., Bogwald, J., Ingebrigtsen, K. and Seljelid, R. (1996). The immunomodulatory effect of laminaran [β(1,3)-D-glucan] on Atlantic salmon, *Salmo salar* L., anterior kidney leucocytes after intraperitoneal, peroral and peranal administration. *J. Fish Dis.* ***19***, 449–457.

[206] Dalmo, R.A., Kjerstad, A.A., Arnesen, S.M., Tobias, P.S. and Bøgwald, J. (2000). Bath exposure of Atlantic halibut (*Hippoglossus hippoglossus* L.) yolk sac larvae to bacterial

lipopolysaccharide (LPS): Absorption and distribution of the LPS and effect on fish survival. *Fish Shellfish Immunol 10*, 107-128.

[207] Söderhäll, K. and Smith, V.J. (1986). The prophenoloxidase activating system: the biochemistry of its activation and role in arthropod cellular with special reference to crustaceans. In: Brehelin, M. (Ed.), *Immunity in Invertebrates, Cells, Molecules and Defence Reactions*. Springer-Verlag, Fair Haven, NJ, pp. 208– 225.

[208] Persson, M., Cerenius, L. and Söderhäll, K. (1987). The influence of haemocyte number on the resistance of the fresh water cray fish *Pacifastacus leniusculus* Dana, to the parasitic fungus *Aphanomyces astaci. J. Fish Dis. 10*, 471– 477.

[209] Söderhäll, K., Cerenius, L. and Johnson, M.W. (1994). The prophenoloxidase activity system and its role in invertebrate defense: foundations for the invertebrate immune system. *Ann. NY Acad. Sci. 712*, 155–166.

[210] Sung, H.H. and Song, Y.L. (1996). Tissue location of Vibrio antigen delivered by immersion to tiger shrimp (*Penaeus monodon*). *Aquaculture 145*, 41– 54.

[211] Cook, M.T., Hayball, P.J., Hutchinson, W., Nowak, B.F. and Hayball, J.D. (2003). Administration of a commercial immunostimulant preparation, EcoActivaTM as a feed supplement enhances macrophage respiratory burst and growth rate of snapper (*Pagrus auratus*, Sparidae (Bloch and Schneider) in winter. *Fish Shellfish Immunol 14*, 333-345.

[212] Misra, C.K., Das, B.K., Mukherjee, S.C. and Pattnaik, P. (2006). Effect of long term administration of dietary ß-gluan on immunity, growth and survival of *Labeo rohita* fingerlings. *Aquculture 255*, 82-94.

[213] Ai, Q., Mai, K., Zhang, L., Tan, B., Zhang, W., Xu, W. and Li, H. (2007). Effects of dietary ß-1,3 glucan on innate immune response of large yellow croaker, *Pseudosciaena crocea. Fish Shellfish Immunol. 22*, 394-402.

[214] Sealey, W.M., Barrows, F.T., Hang, A., Johansen, K.A., Overturf, K., LaPatra, S.E. and Hardy, R.W. (2007). Evaluation of the ability of barley genetypes containing different amounts of ß-glucan to alter growth and disease resistance of rainbow trout *Oncorhynchus mykiss. Animal Feed and Technology xx*:xxx-xxx.

[215] Whittington, R., Lim, C. and Klesius, P.H. (2005). Effect of dietary ß-glucan levels on the growth response and efficacy of Streptococcus iniae vaccine in Nile tilapia. *Orecochromis niloticus. Aquaculture 248*, 217-225.

[216] Bagni, M., Romano, N., Finola, M.G., Abelli, L., Scapigliati, G., Tiscar, P.G., Sarti, M. and Marino, G. (2005). Short- and long-term effects of a dietary yeast beta-glucan (Macrogard) and alginic acid (Ergosan) preparation on immune response in sea bass (*Dicentrarchus labrax*). *Fish Shellfish Immunol.* 18, 311-325.

[217] Welker, T.L., Lim, C., Yildrim-Aksoy, M., Shelby, R. and Klesius, P.H. (2007). Immune response and resistance to stress and *Edwardsiella ictaluri* challenge in channel catfish, *Ictalurus punctatus*, fed diets containing commercial whole-cell or yeast subcomponents. *J. World Aqua. Soc. 38*, 24-35.

[218] Bagni, M., Archetti, L., Amadori, M. and Marino, G. (2000). Effect of long-term administration of an immunostimulant diet on immunity in sea bass (*Dicentrarchus labrax*). *J. Vet. Medicine* B 47, 745-751.

[219] Selvaraj, V., Sampath, K. and Sekar, V. (2005). Administration of yeast glucan enhanches survival and some non-specific and specific immune parameters in carp (*Cyprinus carpio*) infected with *Aeromonas hydrophila. Fish Shellfish Immuno*logy *19,* 293-306.

[220] Sahoo, P.K. and Mukherjee, S.C. (2001). Effect of dietary b-1,3 glucan on immune responses and disease resistance of healthy and aflatoxin B_1-induced immunocompromised rohu (*Labeo rohita* Hamilton). *Fish Shellfish Immunol. 11*, 683-695.

[221] Sahoo, P.K. and Mukherjee, S.C. (2002). The effect of dietary immunomodulation upon *Edwardsiella tarda* vaccination in healthy and immunocompromised Indian major carp (*Labeo rohita*) *Fish Shellfish Immunol. 12*, 1-16.

[222] Kumari, J. and Sahoo, P.K. (2006). Dietary immunostimulants influence specific immune response and resistance of healthy and immunocompromised Asian catfish *Clarias batrachus* to *Aeromonas hydrophila* infection. *Dis. Aqua. Org. 70*, 63-70.

[223] Selvaraj, V., Sampath, K. and Sekar, V. (2006). Adjuvant and immunostimulatory effects of ß-glucan administration in combination with lipopolysaccharide enhances survival and some immune parameters in carp challenged with *Aeromonas hydrophila. Veterinary Immunol. Immunopathol. 114*, 15-24.

[224] O´Garra, A. (2000) T-cell differentiation: Commitment factors for T helper cells. *Current Biol 10:* R492-R494.

[225] Mutoloki, S., Alexandersen, S. and Evensen, Ø. (2004). Sequential study of antigen persistence and concomitant inflammatory reactions relative to side effect and growth of Atlantic salmon (*Salmo salar* L.) following intraperitoneal injection with oil-adjuvanted vaccines. *Fish Shellfish Immunol. 16*, 633-645.

[226] Mutoloki, S., Reite, O.B., Brudeseth, B., Tverdal, A. and Evensen, Ø. (2006). A comparative immunopathological study of injection site reactions in salmonids following intraperitoneal injection with oil-adjuvanted vaccines. *Vaccine 24*, 578-588.

[227] Kamilya, D., Maiti, T.K., Joardar, S.N. and Mal, B.C. (2006). Adjuvant effect of mushroom glucans and bovine lactoferrin upon *Aeromonas hydrophila* vaccination in catla, *Catla catla* (Hamilton). *J. Fish Dis. 29*, 331-337.

[228] Moradali, M-F., Mostafavi, H., Ghods, S. and Hedjaroude, G-A. (2007). Immunomodulating and anticancer agents in the realm of macromycetes fungi (macrofungi). International Immunopharmacology 7: 701-724.

[229] Berg, A., Bergh, Ø., Fjelldal, P.G., Hansen, T., Juell, J.E. and Nerland, A. (2006). Dyrevelferdsmessige konsekvenser av vaksinasjon av fisk – effekter og bivirkninger (Animal welfare and fish vaccination – effects and side-effects). *Fisken og Havet 9,* 1-43.

[230] Rørstad, G., Aasjord, P.M. and Robertsen, B. (1993). Adjuvant effect of a yeast glucans in vaccines against furunculosis in Atlantic salmon (*Salmo salar* L.). *Fish Shellfish Immunol. 3,* 179-190.

[231] [Aakre, R., Wergeland, H.I., Aasjord, P.M., and Endresen, C. (1994). Enhanced antibody response in Atlantic salmon (*Salmo salar* L.) to *Aeromonas salmonicida* cell wall antigens using a bacterin containing ß-1,3-M-glucan as adjuvant. *Fish and Shellfish Immunology 4*, 47-61.

[232] Ashida, T., Okimasu, E., Ui, M., Heguri, M., Oyama, Y. and Amemura, A. (1999). Protection of Japanese flounder *Paralichthys olivaceus* against experimental Edwardsiellosis by formalin-killed Edwardsiella tards in combination with oral adminitration of immunostimulants. *Fish Sci. 65*, 527-530.

[233] Midtlyng, P.J., Reitan, L.J. and Speilberg, L. (1996). Experimental studies on the efficacy and the side-effects of intraperitoneal vaccination of Atlantic salmon (*Salmon salar* L.) against furunculosis, *Fish Shellfish Immunol.* **6,** 335–350.

[234] Romalde, J.L., Magarinos, B. and Toranzo, A.E. (1999). Prevention of streptococcosis in turbot by intraperitoneal vaccination: a review. *J. Applied Ichtyol. 15*, 153-158.

[235] Kawakami, H., Shinohara, N. and Sakai, M. (1998). The non-specific immunostimulation and adjuvant effects of *Vibrio anguillarum* bacterin, M-glucan, chitin and Freund's complete adjuvant against *Pasteurella piscicida* infection in yeallowtail. *Fish. Pathol. 33*, 287-292.

[236] Ara, Y., Saito, T., Takagi, T., Hagiwara, E., Miyagi, Y., Sugiyama, M., Kawamoto, S., Ishii, N., Yoshida, T., Hanashi, D., Koshida, T., Okada, H. and Okuda, K. (2001). Zymosan enhances the immune response to DNA vaccine for human immunodeficiency virus type-1 through the activation of complement system. *Immunol. 103*, 98-105.

[237] Taylor, P.R., Martinez-Pomares, L., Stacey, M., Lin, H-H., Brown, G.D. and Gordon, S. (2005). Macrophage receptors and immune recognition. *Ann. Rev. Immunol. 23*, 901-944.

[238] Dennehy, K.M. and Brown, G.D. (2007). The role of the ß-glucan receptor Dectin-1 in control of fungal infection. *J. Leukocyte Biol. 82*, 253-258.

[239] Taylor, P.R., Brown, G.D., Reid, D.M., Willment, J.A., Martinez-Pomares, L., Gordon, S. and Wong, S.Y. (2002). The ß-glucan receptor, dectin-1, is predominantly expressed on the surface of cells of the monocyte/macrophage and neutrophil lineages. *J. Immunol. 169*, 3876-3882.

[240] Palm, N.W. and Medzhitov, R. (2007). Antifungal turns 17. *Nature Immunol. 8,* 549-551.

[241] Gunimaladevi, I., Savan R. and Sakai, M. (2006). Identification, cloning and characterization of interleukin-17 and its family from zebrafish. *Fish Shellfish Immunol. 21*, 393-403.

[242] Engstad, R.E. and Robertsen, B. (1994). Specificity of a ß-glucan receptor on macrophages from Atlantic salmon (*Salmo salar* L). *Dev Comp Immunol 18*, 397-408.

[243] Ainsworth, A.J. (1994). A beta-glucan inhibitable zymosan receptor on channel catfish neutrophils. *Vet. Immunol. Immunopathol. 41*, 141-152.

[244] Xia, Y. and G. D. Ross. (1999). Generation of recombinant fragments of CD11b expressing the functional beta-glucan-binding lectin site of CR3 (CD11b/CD18). *J. Immunol. 162*, 7285–7293.

[245] Hawlisch, H. and Köhl, J. (2006). Complement and Toll-like receptors: Key regulators of adaptive immune responses. *Molecular Immunology 43,* 13-21.

[246] Nakao, M., Fujiki, K., Kondo, M. and Yano, T. (2003). Detection of complement receptors on head kidney phagocytes of the common carp *Cyprinus carpio*. *Fisheries. Sci. 69*, 929-935.

[247] Zimmerman, J.W., Lindermuth, J., Fish, P.A., Palace, G.P., Stevenson, T.T. and DeMong, D.E. (1998). A novel carbohydrate-glycosphingolipid interaction between a ß-(1-3)-glucan immunomodulator, PGG-glucan, and lactosylceramide of human leukocytes. *J. Biological. Chem. 34*, 22014-22020.

[248] Chen, J. and Seviour, R. (2007). Medicinal importance of fungal ß-(1-3), (1-6)-glucans. *Mycological Res. 111*, 635-652.

[249] Romagne, F. (2007). Current and future drugs targeting one class of innate immunity receptors: the Toll-like receptors. *Drug Discovery Today 12,* 80-87.

[250] Kato, Y., Nakao, M., Mutsuro, J., Zarkadis, I.K. and Yano, T. (2003). The complement component C5 of the common carp (*Cyprinus carpio*): cDNA cloning of two distinct isotypes that differ in a functional site. *Immunogenetics 54*, 807-815.

[251] Løvoll, M., Fischer, Ü., Mathisen, G.S., Bøgwald, J., Ototake, M. and Dalmo, R.A. (2007). The C3 subtypes are differentially regulated after immunostimulation in rainbow trout, but head kidney macrophages do not contribute to C3 transcription. *Vet. Immunol. Immunopathol. 117*, 284-295.

[252] Ellertsen, L.K., Hetland, G., Johnson, E. and Grinde, B. (2006). Effect of medicinal extract from *Agaricus blazei* Murill on gene expression in human monocyte cell line as examined by microarrays and immune assays. *International Immunopharmacol 6,* 133-143.

[253] Somerset, I., Krossøy, B., Biering, E. and Frost, P. (2005). Vaccines for fish in aquaculture. *Expert Rev. Vaccines 4(1),* 89-101.

[254] Duff, D.C.B. (1942). The oral immunization of trout against Bacterium salmonicida. *J. Immunol. 44*, 87-94.

[255] Schaperclaus,W. (1954). Die Verhutung der infectiosen Bauchwasersucht des carpfen duch impfung und Behandlung mit Mdeikamenten. Deutsch *Disch Ztg 1*, 102-108.

[256] Post, G. (1966). Response of rainbow trout (*Salmo gairdneri*) to antigens of *Aeromonas hydrophila*. *J. Fish Res. Bd. Can. 23*, 1487-1494.

[257] Song, Y.L. and Kou, G.H., (1981). Immune response of eel (*Anguilla japonica*) against *Aeromonas hydrophila* and *Edwardsiella anguillimortiferrum* (*E. tarda*) infection. *Nat Sci. Conn. Series No.3*, 107-115.

[258] Thune, R.L. (1980). Immunization of channel catfish (*Ictalurus punctatus*) against *Aeromonas hydrophila* via hyperosmotic infiltration, M.S. Thesis, Auburn University, p. 54.

[259] Thune, R.L. and Plumb, J.A. (1982). Effect of delivery method and antigen preparation on the production of antibodies against *Aeromonas salmonicida* in channel catfish. *Prog. Fish-cult. 44*, 53-54.

[260] Lamers, C.H.J. (1985). The reaction of the immune system to fish vaccination. Thesis, Univeristy of Wageningan, Netherlands. P.256.

[261] Lamers, C.H.J. and De Haas, M.J.M. (1985). Antigen localization in the lymphoid organs of carp (*Cyprinus carpio*). *Cell Tissue Res. 242*, 491-498.

[262] Lamers, C.H.J., De Haas, M.J.M., and Van Muiswinkel, W.B., (1985). Humoral response and memory formation in carp *Cyprinus carpio* after infection of *Aeromonas hydrophila* bacterin. *Develop Comp. Immunol. 9*, 65-76.

[263] Lamers, C.H.J., De Haas, M.J.M., and Van Muiswinkel, W.B. (1985). The reaction of the immune system if fish to vaccination. Development of immunological memory in carp *Cyprinus carpio* following direct immersion in *Aeromonas hydrophila* bacterin. *J. Fish Dis. 8*, 253-262.

[264] Lamers, C.H.J. and Van Muiswinkel, W.B. (1986). Natural and acquired agglutinins to *Aeromonas hydrophila* in carp (*Cyprinus carpio*). *Can. J. Fish Aquat. Sci. 43*, 619-642.

[265] Fryer, J.L., Nelson, J.S. and Garrison, R.L. (1972). Vibriosis in fish. In: *Progress in Fishery and Food Science*, (Ed: R.W.Moore) Seattle: University of Washington, Publications in Fisheries, pp.129-133.

[266] Fryer, J.L., Rohovec, J.S., Tebbit, G.L., Mc Michael , J.S. and Pilcher, K.S. (1976). Vaccination for control of infectious diseases in Pacific salmon. *Fish Path. 10*, 155-164.

[267] Fryer, J.L., Amend, D.F., Harrell, L.W., Novotony, A.S., Plumb, J.A., Rohovec, J.S. and Tebbit, G.L. (1977). *Development of bacterin and vaccines for control of infectious disease of fishes*. Oregon State University, Sea Grant College Program. Publ. No. ORESU-T-77-012.

[268] Fryer, J.L., Rhohvec, J.S. and Garrison, R.L. (1978). Immunization of salmonids for control of vibriosis. *Mar. Fish Review 40*, 20-23.

[269] Dorson, M. (1977). Vaccination trials of rainbow trout fry against infectious pancreatic necrosis. *Bull. Off Int. Epiz. 87(5-6)*, 405-406.

[270] Dorson, M., Castric, J. and Torchy, C. 1978. Infectious pancreatic necrosis of salmonids: biologic and antigenic features of a pathogenic strain and of a non-pathogenic variant selected in RTG2 cells. *J. Fish Dis. 1*, 309-320.

[271] Hill, B.J., Dorson, M. and Dixon, P.F. (1980). Studies on immunization of trout against IPN. In: *Fish Diseases*, Third COPRAQ-Session. Ed. W.Ahne, Berlin, Springer-Verlag.

[272] Dixon, P.F. and Hill, B.J. (1983). Studies on IPN virulence and immunization. *J. Fish. Dis. 6,* 399-409.

[273] de Kinkelin, P and Le Berre, M. (1977). Isolation of a pathogenic rhabdovirus of brown trout (*Salmo trutta* L., 1766). *C R Acad Sci Hebd Seances Acad. Sci. D. 284*, 101-104.

[274] Meier, W., and P. E. V. Jorgensen. (1975). A rapid and specific method for the diagnosis of viral haemorrhagic septicaemia (VHS) of rainbow trout. *Riv. Ital. Piscic. Ittiop. 10*,1115.

[275] Ahne, W. and Thomsen, I. (1985). Occurrence of viral hemorrhagic septicemia virus in wild whitefish *Coregonus* sp. *Zentralbl. Veterinaermed. Reihe* B *32*, 73-75.

[276] Fijan, N., Petrinec, Z., Stancl, Z., Kezic, N. and Teskeredzic, E. (1977). Vaccination of carp against spring viremia: Comparison of intraperitobneal and peroral application of live virus to fish kept in ponds. *Bull. Off Int. Epizoot. 87*, 441-442.

[277] Fijan, N. (1980). Vaccination against spring viremia of carp. In: Ellis, A.E. (Ed.), *Fish Vaccination*. Academic Press, London, pp. 204-215.

[278] Fijan, N., Matasin, Z., Jeney, Z., Oláh, J. and Zwillenberg, L.O., (1984). Isolation of *Rhabdovirus carpio* from sheatfish (*Silurus glanis* L.) fry. *Symposia Biologica Hungarica, 23:*17-25.

[279] Heppell J. and Davis, H.L, (2000). Application of DNA vaccine technology to aquaculture. *Adv. Drug Deliv. Rev. 43(1),* 29-43.

[280] Hazen, T.C., Esch, G.W., Glassman, A.B. and Gibbons, J.W. (1978). Relationship of season, thermal loading and red-sore disease with various haematological parameters in Micropterus salmoides. *J Fish Biol 12*, 491-498.

[281] Azad, I.S., Rajendran, K.V., Rajan, J.J.S., Vijayan, K.K. and Santiago, T.C. (2001). Virulence and histopathology of *Aeromonas hydrophila* (SAH 93) in experimentally infected tilapia, *Oreochromis mossambicus* (L.). *J Aqua Trop 16(3)*, 265-275.

[282] De Fuguerredo, J. and Plumb, J.A. (1977). Virulence of different isolates of *Aeromonas hydrophila* in channel catfish. *Aquaculture 11,* 349-354.

[283] Khalifa, K.A. and Post, G. (1976). Immune response of advanced rainbow trout fry to *Aeromonas liquifaciens*. Prog. Fish-cult., 38: 66-68.

[284] Schachte, J.H. (1978). Immunization of channel catfish, *Ictalurus punctatus* against two bacterial diseases. *Ma. Fish Rev. 40*, 18.

[285] AQUIGRUP (1980). Trial vaccination of rainbow trout against *Aeromonas liquifaciens*. In : *Fish Diseases*; Third COMPAQ-Session, ed. W.Ahne, Berlin: Springer-Verlag, pp 206-211.

[286] Irianto, A., Robertson, P.A.W. and Austin, B. (2003). Oral administration of formalin-inactivated cells of *Aeromonas hydrophila* A3-51 controls infection by atypical *A.salmonicida* in goldfish, *Carrassius auratus* (L.). *J. Fish Dis. 26*,117-120.

[287] Azad, I.S., Shankar, K.M. and Mohan, C.V. (1996). Evaluation of biofilm of *Aeromonas hydrophila* for oral vaccination of carps. In: T.W.Flegel and I.H.Mac Rae (eds.), *Diseases in Asian Aquaculture*, III, Fish Health Section, Asian Fisheries Society, Manila, pp. 181 - 186.

[288] Azad, I.S., Shankar, K.M., Mohan, C.V. and Kalita, B. (1999). Biofilm vaccine of *Aeromonas* hydrophila - Standardization of dose and duration of vaccination of carps. *Fish Shellfish Immunol. 9*, 519 – 528.

[289] Azad, I.S., Shankar, K.M., Mohan, C.V. and Kalita, B. (2000). Protective response of common carp orally vaccinated with biofilm and free cells of *Aeromonas hydrophila* challenged by injection and immersion routes. *J. Aqua Trop. 15(1)*, 65-70.

[290] Azad, I.S., Shankar, K.M., Mohan, C.V. and Kalita., B. (2000). Uptake and processing of biofilm and free-cell vaccines of *Aermonas hydrophila* in Indian major carps and common carp following oral vaccination – antigen localization by monoclonal antibody. *Dis. Aqua Org. 43*, 103 – 108.

[291] Kawai, L., Kusuda, R. and Itami, T. (1981). Mechanisms of protection in ayu orally vaccinated for vibriosis. *Fish Pathol. 15(3/4)*, 257-262.

[292] Rombout, J.H.W.M. and Van Den Berg, A.A., (1989). Immunological response of the second gut segment of carp. I. Uptake and processing of antigens by epithelial cells and macrophages. *J. Fish Biol. 35*, 13-22.

[293] Rombout, J.H.W.M., Bot. H.E. and Taverne-Thiele, J.J. (1989). Immunological importance of the second gut segment of carp. II. Characterization of mucosal leukocytes. *J. Fish Biol. 35*, 167-178.

[294] Rombout, J.H.W.M., Van Den Berg, A.A., Van Den Berg, C.T.G.A., White, P. and Egberts, E. (1989). Immunological response of the second gut segment of carp. III. Systemic and / or mucosal responses after immunization with soluble or particulate antigen. *J. Fish Biol. 35*, 179-186.

[295] Costerton, J.W., Irvin, R.T., and Cheng, K.J. (1981). The bacterial glycocalyx in nature and disease. *Ann. Rev. Microbiol. 35,* 299-324.

[296] Shand, G.H., Anwar, H., Kadurugamuwa, J., Brown, M.R.W., Silverman, S.H. and Melling, J. (1985). *In vivo* evidence that bacteria in urinary tract infections grow under iron-restricted conditions. *Infect. Immun .48,* 35-39

[297] Anwar, H., Shand, G.H., Ward, K.H., Brown, M.R.W., Alpar, K.E. and Gowar, J. (1985). Antibody response to acute *Pseudomoanas aeruginosa* infection in a burn wound. *FEMS Microbiol. Lett. 29*, 225-230.

[298] Nickel, J.C., Ruseska, I., Wright, J.B. and Costerton, J.W., (1985). Tobramycin resistance of *Pseudomonas aeruginosa* cells growing as a biofilm on urinary catheter material. *Antimicrobial Agents and Chemotherapy 27,* 619-624.

[299] Brown, M.R.W., Allison, D.G. and Gilbert, P. (1988). Resistance of bacterial biofilms to antibiotics: a growth-rate related effect ?. *J Antimicrobiol Chemotherapy 22*, 777-780.
[300] Anwar, H., Dasgupta, M., Lam, K. and Costerton, J.W. (1989). Tobramycin resistance of mucoid *Pseudomonas aeruginosa* biofilm grown under iron limitation. *J. Antimicrobial Chemotherapy 24*, 647-655.
[301] Anwar, H., Van Biesen, T., Dasgupta, M., Lam, K. and Costerton, J.W. (1989). Interaction of biofilm bacteria with antibiotics in a novel *in vitro* chemostat systems. *Antimicrobial Agents and Chemotherapy 33*, 1824-1826.
[302] Govan, J.R.W. (1975). Mucoid strains of *Pseudomonas aeruginosa* : the influence of culture medium on the stability of mucus production. *J Med Microbiol 8*, 513-522.
[303] Kreig, D.P., Helmke, R.J., German, V.F. and Mangos, J.A. (1988). Resistance of mucoid *Pseudomonas aeroginosa* to non-opsonic phagocytosis by alveolar macrophages *in vitro*. *Infection Immunity 56*, 3173-3179.
[304] Asha, A., Nayak, D.K., Shankar, K.M. and Mohan, C.V. (2004). Antigen expression in biofilm cells of *Aeromonas hydrophila* employed in oral vaccination of fish. *Fish Shellfish Immunol .16*, 429-436.
[305] Nayak, D.K., Asha, A., Shankar. K.M. and Mohan, C.V. (2004). Evaluation of biofilm of *Aeromonas hygrophila* for oral vaccination of *Clarias batrachus*- a carnivore model. *Fish Shellfish Immunol. 16*, 613-619.
[306] Udey, L.R. and Fryer, J.L. (1978). Immunisation of fish with bacterins of *Aeromonas salmonicida*. *Marine Fish Rev. 40(3),* 12 - 17.
[307] Kay, W.W., Buckley, J.T., Ishiguro, E.E., Phipps, B.M., Monette, J.P.L. and Trust, T.J. (1981). Purification and disposition of a surface protein associated with virulence of *Aeromonas salmonicida*. *J. Bacteriol. 147*, 1077-1084.
[308] Evenberg, D. and Lugtenberg, B. (1982). Cell surface of the fish pathogenic bacterium *Aeromonas salmonicida*. II. Purification and characterization of a major cell envelop protein related to autoagglutination, adhesion and virulence. *Biochim Biophys. Acta 684*, 249-254.
[309] Ellis, A.E., Hastings, T.S. and Munro, A.L.S. (1981). The role of *Aeromonas salmonicida* extracellular products in the pathology of furunculosis. *J. Fish Dis., 4*, 41-51.
[310] Sniezko, S.F. and Friddle, S.B. (1949). Prophylaxis of furunculosis in brook trout (*Salvelinus fontinalis*) by oral immunization and sulphamerazine. *Prog. Fish Cult. 11*, 161-168.
[311] Klontz, G.W. and Anderson, D.F. (1970). Oral immunization of salmonids: a review. In : *Symposium on Diseases of Fishes and Shellfishes*, ed. S.F. Snieszko, American Fisheries Society., Washington DC, Spec. Publ. No. 5, p. 16-20.
[312] Smith, P.D., Mc Carthy, D.H. and Peterson, W.D. (1980). Further studies on furunculosis vaccination, In: Fish Diseases : Third COMPAQ - Session, Ed. W. Ahne, Berlin: Springer-Verlag, pp 113-118.
[313] Austin, B. and Rodgers, C.J. (1981). Preliminary observations on *Aeromonas salmonicida* vaccines. *Develop Biol. Standard 49*, 387-393.
[314] Rodgers, C. J. and Austin, B. (1985). Oral immunisation against furunculosis: an evaluation of two field trials. In: Manning, M. J.. Tatner, M. F. (eds.) *Fish Immunology*. Academic Press, London. p. 185-194

[315] Santos, Y., Pazos, F., Bandin, I. and Toranzo, A.E. (1995). Analysis of antigens present in the extracellular products and cell surface of *Vibrio anguillarum* O1, O2 and O3. *App. Environ. Microbiol. 61*, 2493– 2498.

[316] Toranzo, A.E., Santos, Y. and Barja, J.L. (1997). Immunization with bacterial antigens: Vibrio infections. In: Gudding, R., Lillehaug, A., Midtlyng, P.J., Brown, F. (Eds.), *Fish Vaccinology*. Karger, Basel, Switzeland, pp. 93– 105.

[317] Actis, L.A., Tolmasky, M.E., and Crosa, J.H. Vibriosis. In: Eds Woo PTK, Bruno DW. *Fish Diseases and Disorders*. Volume 3: *Viral, Bacterial and Fungal Infections*. CAB International Publ. (1999); 523-557.

[318] Johnson, K.A. and Amend. D.F. (1983). Efficacy of *Vibrio anguillarum* and *Yersinia ruckeri* bacterin applied by oral and anal intubation of salmonids. *J. Fish Dis. 6*, 473-476.

[319] Rombout, J.H.W.M., Blok, L.J., Lamers, C.H.J. and Egberts, E. (1986). Immunisation of carp with *Vibrio anguillarum* bacterin: indications of a common mucosal immune system. *Dev. Comp. Immunol. 10*, 341-351.

[320] Wong, G., Kaattari, S.L. and Christensen, J.M. (1992). Effectiveness of an oral enteric coated Vibrio vaccines for use in salmonid fish. *Immunol Invest 21(4),* 353-364.

[321] Piganelli, J.D, Wiens, G.D., Zhang, J.A., Christenson, J.M. and Kaattari, S.L. (1999). Evaluation of a whole cell, p57– vaccine againstpathogen *Renibacterium salmoninarum*. *Dis. Aquat. Org. 39*, 37–44.

[322] Polk,A. E., Amsden, B., Scarrat, D.J., Gonzal, A., Okahmafe ,A.O. and Goosen, M.F.A. (1994). Oral delivery in aquaculture: Controlled release of proteins from chitosan-alginate microcapsules. *Aquacul. Engg. 13(4)*, 311-323.

[323] Vervarcke, C., Lescorat, O., Ollevier, F., Kinget, R. and Michoel, A. (2004). Vaccination of African catfish with *Vibrio anguillarum O2*: II. Oral intubation with vaccine-containing pellets. *J. Appl. Ichthyol. 20(2*), 134–138.

[324] Joosten, P.H.M., Avilés-Trigueros, M., Sorgeloos, P. and Rombout, J.H.W.M. (1995). Oral vaccination of juvenile carp (*Cyprinus carpio*) and gilthead seabream (*Sparus aurata*) with bioencapsulated *Vibrio anguillarum* bacterin. *Fish Shellfish Immunol. 5,* 289-299.

[325] Karunasagar I. and Otta S.K. 1996. Biofilm formation by *Vibrio harveyi* on surfaces. *Aquaculture 140*, 241-245.

[326] Toranzo, A.E., Magarinõs, B. and Romalde, L. (2005). A review of the main bacterial fish diseases in mariculture systems. *Aquaculture 246*, 37-61.

[327] Colorni, A., Paperna, I. and Gordin. H. (1981). Bacterial infections in gilt-head sea bream *Sparus aurata* cultured at Elat. *Aquaculture 23*, 257–267.

[328] Balebona, M.C., Zorilla, I. Morinigo, M.A. and Borrego, J.J. (1998). Survey of bacterial pathologies affecting farmed gilt-head sea bream (*Sparus auratus* L.) in southwestern Spain from 1990 to 1996. *Aquaculture 166(1-2)*, 19-35.

[329] Lee, K. K. (1995). Pathogenesis studies on *Vibrio alginolyticus* in the grouper, *Epinephelus malabaricus* Bloch et Schneider. *Microb. Pathog. 19*, 39–48.

[330] Kusuda, R., Yokoyama, J. and Kawai. K. (1986). Bacteriological study on cause of mass mortalities in cultured black seabream fry. *Bull Jap Soc Sci Fish 52*, 1745–1751.

[331] Muroga, K., Higashi, M. and Keitoku, H. (1987). The isolation of intestinal microflora of farmed red seabream (*Pagrus major*) and black seabream (*Acanthopagrus schlegeli*) at larval and juvenile stages. *Aquaculture 65*, 79–88.

[332] [Sedano, J., Zorrilla, I., Morin˜igo, M.A., Balebona, M.C., Vidaurreta, A., Bordas, M.A. and Borrego, J.J. (1996). Microbial origin of the abdominal swelling affecting farmed larvae of gilt-head seabream, *Sparus aurata* L. *Aquacult Res 27*, 323–333.

[333] Park, J.H., Park, W.J. and Jeong, H.D. (2001). Immunological efficacy of *Vibrio vulnificus* bacterins given as an oral vaccine in the flounder, *Paralichthys olivaceus. Aquaculture 201*, 187–197.

[334] Azad, I.S., Al-Marzouk, A., Al-Gharabally, H. and Al-Elah, K.A. (2007). Evaluation of three vibrio isolates for immunogenicity using a test species of mullet, *Liza klunzengeri* (Day). 7th Nordic Symposium on Fish Immunology, 19-22, June, 2007. Sterling University, Sterling, U.K. Abstract No. P-77.

[335] Lin, J.H.Y., Chen, T.Y., Chen, M.S., Chen, H.E., Chou, R.L., Chen, T.I., Su, M.S. and Yang, H.L. (2006). Vaccination with three inactivated pathogens of cobia (*Rachycentron canadum*) stimulates protective immunity. *Aquaculture 255*, 125-132.

[336] Ross, A.J. and Klontz, G.W. (1965). Oral immunization of rainbow trout (*Salmo gairdneri*) against an etiological agent of "Redmouth Disease". *J. Fish Res. Bd. Can. 22*, 713-719.

[337] Anderson, D.P. and Ross, A.J. (1972). Comparative study of Hagerman redmouth disease and oral bacterins. *Prog. Fish. Cult. 34*, 226-228.

[338] Anderson, D.P. and Nelson, J.R. (1974). comparison of protection in rainbow trout (*Salmo gairdneri*) inoculated with and fed Hagerman Redmouth bacterins. *J. Fish Res. Bd. Can. 31*, 214-216.

[339] Fujihara, M.P. and R.E. Nakatani. (1971). Antibody production and immune responses of rainbow trout and coho salmon to *Chondrococcus columnaris. J. Fish Res. Board Can. 28*,1253-1258.

[340] Fukuda, Y. and Kusuda, R. (1981). Efficay of vaccination for Pseudotuberculosis in cultured yellowtail by various routes of administration. *Bull. Jap. Soc. Fish 17*, 147-149.

[341] Plumb, J.A. (1984). Immunization of warm water fish against five important pathogen. In: *Symposium of Fish Vaccination*, (Ed. by P. de Kinkelin and C. Michel). O.I.E., Paris, p 119-222.

[342] Bakopoulos, V., Volpatti, D., Gusmani, L., Galeotti, M., Adams. A. and Dimitriadis, G.J. (2003). Vaccination trials of sea bass, *Dicentrarchus labrax* (L.), against *Photobacterium damselae* subsp. *piscicida*, using novel vaccine mixtures. *J. Fish Dis.* 26**,** 77–90.

[343] Griffiths, S.G., Melville, K.J. and Salonius, K. (1998). Reduction of *Renibacterium salmoninarum* culture activity in Atlantic salmon following vaccination with avirulent strains. Fish Shellfish Immunol 8:607–619.

[344] Kitao, T. (1993). Streptococcal infections. In: *Bacterial Diseases of Fish* (ed. by V. Inglis, R. J. Roberts and N. R. Bromage), Blackwell Scientific Publications, Oxford. pp. 196-210.

[345] Plumb, J.A., J.H. Schachte, J.L. Gaines, W. Peltier and B. Carrol. (1974). *Streptococcus* sp. from marine fishes along the Alabama and northwest Florida coast of the Gulf of Mexico. *Transactions of the American Fisheries Society 103*, 358-361.

[346] Evans, J.J., Klesius, P. H., Glibert, P. M., Shoemaker, C. A.,. Al Sarawi, M ., Landsberg, J., Duremdez, R., Al-Marzouk, A. and Al-Zenki, S. (2002). Characterization

of ß- haemolytic Group *Streptococcus agalactiae* in cultured sea bream, *Sparus auratus* L., and wild mullet, *Liza klunzingeri* (Day), in Kuwait. *J. Fish Dis. 25*, 505-513.

[347] Akhlaghi, M., Munday, B.L. and Wittington, R.J. (1996). Comparison of passive and active immunization of fish against streptococcosis (enterococcosis). *J. Fish Dis. 19*, 251-258.

[348] Shoemaker, C.A. and Klesius, P.H. (1997). Protective immunity against enteric septicemia in channel catfish, *Ictalurus punctatus* (Rafinesque), following controlled exposure to *Edwardsiella ictaluri*. *J. Fish Dis. 20*: 101-108.

[349] Eldar, A. and Ghittino, C. (1999). *Lactococcus graviae* and *Streptococcus iniae* infection in rainbow trout *Oncorhynchus mykiss*: similar, but different diseases. *Dis. Aquat. Org. 36,* 227-231.

[350] Kusuda, R., Kawai, K., Salati, F., Banner, C.R. and Fryer, J.L. (1991). *Enterococcus seriolicida* sp. Nov., a fish pathogen. *Int. J. Syst. Bacterio.l 41*, 406-409.

[351] Romalde, J.L., Alvárez, A.L., Ravelo, C., Toranzo, A.E. and Méndez, J.B. (2004). Oral immunization using alginate microparticles as a useful strategy for booster vaccination against fish lactococcosis. *Aquaculture 236*, 119-129.

[352] Beckert, H., and Allison, R. (1964). Some host responses of the white catfish to *Ichthyophthirius multifiliis* (Fouquet). *Proc. S. E. Assoc. Game Fish Comm.* 18, 438–441.

[353] Clark, T. G., Dickerson, H. W. Gratzek, J. B. and Findley, R. C. (1987). *In vitro* response of *Ichthyophthirius multifiliis* to sera from immune catfish. *J. Fish Biol. 31*, 203–208.

[354] Clark, T. G., Lin, T. L. and Dickerson, H. W. 1995. Surface immobilization antigens of *Ichthyophthirius multifiliis*: their role in protective immunity. *Annu. Rev. Fish Dis. 5*, 113–131.

[355] Maki, J.L. and Dickerson, H.W. (2003). Systemic and cutaneous mucus antibody responses of Channel catfish immunized against the protozoan parasite *Ichthyophthirius multifiliis*. *Clin. Diag. Lab. Immunol. 10*, 876–881.

[356] Bower, S.M. and Thompson, A.B. (1987). Hatching of the Pacific salmon leech (*Piscicola salmositica*) from cocoons exposed to various treatments. *Aquaculture 66*, 1-8

[357] Woo, P.T.K. and Li, S. (1990). In vitro attenuation of *Cryptobia salmositica* and its use as a live vaccine against cryptobiosis in *Oncorhynchus mykiss*. *J. Parasitol. 76*, 752-755

[358] Woo, P.T.K. (2003). *Cryptobia* (*Trypanoplasma*) *salmositica* and salmonid cryptobiosis. *J. Fish Dis. 26*, 627-646

[359] Li, S. and Woo, P.T.K. (1995). Efficacy of a live *Cryptobia salmositica* vaccine, and the mechanism of protection in vaccinated *Oncorhynchus mykiss* (Walbaum) against cryptobiosis. *Vet Immunol. Immunopathol. 48,*343-353

[360] Chin, A. and Woo, P.T.K. (2005). Innate cell-mediated immune response and peripheral leukocyte populations in Atlantic salmon, *Salmo salar* L., to a live *Cryptobia salmositica* vaccine. Parasitol. Res. *95,* 10.1007/s00436-004-1270-x

[361] Christie, K.E. (1997). Immunization with viral antigens: infectious pancreatic necrosis. In *Fish Vaccinology* (Eds. R. Gudding, A. Lillehaug, P.J. Midtlyng and F. Brown). *Dev. Biol. Standard 90*, 191-199.

[362] Jorgensen, P. E. V. (1976). Partial resistance of rainbow trout *(Salmo gairdneri)* to viral haemorrhagic septicaemia (VHS) following exposure to nonvirulent Egtved virus. *Nord Veterinaermed. 28,* 570-571.

[363] Lorenzen, N. and Olesen, N.J. (1997). Immunization with viral haemorrhagic septicaemia antigens. In Fish Vaccinology (R. Gudding, A. Lillehaug, P.J. Midtlyng and F. Brown eds.). *Dev. Biol. Standard. 90*, 201-209.

[364] Nakajima K., Maeno, Y., Kurita, J. and Inui, Y. (1997). Vacclnation against red sea bream iridoviral disease in red sea bream *Fish Pathol. 32*, 205-209.

[365] Nakajima, K., Maeno, Y., Honda, A., Yokoyama, K., Turiyama, T. and Manabe, S. (1999). Effectiveness of a vaccine against red sea bream iridoviral disease in a field trial test. *Dis. Aquat. Org. 36*, 73-75.

[366] Cho, M.Y., Oh, M.J., Jung, S.J., Sohn, S.G., Kim, J.W. and Park, S.I. (2003). Immune responses to the vaccines of viral systemic necrosis of carp virus (VSNCV) of common carp, *Cyprinus carpio* L. *J. Fish Pathol. 16*, 175-181

[367] Lorenzen, N. and LePatra, S.E. (2005). DNA vaccines for aquacultured fish. *Rev. Sci. Tech. Off Int. Epiz. 24*, 201-213.

[368] Lin, J.H., Yu, C.C., Lin, C.C. and Yang, H.L. (2005). An oral delivery system for recombinant subunit vaccine to fish. *Dev. Biol. (Basel) 121*, 175-180.

[369] Lorenzen, N., Olesen, N.J. and Jørgensen, P.E.V. (1990). Neutralization of Egtved virus pathogenicity to cell cultures and fish by monoclonal antibodies to the viral G protein. *J. gen. Virol. 71*, 561-567.

[370] Huang, C., Chien, M., Landolt, M., Batts, W. and Winton, J. (1995). Mapping the neutralizing epitopes on the glycoprotein of infectious hematopoietic necrosis virus, a fish rhabdovirus . *J. gen. Virol. 77*, 3033-3040.

[371] Yasumoto, S., Kuzuya, Y., Yasuda, M., Yoshimura, T. and Miyazaki, T. (2006). Oral immunization of common carp with a liposome vaccine fusing koi herpesvirus antigen. *Fish Pathol. 41*, 141-145.

[372] [372] Conteras, M.L.M., Murillo, M.A.G., Bustos, E.R., Romero, M.J., Roa, M.A.C. and Ascencio, F. (2001). Mucosal immune response of spotted sand bass *Paralabrax maculatofasciatus* (Steindachner, 1868) orally immunised with an extracellular lectin of *Aeromonas veronii*. *Fish Shellfish Immunol. 11*, 115-126.

[373] Hara, T., Inoue, K., Morikawa, S. and Tashiro, R. (1976). Vaccination trials for control of furunculosis in salmonids in Japan. *Fish. Pathol. 10,* 227-235.

[374] Michel, C. (1979). Furunculosis of salmonids: vaccination attempts in rainbow trout (*Salmo gairdneri*) by formalin killed germs. *Ann. Rech. Vet. 10*, 33-40.

[375] Spence, K. D., Fryer, J. L. and Pilcher, K. S. (1965). Active and passive immunization of certain salmonid fishes against *Aeromonas salmonicida. Can. J. Microbial. 11*, 397-405.

[376] Rohovec, J.S., Garrison, R.L. and Fryer, J.L. (1975). Immunization of fish for the control of vibriosis. In: Proc. 3rd U.S.-Japanese meeting on aquaculture, Spec. Publ. of Japan Fisheries Agencies and Sea Regional Fisheries Research Laboratory, Niigato, Japan, p 105.

[377] Hastein. T. and Refsti, T. (1986). Vaccination of rainbow trout against vibriosis by injection, dip, and bath. *Bull. Eur. Assoc. Fish Pathol. 6*, 45-49.

[378] Anders, E. (1978). Experimental oral immunization of rainbow trout and eel against *Vibrio anguillarum. Fish.-Forsh., Rostock, 16*, 53-54.

[379] Kusuda, R., Kawai, K., Jo, Y., Akijzuki, T., Fukunaga, M. and Koteke, M. (1978). Efficacy of oral vaccination for vibriosis in cultured ayu. *Bull Jap. Soc. Sci. Fish 44*, 21-25.

[380] Horne, M.T., Tatner, M., McDerment, S. and Agius, C. (1982). Vaccination of rainbow trout, *Salmo gairdneri* Richardson, at low temperatures and the long-term persistence of protection. *J. Fish Dis. 5*, 343-345.

[381] Agius, C., Horne, M.T., and Ward, P.L.D. (1983). Immunization of rainbow trout *Salmo gairdneri*, Richardson, against vibriosis: comparison of an extract antigen with whole cell bacterins by oral and intraperitoneal routes. *J.Fish. Dis. 6*, 129-134.

[382] Llillehaug, A. (1989). A cost-effective study of three different methods of vaccination against vibriosis in salmonids. *Aquaculture 83*: 227-236.

[383] Kawai, L. and Kusuda, R. (1985). Field testing of *Vibrio anguillarum* bacterin in pond cultured ayu. *Fish Pathol. 20*, 413-415.

[384] Gassent, M.D.E., Fouz, B. and Amaro, C. (2004). Efficacy of bivalent vaccine against eel diseases caused by *Vibrio vulnificus* after its administration by four different routes. *Fish Shellfish Immunol 16*, 93-105.

[385] Shoemaker, C.A., Vandenberg, G.W., D sormeaux, A., Klesius, P.H. and Evans, J.J. (2006). Efficcay of *Streptococcus iniae* modified bacterin delivered using Oralject™ technology in Nile tilapia (*Oreochromis niloticus*). *Aquaculture 255*, 151-156.

[386] Kolbl, D. (1975). Neue Erkenntnisse uber die Bauchwassersucht bei Karpfen. *Osterr Fisch 28*, 69-72

[387] Lin, C.C., Lin, J.H.Y., Chen, M.S. and Yang, H.L. (2007). An oral nervous necrosis virus vaccine that induces protective immunity in larvae of grouper (*Epinephelus coides*). *Aquaculture* 268, 265-273.

[388] Sandnes, K., Hansen, T., Killie, J.E.A. and Wagbø, R. (1990). Ascorbate 2-sulfate as dietary vitamin C source for Atlantic salmon (*Salmo salar*). I. Growth bioactivity, hematology and humoral immune response.

[389] Halver, J.E. (1972). The role of ascorbic acid in fish diseases and tissue repair. *Bull. Jap. Soc. Sci. Fish 38*, 79-92.

[390] Roberts, M.L., Davies, S.J. and Pulsford, A.L. (1995). The influence of ascorbic acid (vitamin C) on non-specific immunity in the turbot (*Scophthalmus maximus* L.) *Fish Shellfish Immunol. 5*, 27-38.

[391] Montero, D., Morrero, M., Izquierdo, M.S., Robiana, L., Vergara, J.M. and Tort, L. (1999). Effect of vitamin E and C dietary supplementation on some immune parameters of gilthead sea bream (*Sparus aurata*) juveniles subjected to crowding stress. *Aquaculture 171*, 269-278.

[392] Gallindo-Villegas, J., Masumoto, T. and Hosokawa, H. (2002). Immunostimulants and disease resistance in Japanese flounder *Paralichthys olivaceus*. S-26, In: Proceedings of the 5th Korea-Japan Biannaul Joint Meeting. Kunsan National University, Kunsan, Korea.

[393] Yano, T., Furuichi, M., Nakao, M. and Ito, S. (1990). Effect of L-ascorbyl-2-sulphate Mg on the growth and non-specific immune system of red sea bream *Pagrus major*. Abstract T29.12. Worl Aquaculture Society Meeting. June 10-14, Halifax, Canada.

[394] Pulsford, A.L., Crampe, M., Longston, A. and Glynn, P.J. (1995). Modulatory effects of disease, stress, copper, TBT and vitamin E on the immune system of flatfish. *Fish Shellfishh Immunol. 5,* 631-643.

[395] Hari, B. and Kurup, B.M. (2002). Vitamin C (ascorbyl 2 polyphosphate) requirement of freshwater prawn *Macrobrachium rosenbergii* (de Man) 2002. *Asian Fish Sci. 15(2)*, 145-154.
[396] Kumari, J. and Sahoo, P.K. (2006). Non-specific immune response of healthy and immunocompromised Asian catfish (*Clarias batrachus*) to several immunostimulants. *Aquaculture 255*, 133-141.
[397] Cuesta, A., Orunto, J., Rodriguez, A., Esteban, M.A. and Meseguer, J. (2002). Changes in some innate defence parameters of sea bream (*Sparus aurata* L.) induced by retinol acetate. *Fish Shellfish immunol 13*, 279-291.
[398] Sahu, S. K., Paul, B. N., Mohanta, K. N. and Mohanty, S. N. (2004). Dietary vitamin E requirement, fish performance and carcass composition of rohu (*Labeo rohita*) fry. *Aquaculture 240,* 359-368.
[399] Lall, S.P. and and Weerakoon, D.E.M. (1990). Vitamin B6 requirement of Atlantic salmon (*Salmo salar*). *J. Federa. Am. Soc. Exp. Biol. 4*, 9-12.
[400] Cuesta, A., Esteban, M.A., Ortunõ, J. and Meseguer, J. (2001). Vitamin E increasing natural cytotoxic activity in sea bream (*Sparus aurata* L.). *Fish Shellfish Immunol. 11,* 293-302.
[401] Puangkaew, J., Kiron, V., Somamoto, T., Okamoto, N., Satoh, S., Takeuchi, T. and Watanabe, T. (2004). Nonspecific immune response of rainbow trout (*Oncorhynchus mykiss* Walbaum) in relation to different status of vitamin E and highly unsaturated fatty acids. *Fish. Shellfish Immunol.q 16*, 25-39.
[402] Lee, M.H. and Shiau, S.Y. (2004). Vitamin E requirements of juvenile grass shrimp *Penaeus monodon*, and effects on non-specific immune responses. *Fish Shellfish Immunol. 16,* 475-485.
[403] Hardy, R.W., Halver, J.L. and Brannon, E.L. (1979). Effects of dietary pyridoxine levels on growth and disease resistance of Chinook salmon. In J.E.Halver and K. Tiews (Eds.), Finfish Nutrition and Fish Feed Technology, Berlin, H. Heenemann GmbH and Co. pp. 253-260.
[404] Esteban, M.A., Cuesta, A., Orunto, J. and Meseguer, J. (2001). Immunomodulatory effects of dietary intake of chitin on gilthead sea bream (*Sparus aurata* L.) innate immune system. *Fish Shellfish Immunol. 11*, 303-315.
[405] Shiau, S.Y. and Chen, Y. (2000). Estimation of the dietary vitamin A requirement of juvenile grass shrimp *Penaeus monodon*. *J. Nutr. 200*: 339-347.
[406] Molina-Aja, A., García-Gasca, A., Abreu-Grobois, A., Bolán-Mejía, C., Roque A. and Gomez-Gil B. (2002). Plasmid profiling and antibiotic resistance of Vibrio strains isolated from cultured penaeid shrimp. *FEMS Microbiol Lett, 213*, 7-12.
[407] Vivekanandhan, G., Savithamani, K., Hatha, A.A.M. and Lakshmanaperumalsamy, P. (2002). Antibiotic resistance of *Aeromonas hydrophila* isolated from marketed fish and prawn of South India. *Int. J. Food Microbiol. 76*, 165–168.
[408] Vattanaviboon, P., Panmanee, W. and Mongkolsuk, S. (2003). Induction of peroxide and superoxide protective enzymes and physiological cross-protection against peroxide killing by a superoxide generator in Vibrio harveyi. *FEMS Microbiol Lett 221*, 89–95.
[409] Salminen, S., A. C.Ouwehand, Y. Benno , and Y. K. Lee (1999). Probiotics how should they be defined. Trends food science technol.10:107-110.
[410] FAO/WHO (2002). Guidelines for the Evaluation of Probiotics in Food. London, Ontario, Canada, p. 1–11.

[411] [Vershuere, L., Rombaut, G., Sorgeloos, P. and Verstraete, W. (2000) Probiotic bacteria as biological control agents in Aquaculture. *Microbiol Mol Biol Rev 64*, 655-671
[412] Gatesoupe, F.J. (1994) Lactic acid bacteria increase the resistance of turbot larvae, *Scophthalmus maximus*, against pathogenic *Vibrio. Aqu. Living Res. 7*, 277-282.
[413] Robertson, P.A.W., Dowd, C.O. and Burrells, C. (2000). Use of *Carnobacterium* sp. as a probiotic for Atlantic salmon *Salmo salar* L. and rainbow trout *Oncorhynchus mykiss. Aquaculture 185*, 235-243
[414] Nikoskelainen, S., Ouwehand, A. C., Bylund, G. and Salminen, S. (2001). Protection of rainbow trout (*Oncorhynchus mykiss*) from furunculosis by *Lactobacillus rhamnosus. Aquaculture 198*, 229-236.
[415] Nikoskelainen, S., Ouwehand A.C., Bylund, G., Salminen, S. and Lilius, E. (2003). Immune enhancement in rainbow trout (*Oncorhynchus mykiss*) by potential probiotic bacteria *(Lactobacillus rhamnosus). Fish Shellfish Immunol. 15*, 443-452.
[416] Vadstein, O., Oie, G., Olsen, Y., Salvesen, I., Skjermo, J. and Skjak-Braek, G. (1993). A strategy to obtain microbial control during larval development of marine fish, In Reinertsen, H., Dahle, L. A., Jørgensen, L., Tvinnereim, K. (Eds.), *Proceedings of the First International Conference on Fish Farming Technology*. Balkema, Rotterdam, The Netherlands. pp. 69–75.
[417] Skjermo, J. and Vadstein, O. (1999). Techniques for microbial control in the intensive rearing of marine larvae. *Aquaculture 177,* 333-343.
[418] Ronnestad, I., Thorsen, A. and Finn, R. N. (1999). Fish larval nutrition: a review of recent advances in the roles of amino acids. *Aquaculture 177*, 201-216.
[419] Gildberg, A. and Mikkelsen H. (1998). Effects of supplementing the feed to Atlantic cod (*Gadus morhua*) fry with lactic acid bacteria and immuno-stimulating peptides during a challenge trial with *Vibrio anguillarum. Aquaculture 167*, 103–113.
[420] Gibson, L.F. (1999). Bacteriocin activity and probiotic activity of *Aeromonas media. J Appl Microbiol 85*, 243–248.
[421] [Garriques, D. and Arevalo, G. (1995). An evaluation of the production and use of a live bacterial isolate to manipulate the microbial flora in the commercial production of *Penaeus vannamei* postlarvae in Ecuador. In: *Swimming Through Troubled Waters.* Proceedings of the Special Session on Shrimp Farming (ed. by C.L. Browd and J.S. Hopkins), pp. 53–59. World Aquaculture Society, Baton Rouge, LA, USA.
[422] Gildberg, A., Mikkelsen, H., Sandaker, E. and Ringo, E. (1997). Probiotic effect of lactic acid bacteria in the feed on growth and survival of fry of Atlantic cod (*Gadus morhua*). *Hydrobiologia 352*, 279–285.
[423] [Kennedy, S.B., Tucker, J.W., Neidic, C.L., Vermeer, G.K., Cooper, V.R., Jarrell, J.L. and Sennett, D.G. (1998). Bacterial management strategies for stock enhancement of warmwater marine fish: a case study with common snook (*Centropomus undecimalis*). *Bull. Mar. Sci. 62*, 573–588.
[424] Suyanandana, P., Budhaka, P., Sassanarakkit, S., Saman P., Disayaboot, P., Cai Y. and Benno, Y. (1998). New probiotic lactobacilli and enterococci from fish intestine and their effect on fish production. In: Proceedings of the International Conference on Asian Network on Microbial Researches, pp. 23–25. Yogyakarta, Indonesia.
[425] De Schrijver R. and Ollevier F. (2000). Protein digestion in juvenile turbot (*Scophthalmus maximus*) and effects of dietary administration of *Vibrio proteolyticus. Aquaculture 186*, 107-116.

[426] Queiroz, J. F. and Boyd, C. E. (1998). Effects of a Bacterial Inoculum in Channel Catfish Ponds. *J. World Aqua. Soc. 29(1)*, 67-73.

[427] Austin B, Stuckey L F, Robertson, P.A.W, Effendi, I,Griffith, D R W (1995) A probiotic strain of *Vibrio alginolyticus* effective in reducing diseases caused by *Aeromonas salmonicida, Vibrio anguillarum* and *Vibrio ordalii. J. Fish. Diseases 18*, 93-96.

[428] Rietschel, E. T., Brade, L., Schade, U., Seydel, U., Zähringer, U., Kusumoto, S. and Brade, H. (1988). Bacterial endotoxin: Properties and structure of biologically active domains. In: Schriner, E., Richmon, M.H., Seibert, G., Schwarz, U. (Eds.), *Surface Structure of Microorganisms and Their Interaction with Mammalian Host.* Verlag Chemie, Weinheim, pp. 1-41.

[429] Panigrahi A., Kiron V., Kobayashi T., Puangkaew J., Satoh S. and Sugita H. (2004). Immune responses in rainbow trout *Oncorhynchus mykiss* induced by a potential probiotic bacteria *Lactobacillus rhamnosus* JCM 1136. *Vet. Immunol. Immunopathol. 102,* 379–88.

[430] Panigrahi A., Kiron V., Puangkaew J., Kobayashi T., Satoh S. and Sugita H. (2005). The viability of probiotic bacteria as a factor influencing the immune response in rainbow trout *Oncorhynchus mykiss. Aquaculture 243*, 241-54.

[431] Panigrahi A., Kiron V., Satoh S., Hirono I., Kobayashi T., Sugita H., Puangkaew J. and Aoki T. (2006). Immune modulation and expression of cytokine genes in rainbow trout *Oncorhynchus mykiss* upon probiotic feeding. Development and Comparative Immunology. *Dev. Comp. Immunol. 31(4)*, 372-82

[432] Sakata, T., Ichikawa, H. and Inagaki, A. (1999). Influence of lactic acid, uric acid and ammonia on epithelial cell proliferation and motility of the large bowel. *Asia Pacific J. Clin. Nutr. 8*, S9- S13.

[433] Picchietti, S., Mazzini, M., Taddei, A.R., Renna, R., Fausto, A.M., Mulero, V., Carnevali, O., Cresci, A. and Abelli, A. (2007). Effect of administration of probiotics strains on GALT of larval gilthead seabream: Immunohistochemical and ultrastructural studies. *Fish. Shellfish Immunol. 22*, 57-67.

In: Aquaculture Research Progress
Editor: Takumi K. Nakamura
ISBN 978-1-60456-247-7

Chapter 2

TRENDS AND PERSPECTIVES IN PREVENTING AND CONTROLLING INFECTIOUS DISEASES IN MOLLUSCS

Tristan Renault[1]

Institut Français de Recherche pour l'Exploitation de la Mer (Ifremer), Laboratoire de Génétique et Pathologie, 17390 La Tremblade, France

ABSTRACT

Global molluscan production is continuously increasing, and now becomes one of the large aquaculture activity around the world. With the rapid production and development of molluscs worldwide, diseases due to an extended variety of pathogens have become severe and threaten the mollusc industry. Infectious diseases are a major concern for aquaculture and may result in massive losses and sometimes shutdown entirely a particular activity. In this context, preventing and controlling diseases has become a priority for aquaculture sustainability.

Molluscs including bivalves are unique in terms of health management and very few ways to reduce the detrimental effect of pathogens have been identified. Molluscs, like other invertebrates, lacks a true adaptative immune system and an immunological memory relying totally on their innate immune system to overcome diseases. Vaccination can not be used to protect them against pathogens. Moreover, molluscs are usually reared in the open sea which strongly limits the use of drugs. Finally, pathogen transfers via movements of aquatic organisms appear to be a major cause of epizootics. However, great effort has been made in preventing and controlling infectious diseases based on different approaches:

IV) Control of mollusc movements. Transfer regulations have been developed in order to avoid the introduction of animals from an enzootic area to a pathogen-free area. A such approach needs well suited tools for diagnosing pathogens. Molecular techniques for diagnosing bivalve infectious agents have been developed during the last decade and are now moving from a developmental phase in specialized laboratories for research purposes, to routine application.

---

[1] Phone: 33 5 46 76 26 49, Fax: 33 5 46 76 26 11, E-mail: trenault@ifremer.fr.

V) Understanding complex interactions between animal, environment and pathogen - Marine invertebrates, especially bivalves, are particularly challenged by their environment. Molluscs are typically reared in estuarine environments frequently subjected to fluctuations of environmental factors including temperature, salinity and pollution. Reducing the impact of pathogens is likely to rely on knowledge of their biology.

VI) Genetic improvement for disease resistance - Enhancing the mollusc defense ability through their immunity is one of the important approach to prevent and control infectious diseases. Researchs were recenlty developed in order to provide valuable information for further understanding of the defense mechanisms in bivalves and to define genetic markers of interest.

VII) Alternative methods of prevention and treatment - The preventive or therapeutical use of chemotherapeutants is highly restricted since mollusc farming is carried out in the natural marine environment. Moreover, efficiency of antibiotics treatemts for ubiquitous pathogens is seriously questioned. Alternative exist, including use of probiotics, although these were not extensively applied until recently.

INTRODUCTION

The potential of aquaculture to meet the challenge of food security and to generate employment and international trade is clearly demonstrated by the rapid expansion of this sector, which has grown during the last fifty years from a production of less than one million tonnes in the early 1950s to 59.4 million tonnes by 2004 valued at US$70.3 billions (FAO 2006). World aquaculture has grown at an average annual rate of 8.8 percent from 1950 to 2004 and its drastic growth has been most notable in Asia and the Pacific region, mainly in China. Although aquaculture accounts now for almost 50 percent of the worl aquatic food, the global population demand is always increasing. An additional 40 million tonnes of aquatic food will be required by 2030 to maintain the current consumption.

Global molluscan production is continuously increasing, and now becomes one of the large aquaculture activity around the world. Molluscs are the third most important group by quantity and by value, whereas crustaceans are fourth (FAO, 2006). The world aquaculture production of molluscs for 2004 was estimated at 12.8 million tonnes, representing 28.2 percent of the total world production valued at US$ 9.4 billions. This production is mainly composed of oysters, clams, cockles, mussels, and scallops. By volume, oysters (Ostreidae) are the second most important aquacultured taxonomic group at 4.6 million tonnes following the cyprinids (18.2 million tonnes, US$ 16.3 billions) (FAO, 2006). The Pacific cupped oyster, *Crassostrea gigas*, itself had the greatest contribution with a production volume at 4.4 million tonnes.

Although mollusc culture is steadily growing in importance in the aquaculture sector, populations of cultivated molluscs may suffer from severe mortality outbreaks. Among the possible causes are the occurrence of infectious diseases due to a variety of pathogens including protozoan parasites, bacteria and viruses. Disease outbreaks can be related to (i) an increase of population density that may favours disease transmission, (ii) an increase of animal movements, (iii) the weakening of aquacultured species under environmental stress, and (iv) the diversity in host genetic and immune-competency.. Major disease outbreaks result in loss of income and sometimes the complete shutdown of operations. Preventing and controlling diseases has become a priority for aquaculture sustainability. In this context, attempts must be made to develop new approaches and technologies suited to health improvement of aquacultured species and that are not harmful for the environment.

However, molluscs including bivalves are unique in terms of health management and very few ways to reduce the detrimental effect of pathogens have been identified. Molluscs, like other invertebrates, lack an adaptative immune system and an immunological memory relying totally on their innate immune system to overcome diseases. Vaccination can not be used to protect them against pathogens. Moreover, molluscs are usually reared in the open sea which strongly limits the use of drugs.

Great effort has already been made, but needs to be pursued, in preventing and controlling infectious diseases based on different approaches:

a) control of mollusc movements;
b) understanding complex interactions between hostl, environment and pathogen;
c) genetic improvement for disease resistance;
d) alternative methods of prevention and treatment.

Pathogens Affecting Molluscs

Molluscs may suffer from mortality outbreaks due to various pathogens. Although a large number of pathogens including viruses, bacteria, protozoan and metazoan parasites have been reported in molluscs (reviewed by Bower et al 1994), protozoan diseases are the most frequently described diseases.

Protistan parasites belonging to the genus *Perkinsus* affect various marine molluscs in the five continents and some of them are responsible of massive mortality outbreaks. The most severe economic losses due to mollusc diseases have been related to perkinsosis and considerable efforts have been made by researchers in studying these parasites. Based on molecular data, *Perkinsus* spp were found to be close from dinoflagellates. The type species of the genus *Perkinsus*, *Perkinsus marinus*, first referred as *Dermocystidium marinum*, has been identified in eastern oysters, *Crassostrea virginica*, along the Atlantic coast of the United States and Mexico in association with massive mortalities (Ray 1996). Several new species have then been described. The parasite *Perkinsus olseni* has first been reported in the abalone *Haliotis ruber* in Australia (Lester and Davis 1981) and the species *P. atlanticus* in the carpet shell clam *Ruditapes decussatus* in southern Portugal (Azevedo 1989). Based on molecular analyses, both species were recently proposed to be conspecific, with *P. olseni* having priority (Murrell et al 2002).

Paramyxeans constitute also an important group of protistan parasites affecting molluscs world-wide. The disease caused by the protozoan parasite *Marteilia refringens* spread in the late 60's and mainly reduced the production of the European flat oyster, *Ostrea edulis,* in Europe (Comps 1970). *Marteilia sydneyi*, the aetiological agent of the QX disease of Sydney rock oyster, *Saccostrea glomerata*, is of particular concern for oyster aquaculture in Australia (Perkins and Wolf 1976). Adult Pacific oysters, *C. gigas*, have also been reported affected by a paramyxean parasite, *Marteilioides chungmuensis*, in Japan and Korea (Comps et al 1986).

Several protozoan parasites belonging to the group of Haplospridia are responsible of serious diseases among commercially important molluscs. *Haplosporidium nelsoni* is the causative agent of MSX disease and it affects the eastern oyster, *C. virginica*, on the east coast of North America (Ford and Haskin 1987, Oliver et al 1998). A closely related species, *H. costale*, causes SSO disease, also in the eastern oyster in the USA (Wood and Andrew 1962).

Two genera (*Bonamia* and *Mikrocytos*) and four species of microcells have been described in association with mollusc mortility outbreaks and are currently considered valid. *Bonamia ostreae* (Pichot et al 1979) has been reported in flat oysters, *O. edulis*, in Europe and in Washington, California and Maine, USA (Barber and Davis 1984, Elston et al 1986, Friedman and Perkins 1994, Friedman et al 1989). In the Southern hemisphere, *Bonamia exitiosia* (Hine et al 2001) causes mortality outbreaks in flat oysters, *Tiostrea chilensis* (Dinamani et al 1987) and *Ostrea angasi*. *Mikrocytos roughleyi* is the aetiological agent of winter mortality in *S. glomerata* in Australia (Farley et al 1988). Based on molecular analysis, it has recently been proposed to reassign this parasite to the genus *Bonamia* as *B. roughleyi* (Cochennec-Laureau et al 2003). *Mikrocytos mackini* is a small protistan parasite that causes Denman Island disease in Pacific oyster, *C. gigas*, in British Columbia in Canada (Bower 1988). It is at present the single member of the genus *Mikrocytos*.

Populations of *C. virginica* may be affected by juvenile-oyster-disease (JOD) in eastern USA (Ford and Borrero 2001). JOD is a syndrome recognized in 1988 and characterised by an anomalous conchiolin deposit on the inner surface of the shell. The aetiological agent of JOD has recently been identified as a bacterium, *Roseovarius crassostreae* (Boettcher et al 2005). The Manila clam, *R. philippinarum*, is suffering from *Vibrio tapetis*, the causative agent of brown ring disease (BRD) in France (Paillard and Maes 1989, 1990). BRD is a shell disease, first reported in 1987 among cultured French clams in Brittany (France). High mortality rates were associated with the detection of a charactristic feature, a brown deposit on the inner surface of the shell. Scallops, *Pecten maximus*, may be affected by prokaryotic rickettsia in North Eastern USA and France (Le Gall et al 1988), and by bacteria in NW Spain (Riquelme et al 1996) and France (Lambert et al 1998). The intracellular bacterium *Candidatus* xenohaliotis californiensis has been described as the aetiological agent of the whitering syndrom affecting abalone in California (Friedaman et al 2000) and in other locations all around the world. The Pacific oyster, *C. gigas* suffers from several bacteria belonging to the *Vibrio* genus (Sutton and Garrick 1993, Lacoste et al 2001, Waechter et al 2003, Gay et al 2004, Le Roux et al 2005).

High mortality rates of adult Portuguese oysters, *C. angulata,* have been reported in French livestocks between 1967 and 1973 and were associated with the detection of irido-like viruses (Comps et al 1976, Comps and Bonami, 1977). Herpes-like viruses have also been identified in various marine mollusc species in different countries including the USA (Farley et al 1972, Friedman et al 2005), New Zealand (Hine et al 1992), France (Nicolas et al 1992, Arzul et al 2001, Renault and Arzul 2001, Renault et al. 2000, Renault et al 2001), Australia (Hine and Thorne 1997), and Mexico (Vásquez-Yeomans et al 2004). Herpes-like viruses infections are commonly associated with mortality outbreaks in bivalves. More recently, herpes-like viruses were detected in marine gastropods. Highly pathogenic herpes-like viruses were reported in several abalone species including *Haliotis diversicolor supertexta*, *H. laevigta* and *H. ruber*. (Chang et al 2005, Hooper et al 2007). Other viruses described in bivalves are interpreted as members of the *Papovaviridae*, *Togaviridae*, *Retroviridae*, *Reoviridae, Birnaviridae* and *Picornaviridae* (Farley 1976, Farley 1978, Oprandry et al 1981, Ramussen 1986, Bower 2001).

An increase in both, the number and types of tumors found worldwide in molluscs has been noted over the past decades. Proliferate disorders have been described in more than 20 bivalve species (Elston et al 1988, Elston et al 1990, Villaba et al 2001, Smolarz et al 2005). Although several studies have been conducted in order to define the causative factors involved in the development of such neoplasia in molluscs, no clear response has been obtained. Disseminated neoplasia has been induced through proximity experiments or injection of neoplastic cells (Elston et al 1988, Kent et al 1991, Weinberg et al 1997) and a viral aetiology (i.e. retrovirus) has been suspected in several bivalve species based on transmission trials, detection of a reverse transcriptase activity and viral induction using 5-bromodeoxyuridine (Oprandry et al 1981, Oprandy and Chang 1983, House et al 1998). However, no obvious viral agent has yet been identified.

Infectious diseases may induce a bottle neck in the development of mollusc production of new species, and may induce significant losses when new pathogens are isolated from established aquaculture species. Viral pathogens are difficult to diagnose, prevent and control in open systems. Parasites and bacteria, although easier to diagnose and control, may often acquire greater economic importance due to cumulative mortality rates and effects on the condition of the final products.

SPECIFIC ISSUES RELATED TO MOLLUSCS DISEASES: LIMITATION OF CURRENT MANAGEMENT TECHNIQUES

Molluscs, including wild and cultivated ones, live in the open water. Most of the aquacultured mollusc species are sessile filter feeders. It means that they can get infected with all kinds of pathogens, as they filter large volumes of water. They are also directly subjected to the fluctuations of their environment. Cultured and wild molluscs tend to share same ecological habitats and there is no real distinction between farming and harvest from natural populations regarding risk of disease emergence and spread. Published data suggests that most diseases affecting cultured mollucs are also affecting wild populations and cross transfers are possible (Mortensen et al 2007).

The behaviour of diseased individuals is usually extremely difficult to observe in relation with the presence of the shell. Moreover, shellfish production is an economical activity occurring usually in open-water (with an exception for hatchery-reared animals). No close observation can be carried out. Classical production techniques do not offer the conspicuous advantage of daily monitoring and early warnings including feeding behaviour and signs of weakening. Frequently, shellfish farmers only perform "after-the-fact" monitoring. In these conditions, it is difficult to collect moribund or recently deceased specimens for pathogen detection and diagnosis. Decomposition is rapid in aquatic environments and there may be little tissue in an appropriate state for preservation, processing and examination. In addition, farmers rearing molluscs have little or no control of environmental factors that could potentially predispose animals to disease.

Molluscs, like other invertebrates, lack a true adaptive immune system and an immunological memory, relying on a innate immune system to overcome diseases (Roch 1999). It relies on cells (i.e. phagocytes) and blood-borne molecules (Marchalonis and Schluter 1994, Kuby 1997). Molluscs do not possess lymphocytes and do not produce antibodies. Hence, vaccination can not be used to protect them against pathogens and indirect diagnosis tests including serology are not available. The direct detection of infectious agents remains the only possible approach.

As a consequence of traditional techniques used for shellfish production, there are many transfers of different mollusc species throughout the countries culturing them. Moreover, because of the success stories of certain species or groups, there are many introductions of non native mollusc species (alien species) in order to start new farming activities (Grizel and Heral 1991, Goulletquer and Heral 1997, Calvo et al 2001, Anderson et al 2004). Although the use of introduced species in aquaculture is not new, with air transport and increased global trade, the rate of introductions has increased in last decades. Non native species have successfully been used to generate income and social benefits. Animal movements and transfers have had positive impact on the development of mollusc aquaculture sector. They have, however, also been identified as a major threat to biodiversity and as vectors for pathogens. Pathogen transfers via movements and introductions of live molluscs have shown to be a major cause of epizootics.

The use of currently available drugs in aquaculture is highly regulated and restricted to avoid risks to public safety. Many countries are thus developing and implementing regulations on the use of antimicrobials, veterinary drugs and chemicals in aquaculture. Moreover, molluscs are usually reared in the open sea which strongly limits the use of drugs because of the quantity to be used and impact on the environment. In addition, animals are continuously at risk of exposure to pathogens present in their environment including those introduced by native fauna. Chemotherapy may, however, be successfully applied in mollusc hatcheries and nurseries because of confined facilities.

Despite the impact that infectious diseases may have on molluscs, relatively little is known about what farmers should do to prevent and treat infectious diseases and how molluscs fight against pathogens. Consequently, farmers are left with few resources other than the use of preventive measures. Prevention and control of infectious diseases pass through control of mollusc movements, understanding complex interactions between animal, environment and pathogen (*i.e.* temperature and salinity tolerance, life-cycles including the presence of intermediate hosts or reservoirs, pathogen transmission, etc) and genetic improvement towards disease resistance. In all cases, exact identification of pathogens (taxonomy) and specific diagnosis tools are needed.

Control of Animal Movements and Transfers

Mollusc trade is associated with an inherent risk of moving and spreading infectious diseases. Introduction of known exotic pathogens, changes in the pattern of existing endemic diseases or the appearance of previously unrecognized pathogens may be the basic causes of disease emergencies. The effective management of such disease emergencies relies on a rapid reaction including early warning and contingency planning. Surveillance programmes to assess the presence and prevalence of pathogens in molluscs are needed to protect aquatic resources from infectious diseases where early detection and intervention can significantly reduce impact and losses. In this context, a comprehensive approach for animal health control in mollusc cultures requires: (i) the assessement of health status of mollusc populations, (ii) the constraint of restocking open waters and farming facilities only with molluscs having a health status higher than or equal to that of molluscs already living in the considered areas, and (iii) the eradication of disease when possible by slaughtering infected stocks, disinfecting facilities and restocking with molluscs from approved disease-free sources.

As trade of live molluscs is a major cause of epizootics, restriction of animal movements is a basic way to control infectious diseases. Various codes of practices, guidelines and rules exist in order to minimize risk due to infectious diseases associated with aquatic animal movements. Transfer regulations at international and European levels have thus been developed in order to avoid the introduction of animals from an enzootic area to a pathogen-free area. Minimum measures for the control of certain diseases affecting bivalve molluscs were established by Council Directive 95/70/EC of 22 December 1995. More recently, the Council of the European Union adopted on 24 October 2006 a new directive 'on animal health requirements for aquaculture animals and products thereof, and on the prevention and control of certain diseases in aquatic animals' (2006/88/EC). Directive 2006/88/EU lays down harmonised animal health provisions for placing on the market of aquaculture animals and products thereof, including specific provisions applicable to species susceptible to certain exotic and non-exotic diseases. Annex IV Part II contains a list of cexotic and non-exotic diseases and the species susceptible to these diseases. Directive 2006/88/EC will be in force in August 2008.

Specific infectious agents have thus been or are listed as notifiable pathogens by the EU legislation (Directives 91/67/EEC, Annex A and 2006/88/EC, Annex IV) (Tables 1A and 1B) and by the OIE (Manual of Diagnostic Tests for Aquatic Animals, 2006 and the Aquatic Animal Health Code, 2007) (Table 2). Target pathogens are listed as notifiable according to the following basic considerations: (i) they resist or respond poorly to therapy, (ii) they have a restricted geographical range, (iii) they are of high socio-economic importance, and (iv) they occur in species involved in international trade. In this context, defining the health status (free from infection or infected) of mollusc stocks from a production site, a geographical zone or an entire country through health surveillance programmes is an essential prerequiste to allow or not animal movements and transfers. According to Directive 2006/88/EC (Article 16 (1)) aquaculture animals of species susceptible to diseases listed in Annex IV Part II may be introduced for farming or restocking purpose into a Member State, zone or compartment declared free of a specific disease, only if molluscs come from Member States, zones or compartments also declared free of that disease. Where it can be scientifically justified that a species susceptible to the specific disease at certain life stages do not transmit that disease, Article 16 (1) shall however not apply (Article 16(2)). Moreover, a list of vector species (vector species: 'species that are not susceptible to a disease but which is capable of spreading infection by conveying pathogens from one host to another', 2006/88/EC, Annex I), other than those referred as susceptile species in Part II of Annex IV will be established on the basis of an unrestricted risk assessment. Authorized movements of such vector species should depend on the health status of the Member State, zone or compartment from which they originate (Article 17).

The requirements for international recongnition of freedom from infection depend on the previous infection status of the country, zone or compartiment, namely: (i) absence of susceptible species, (ii) historically free, (iii) last known occurrence within the previous 25 years, and (iv) previously unknown infectious status (OIE, 2007).

Table 1A. list of exotic mollus pathogens and susceptible species

Exotic diseases		
	Disease	Susceptible species
MOLLUSCS	Infection with *Bonamia exitiosa*	Chilean flat oyster (*Ostrea chilensis)* and Australian mud oyster *(O.strea angasi).*
	Infection with *Perkinsus marinus*	Pacific oyster (*Crassostrea gigas*) and Eastern oyster (*C. virginica*)
	Infection with *Microcytos mackini*	Pacific oyster (*Crassostrea gigas*), and Eastern oyster (*C. virginica*), Olympia flat oyster (*Ostrea conchaphila*), and European flat oyster (*O. edulis*)

Table 1B. list of non exotic mollus pathogens and susceptible species

NON Exotic diseases		
	Disease	Susceptible species
MOLLUSCS	Infection with *Mateilia refringens*	Australian mud oyster (*Ostrea angasi*), Chilean flat oyster (*O. chilensis*), European flat oyster (*O. edulis*), Argentinian oyster (*O. puelchana*), blue mussel (*Mytilus edulis*) and Mediterranean mussel (*M. galloprovincialis*)
	Infection with *Bonamia ostreae*	Australian mud oyster (*Ostrea angasi*), Chilean flat oyster (*O. chilensis*), Olympia flat oyster (*O. conchaphila*), Asiatic oyster (*O. denselamellosa*), European flat oyster (*O. edulis*), and Argentinian oyster (*O. puelchana*)

Methodologies to demonstrate freedom from infection must be able to accommodate the variety and the complexity of real life situations. In practice, demonstrating freedom from a specific infection is based on an adequate evidence (to an acceptable level of confidence), that infection, if present, is present in less than a specified proportion of the target population. In addition, risk analysis is increasingly being used as a decision making tool to determine risk associated with the movement of live aquatic animals and trade in aquatic products. International (e.g. OIE) and regional (e.g. ICES and NACA) disease reports provide regular and updated information on major infectious diseases to facilitatae risk analysis and minimize the introduction of pathogens as a result of trade. However, research is required to allow effective development of risk analysis procedures pertaining to mollusc diseases.

It is also important to keep in mind that over-regulation could place unnecessary restrictions on free trade. Finally, at the administrative levels the impact of infectious diseases, especially non-notifiable diseases, are generally misunderstood or seriously underestimated, and it is not uncommon to find a lack of interest in their assessement.

Table 2. list of notifiable diseases and susceptible species

PATHOGENS	SUSCEPTIBLE SPECIES
Bonamia ostreae	*Ostrea edulis, O. puelchana, O. angasi, O. chilensis, O. densellammellosa, C. ariakensis*
Bonamia existiosa	*O. chilensis, O. angasi*
Haplosporidium nelsoni	*C. virginica, C. gigas*
Marteilia refringens	*O. edulis, O. chilensis, O. puelchana, O. angasi* and *Mytilus edulis, M. galloprovincialis*
Mikrocytos mackini	*C. gigas, C. virginica, O. edulis, O. conchaphila (= O. lurida)*
Perkinsus marinus	*C. virginica, C. gigas, C. ariakensis, C. rhizophorae, Mya arenaria, Macoma balthica, Mercenaria mercenariea* + all susceptible species OIE Manual
Perkinsus olseni	*Austrovenus stutchburyi, Venerupis pullastra, V. aurea, Ruditapes decussatus, Ruditapes philippinarum* *Anadara trapezia, Barbatia novazlandiae, Macoma liliana, Paphies australis, C. gigas, C. ariakensis* *Haliotis rubra, H. laevigata, H. scalaris, H. cyclobates*
Candidatus xenohaliotis californiensis	*Haliotis cracherodii, H. rufescens, H. corrugata, H. fulgens, H. sorenseni; H. wallalensis* and *H. discus-hannai*

In mollusc hatcheries/nurseries, infectious diseases are frequently reported in larvae and spat and are involved in mortality outbreaks. These diseases are not currently notifiable diseases subjected to specific control measures under EU or OIE legislation. However, pathogens associated with these mortalities, mainly virus and bacteria, generate important economic loss and may jeopardize the sustainable development of this important socio-economic activity in coastal regions.

KEY DRIVERS TO IMPROVE DISEASE DIAGNOSIS

Effective health management measures are based on diagnosis procedures such as: (i) a systematic approach to disease monitoring including the detection of early infected animals, (ii) the use of sensitive and rapid diagnosis techniques, (iii) the collection of diagnosis data, and (iv) the establishement of diagnositic laboratories using standardised procedures with samples collected according to defined rules. Mollusc health surveillance/control programmes require both adequate legislation and standardised procedures for pathogen diagnosis. Valid laboratory results are essential for diagnosis, surveillance and trade.

However, for many pathogens affecting molluscs, available diagnostic techniques are rather limited, and investigation restricted to histological and ultrastructural examination.

Histopathology, cytology (tissue imprints) and transmission electron microscopy are still widely used in diagnostic laboratories for mollusc diseases. Moreover, the lack of marine mollusc cell lines does not allow carrying out diagnosis based on the detection of cytopathogenic effects in cell cultures. Some infectious agents can be tentatively diagnosed by applying simple methodologies, such as stained tissue imprints usually in association with gross pathology. Histopathology provides general information including physiological aspects and general health condition and allows screening the presence of several pathogens in all the organs present in the slide. In addition, histopathological examination provides valuable information on the intensity and severity of infection at the individual level, co-infections with other notifiable pathogens as well as potential emerging pathogens and non-infectious conditions. However, these techniques are not well suited for epidemilogical surveys. They are time consuming and require experienced staff. The diagnosis based on these methods may be difficult when clinical signs of disease are absent as it is frequently observed with mollusc infectious diseases. Finally, these techniques including histology do not allow identification to the species level for many pathogens.

Recognising the need for diagnostic tools to discriminate between genera and species, efforts have been made to overcome limitations of microscopy. Molecular detection assays for pathogens infecting molluscs appear clearly as suitable tools. Molecular techniques - including PCR and *in situ* hybridization - have thus been developed during the last two decades. They are being developed at an increasingly rate (Walker and Subasinghe, 2000). Molecular techniques for diagnosing pathogens in molluscs are now moving from development in specialised laboratories for research purposes, to routine application and are expected to be increasingly used in pathogen monitoring programmes for the detection of specific pathogens. International standards proposed by the OIE are including molecular techniques for the detection and identification of listed pathogens (OIE 2007).

The routine use of DNA based diagnosis tools is however hampered by a number of major concerns:

(i) Not all regions of the pathogen genome are equally useful as targets for molecular detection. Closely related pathogens may present high sequence similarities.
(ii) Moreover, molecular tools detect the genome and not a viable pathogen.
(iii) The assays often have not been thoroughly tested for inclusivity (detection of all strains of the pathogen) or specificity (cross reaction with any other organism). The main concern is that molecular tools too often are developed from a few sequences without a good understanding of the overall sequence variability within the species.
(iv) PCR - despite its sensitivity - is still subject to the same problems that other diagnostic methods are: the tiny piece of tissue used for amplification may simply not contain the pathogen.

Molecular tools need also formal validation against traditional techniques and testing for their specificity. Studies conducted in parallel with the same isolates in several laboratories would be ideal. It will also be necessary to identify regions of the pathogen genome that can be utilised for species differentiation. Moreover, all molecular assays specific for a pathogen should be tested in parallel and validated, and further sensitive confirmatory tests that will clearly discriminate between all "valid" species should be developed.

As recommended by OIE, validation activities might include:

1) comparison with other methods (standard methods);
2) comparison with reference standards (if available);
3) field and/or epidemiological studies;
4) collaborative studies with other laboratories using the same documented method and including the exchange of samples;
5) reproduction of data from an accepted standard method, or from a reputable publication;
6) experimental infection studies;
7) analysis of internal quality control data.

Validation needs also to be a balance between costs, risks and technical possibilities.

Dagnostics tools are generally not standardized and differences could exist in reagents quality and preparation, in controls, as well as in the interpretation of results. Obviously, the use of a "standardized" diagnostic tool for routine analysis should allow the implementation of a calibrated and controlled process in laboratories but also in sites of production. The development of commercial kits for the detection of mollusc pathogens appears as a promsing approach. Such diagnostic tools may allow mollusc producers to stand sentinel, become a proactive player in the health management of their production and implement optimized bio-security procedures. However, to date there is no commercial kit available for the detection of any mollusc pathogen. The development of such commercial kits is particularly needed.

To confirm the presence of a viable pathogen, molecular approaches should be used in conjunction with other methods including histology in order to allow pathogen visualization. In known susceptible species within the known geographical range of *Bonamia ostreae*, a suspect case of infection with *B. ostreae* is a positive result by one of the following methods: histopathology, tissue imprints, or PCR. In other host species or outside the known range of *B. ostreae*, a suspect case is detection by histopathology, tissue imprints, PCR or *in situ* hybridisation. A presumptive case of *B. ostreae* is detection by tissue imprints or histology combined with a positive result with PCR or *in situ* hybridisation. PCR-RFLP and possibly, transmission electron microscopy are recommended for a confirmatory diagnosis.

As reported in the The Manual of Diagnostic Tests for Aquatic Animals (OIE, 2007) pathogen diagnosis 'may be achieved by the use of good management practices, valid test and calibration methods, proper technique, quality control and quality insurance, all working together within a quality system'. Scientifically sound and uniform testing is main importance for appropriate disease diagnosis and for the application of the necessary control.

Disease Resistance by Selective Breeding

Selective breeding of mollusc stocks appears as one of the most promising approaches for aquaculture development. Examples have been reported for *H. nelsoni*, *P. marinus* and Juvenile Oyster Disease in *C. virginica* (Guo et al 2003) and for *B. ostreae* in *O. edulis* (Naciri-Graven et al 1998).

A selective breeding programme for developing eastern oyster strains tolerant to the parasite *H. nelsoni* has succesfully been conducted in the USA (Haskin and Ford 1979, Ford and Haskin 1987). Traditional selective breeding approaches have also been employed to obtain strains of the same oyster species resistant to the parasite *P. marinus*. Although previously selected *H. nelsoni*-tolerant strains were highly susceptible to *P. marinus* (Burreson 1991), the more recenlty developed strain 'DEBY' demonstrated promising performances showing an enhanced tolerance to both parasites *H. nesloni* and *P. marinus* (Guo et al 2003, Ragone Calvo et al 2003). It has also been reported that natural selection might occur in the natural enviromnent and might lead to the development of oyster populations more tolerant to *P. marinus* (Bushek and Allen, 1996). Moreover, a increased resistance to *P. qugwadi* has been repored in surviving Japanese scallops, *Patinpecten yessoensis*, after an epizootic outbreak (Bower et al 1999). Since 1985, European flat oysters have selectively been bred in France for resistance to *B. ostreae*. Third generation selected oysters showed significantly higher survival rates than non-selected oysters. Moreover, survival rates were correlated with *B. ostreae* prevalence. Resistance (or tolerance) is clearly heritable and mainly additive (Naciri-Graven et al 1998). Selection for bonamiosis-resistant oysters has also been achieved in Ireland (Hugh Jones 1999, Culloty et al 2001, Culloty et al 2004). Genetic improvement has been conducted in order to select resistant animals to JOD in eastern oysters (Farley et al 1997, Farley et al 1998). The evaluation of Sydney rock oyster *S. glomerata* selected lines for resistance to QX disease *M. sydneyi* and winter mortality *B. roughleyi* showed also promising results (Nell et al 2000, Nell and Perkins 2006). Based on recent data, research has been iniated for developing Pacific oyster strains resistant or tolerant to bacterial and viral diseases in France (T. Renault, pers. comm.).

Only mollusc hatcheries are able to supply selected animals. They may contribute to the development of a substantial international trade in mollusc gametes and larvae, allowing for the distribution of seed stocks improved through selective breeding. Animal transfers are currently recognised as a major cause underlying outbreaks of mortality, epidemics and spread of diseases. Although hatchery technology is constantly being improved, significant production problems including infectious diseases must be solved. Current practise in the commercial shellfish hatcheries takes into account basic research findings about food provision and avoidance of *Vibrio* infection but uncontrolled variables are still damaging the industry. Among these uncontrolled variables, virus infections may play a key role.

With this in mind, hatcheries should be underpinned as a risky segment of the mollusc aquaculture sector. However, mollusc hatcheries are unique in having closed facilities that enhance disease control capabilities. Hatcheries may produce checked progeny, namely specific pathogen free, and reduce circulation of infected stocks. Moreover, they may have a pivotal role in the development and implementation of health management strategies based on improved resistance to infectious diseases.

Advances in Understanding of Immunity of Aquacultured Mollus Species

Bivalve molluscs do not possess acquired immunity. Their defence capabilities rely on an innate, non-adaptive immune system involving circulating cells, the haemocytes, and various effector molecules (Canesi et al. 2002) including lysozyme (McDade and Tripp 1967, Cheng and Rodrick 1974, Ito et al 1999, Bachali et al 2002, Xue et al 2007), lysosomal enzymes (Cheng 1986, Oubella et al 1994, Xue et al 2001), phenol oxidase (Coles and Pipe 1994,

Hellio et al 2007), protease inhibitors (Faisal et al 1998), heat shock proteins (Song et al 2006) and antimicrobial peptides (AMP) (Charlet et al., 1996, Hubert et al 1996, Mitta et al., 2000, Seo et al 2005, Gueguen et al 2006, Gonzalez et al 2007). Besides the cell-free response, haemocytes are primarily responsible for immune defence. Hemocytes are responsible for recognition, phagocytosis, and elimination by oxidation of non-self particles (Cheng 1981). They mediate phagocytosis, a primordial mechanism for the elimination of foreign materials including inorganic particles, living organisms (pathogenic or non-pathogenic) and modified self-cells. The intracellular degradation of the phagocytised material involves free radicals such as reactive oxygen intermediates (ROIs) (Pipe 1992), and lysosomal enzymes.

Although various data has already been published on the topic, a major concern in bivalve immunity is the fragmentary or lacking knowledge on effector proteins, regulatory pathways and related genes. The early appearance and maintenance of defence capabilities against pathogens involve gene expression and regulation including signalling pathways, cell cycle regulation, protein turnover, and immune functions. The mechanisms underlying innate immunity critically depend on the presence of functional proteins. These proteins are encoded by genes whose transcription involved complex programmes. Altered or defective gene expression and compromised regulation may increase host susceptibility to pathogens and lead to disease outbreaks.

Last decade, various molecular technologies including mRNA differential display, Suppressive Subtractive Hybridization (SSH) and expressed sequence tag (EST) libraries have been applied in molluscs in order to identify immune genes and to assess the level of gene expression. About 50,000 ESTs from bivalves are now available in databases and other thousands will be released soon.

Various genes have recently been identified in bivalves using EST libraries. EST libraries were constructed from the eastern oyster, *C. virginica* (Jenny et al 2002). Among the identified sequences, less than 5% were related to putative immune functions. EST from gill and gonad tissues of adult eastern oyster *C. virginica* revealed several heat shock proteins, stress-induced proteins and the AMP defensin (Peatman et al 2004). A recent study has also been carried out in order to analyse at a molecular level the response to the parasite *P. marinus* in two oyster species, *C. gigas* and *C. virginica*, the latter presenting the highest susceptibility to the pathogen (Tanguy et al 2004). Five hundred singleton ESTs have been obtained and 19 of them were related to genes involved in immunity and cell communication, such as toll-like receptors and TNF receptor associated proteins (Tanguy et al 2004). Libraries have also been constructed from mussels *M. galloprovincialis* using multiple tissues extracts. They allowed the identifiation of AMP myticin, methallothionein and heat shock protein genes (Venier et al 2003).

The EST technology has been applied to *C. gigas* by Guegen et al (2003) and led to the identification of 20 sequences presenting homologies with known genes involved in immunity. Huvet et al (2004) have then constructed EST libraries using selected *C. gigas* populations presenting different susceptibility to summer mortality syndrom. They also identified ESTs more and less related to immune functions (16%) (Huvet et al 2004). The construction of an EST library from *C. gigas* mantle allowed the identification of the first oyster AMP belonging to the mussel defensin family (Gueguen et al 2006). In addition, during the course of an EU funded project (AVINSI, QLK2-CT-2002-01691) SSH libraries have been constructed in order to study the antiviral response of adult Pacific oyster,

Crassostrea gigas, to Ostreid Herpes virus 1 (OsHV 1) by identifying and studying virus-induced genes. Among the sequences that matched with the product of known genes, 9% were related to putative immune functions (laccase, macropahge-expressed protein, molluscan defence protein, IK cytokine, myeloid differenciation factor 88, …) and the related oyster genes were totally sequenced by RACE PCR (T. Renault, pers comm).At least, a EST library constructed from the bay scallop *Argopecten irradians* revealed 131 host-defense-related gene sequences including proteases, protease-inhibitors, heat shock proteins, lectins, defensins, antioxidants, and Toll-like receptors (Song et al 2006).

Such data provides the basis for understanding the role of the innate immune system against various pathogens and may be useful in defining selection markers. The immune relevant genes identified in bivalves may serve to construct genetic maps in different mollusc species on the basis of a QTL approach. Promising results have been obtained in the eastern oyster, *C. virginca* (Yu and Guo 2006).

The innate immune responses of molluscs remain a vast domain to be explored and is very likely to present potential applications by the use of encoding genes as selection markers for improving resistance to infections and by the use of products of identified genes as therapeutic agents. Moreover, the fact that invertebrates make up greater than 90% of all species on earth attests to the efficiency of their 'primitive' host defence systems, at least at the population level. It has become more and more evident that some of the innate mechanisms are very conserved in evolution. It has been clearly demonstrated that some pathways in non-specific defence systems are conserved in *Drosophila* and mammals (Medzhitov et al 1997, Means et al 2000). In addition, it has been shown the non-specific response is required to initiate the specific immune response in mammals. Thus, non-specific immunity increasingly perceived as being of fundamental importance.

Studying Complex Interactions between Molluscs, Environment and Pathogens: A Key for a Sustainable Aquaculture

Shellfish production is an economical activity occurring usually in open-water and farmers rearing molluscs have no real control on environmental factors that could potentially predispose animals to disease. Marine molluscs are particularly challenged by their environment. Due to their filter-feeding habits, bivalves gather large number of micro-organisms Pollutants or xenobiotics can interact with immune system components and interfere with protection functions. According to Snieszko (1974), the development of an infectious disease results from an unbalance between the host and the pathogen due to external factors including pollutants and/or internal factors of both protagonists (virulence of the pathogen, susceptibility of the host). Although , in the case of molluscs, this definition appears particularly adapted, environmental suppression/exacerbation factors associated with disease expression are notably lacking in most mollusc health literature. The impact of disease under 'changing' habitats or environmental conditions requires more detailed examination. The environment of pathogens is made of many different types of factors. The genotype of the host itself produces the background of the pathogen development and interactions between host and pathogen genotypes can result in the shared control of their traits. Abiotic and biotic factors, occurring inside and outside molluscs can affect the translation of genotypic variations into phenotypes. Such environmental factors add another layer of complexity to the determinism of phenotypes as they can affect the interactions between host and pathogen genotypes leading to highly complex host/pathogen/environment interactions.

The elimination of pathogens can be regarded as a dynamic process. The pressure of different stressors including variations of salinity, and temperature and chemical pollutants can disrupt homeostasis and define irreversible alterations rendering molluscs more susceptible to pathogens. Pollutants may be partly responsible by depressing defence mechanisms. Demonstration of the relationship between pollution and increase of susceptibility to infectious diseases have been carried out in marine vertebrates (Jepson et al 2005). Some studies have also been carried out to link pollutant presence and susceptibility to infectious diseases in bivalves. The intensity of *P. marinus* infection increased in eastern oysters, *C. virginica* after contamination by polluted sediment and tributyltin (Chu et al 2002). Cadmium contamination of *M. edulis* followed by a *V. tubiashii* challenge resulted in an increased hemocyte count (Pipe and Coles 1995). Mortality rates of *Meretrix lusoria* contaminated with heavy metals and experimentally exposed to a birnavirus were higher than those of animals only contaminated or only infected (Chou et al 1998). Summer seed mortality reported in Tomales Bay, California has been associated with extreme temperature, phytoplankton blooms, and a herpesvirus (Burge et al 2007).

Although animals are exposed to complex mixtures of pollutants in the environment, most of laboratory experiments are based on the use of a unique pollutant (Banerjee et al 2001). Studies on herbicides (Faust et al 2001) have shown some positive interactions: toxicity of a mixture was more important than toxicity of molecules individually tested. A pesticide mixture modulated *C. gigas* hemocyte phagocytosis *in vitro* whereas no modulation was reported for each molecule tested individually (Gagnaire et al 2006a). Antagonist effects have also been shown for heavy metals (Breittmayer et al 1984).

A pesticide mixture used at realistic concentrations on Pacific oyster, *C. gigas*, in laboratory conditions (i) decreased hemocyte phagocytosis, (ii) down-regulated genes involved in haemocyte functions and particularly in phagocytosis and (iii) increase susceptibility to a bacterial challenge (Gagnaire et al 2007). Oysters were exposed to a mixture of eight pesticides (atrazine, glyphosate, alachlor, metolachlor, fosetyl-alumimium, terbuthylazine, diuron and carbaryl) at environmentally relevant concentrations over a seven day period. Hemocyte parameters (cell mortality, enzyme activities and phagocytosis) were monitored using flow cytometry. The expression of 19 genes interpreted as potentially involved in *C. gigas* defence mechanisms was also analysed using real-time RT-PCR. Phagocytosis was significantly reduced and the 19 selected genes were down-regulated in contaminated animals. After pesticide contamination, two pathogenic *V. splendidus*-related strains were simultaneously injected into the adductor muscle of oysters in order to study interactions between pesticide exposure and a bacterial challenge. Oyster mortality and expression of selected oyster genes were followed 4h and 24h post-injection. Oyster mortality rates were significantly higher in pesticide-treated oysters compared to untreated oysters 48h and 72h after the bacterial challenge. Genes were over-expressed in pesticide-treated oysters compared to untreated oysters 24h post bacterial injection. Gene over-expression could lead to an injury of host tissues, resulting in pathogenesis and oyster mortality (Gagnaire et al 2007). Moreover, pollutants could also have an effect on pathogen virulence. Several genes potentially involved in the virulence of these bacterial strains have been identified (D. Saulnier, pers comm). The joint study of host and pathogen response to pollutants would be of interest to identify the conditions needed for a disease to be established.

EMERGING METHODS TO CONTROL PATHOGENS

The preventive or therapeutical use of chemotherapeutants is highly restricted since mollusc farming is carried out in the natural marine environment. Moreover, efficiency of antibiotics treatments for ubiquitous pathogens is seriously questioned. Alternatives exist, including use of probiotics.

Probiotics are however not extensively applied in mollusc aquaculture and research effort on the topic in bivalve molluscs is moderate. Probiotics can only be used in confined facilities, such as hatcheries or land based operations. Douillet and Langdon (1993) conducted the first work using axenic Pacific oyster, *C. gigas*, larvae and reported that a putative *Alteromonas* sp. has had a positive impact on larval growth and survival. Other potential probiotics for *C. gigas* have also been determined by Gibson et al (1998) and more recently by Elston et al (2004). Riquelme et al (1996) has identified a bacterium, *Alteromonas haloplanktis*, capable of reducing larval mortality due to *V. anguillarum* in Chilean scallop, *Argopecten pupuratus*. The same research group later identified other promising bacteria (Riquelme et al 2000). Based on their reserach work, three selected bacteria (*Vibrio* sp., *Pseudmomoanas* sp. and *Bacillus* sp.) have been used in a commercial scale hatchery production (Riquelme et al 2001) and allowed larval production without antibiotics. A strain of *Roseobacter* sp. presenting an antibacterial activity has been isolated from scallop, *Pecten maximus* (Ruiz-Ponte et al 1999). Although the bacterial extracts were capable to enhance larval survival in normal culture, they did not demonstrate any positive effect when scallop larvae were challenged with a pathogen. More recently, Castro et al (2002) reported an inhibitory effect of five strains of *V. tibiashii* or *V. tubiashii*-like bacteria on the growth of *V. tapetis*, the aetiological agent of brown ring disease in clams. As a conclusion, we can keep in mind that the use of probiotics is not today a current pratice in mollusc aquaculture and is still an emerging method.

Certain diatoms, e. g. *Skeletonema costatum*, present negative effects on marine bacteria belonging to the genus *Vibrio* and may be applied in controlling bacterial infections. However, the development of a such approach needs further studies in order to understand the modes of action.

On the contrary, the use of polypoid animals has clearly emerged last decades as a possible way to fight against infectious diseases. Triploids contribute now to a large part of the oyster production and demonstrate a faster growth and a better test (higher glycogen content). We may assume that growing triploid oysters, market size due to faster grow could be reached before epizootic outbreaks occur. Morever, reduced gametogenesis could be associated to a better use of energy for defence capabilities rendering triploids less susceptible to infectious diseases. Recent data demonstrated that triploid Pacific oysters showed higher haemocyte activites than diploids and were less susceptible to summer mortality after deployment in the field (Gagnaire et al 2006b). Triploid *S. glomerata* showed reduced winter mortality due to QX disease relative to diploid oysters (Hand et al 1998). However, triploid *C.*

virginica did not demonstrate enhanced resistance to *P. marinus* (Barber and Mann 1991, Meyers et al 1991).

Testing of non native mollusc species (allochtonous species) for resistance to pathogens is also a potential option. This option has been selected in the Chesapeake Bay Region of the USA in order to overcome the drastic decline in the production of *C. virginica* related to *H. nelsoni* and *P. marinus* epizootics (United States National Marine Fisheries Service 2003). In this context, the oyster species *C. ariakensis* originating from Asian Pacific has been tested in field experiments (triploids) and its susceptibility to *H. nelsoni* analysed. This oyster species demonstrated higher survival, lower disease susceptibility and faster growth than native oysters (Calvo et al 2001). However, the introduction of *C. ariakensis* in American water is still under debate. Pathological analysis demonstrated that *C. ariakensis* could be infected by various pathogens (Moss et al 2007). Although existing quarantine procedures (ICES 1985, 1995) may be applied in order to minimize the risk, such introductions must be considered as risky considering the putative presence of pathogens in introduced animals. Moreover, a non native mollusc species could be found after its introduction highly susceptible to previously unrecognized pathogens present in native species, those species acting as vectors (no clinical sign or mortality) (Carnegie et al 2006).

It is necessary to keep in mind that genetic engineering (gene transfer) could be a way to improve infectious disease resistance in molluscs. Although some tools are now available (Cadoret et al 2000, Buchanan et al 2001a and b), this approach must be analysed in the context of the general evolution of the society with a greater demand for environmentally responsible practices in agriculture and food safety.

CONCLUSION

Aquaculture is developing, expending in new areas, intensifying and diversifying world-wide. Driving forces in shaping aquaculture rely now on demand market and marketing aspects. Consumers are becoming more demanding on socially and environmentally responsible aquaculture practices, food safety and traceability. In this context, primary producers need to be equipped with adapted skills and knowledge so that the commodities they produce meet the requirments demanded by the market and consumers.

No doubt, an efficient management of the sanitary status of mollusc production implies a significant involvement of the farmers who -unavoidably- are in the front line in the fight against the diseases and that can become key players in the control of the pathogens that threaten their livestock. This outlook implies first, that farmers efforts are supported by the Authorities and the National laboratories involved in control or disease policy, and second that they can have at their disposal validated, accurate, easy to use in the field and affordable screening tools, that can allow an efficient monitoring of the sanitary status of their production. However, to date there is no commercial kit available for the detection of any mollusc pathogen. In this context, the promotion of an integrated bio-security approach and the development of rapid, reliable and sensitive diagnosis tools including commercial kits that meet the time constraints experienced under hatching conditions, are of the utmost importance for efficient control of diseases in mollusc aquaculture.

Mollusc hatcheries are to face an increasing demand and may therefore play a key role in mollusc aquaculture development in next decades contributing to movements, transfers and introductions of live animals all around the world. Although serious infectious diseases may affect molluscs during the course of hatchery/nursery processes, based on their closed facilities hatcheries are capable to produce checked molluscs progeny and therefore play a key role in controlling the circulaton of infected animal.

There is no real distinction between farming and harvest from natural populations regarding risk of disease emergence and spread. Published data suggests that most diseases affecting cultured molluscs are also affecting wild populations. As a consequent, we can hypothesise that several mollusc species may act as vectors for pathogens. A recent review reports that numerous pathogens are strongly suspected to be exchange from farmed shellfish to the wild and vice versa (Mortensen et al 2007). Since mollusc culture is commonly practised in close contact from wild stocks, the introduction of a new infectious agent into open-water can impact sympatric wild resources. Also, transplanted wild shellfish can be asymptomatic carriers of pathogens that may infect cultured populations. Impacts on mollusc fom accidental introduction of an exotic pathogen could present consequences for both cultured and wild stocks. However, the hard evidences of transfer of pathogens between wild and farm populations are missing due mainly to lack of relevant data on life cycles and important ecological aspect of the life of pathogens. Lack of knowledcgc on host range for mollusc pathogens seriously impedes the reliability of risk analysis results.

There is recent evidence that global change including climat change has an effect on the aquatic environments. Alterations in salinity, temperature, current flow rate and direction, and weather conditions are likely to occur. As a consequence, the health status of aquatic organisms, such as host physiology, virulence of pathogens and their abundance and distribution, may be affected (Harvell et al 1999, Harvell et al 2004, Lafferty et al 2004). Increase or decrease disease prevalence or intensity and distribution of diseases could be oberved as a result in temperature and salinity shifts related to climate change. During periods of winter warming, an extension of the geograpical range of the parasite *P. marinus* (a warm water parasite) from the southern USA to the northeastern USA has been reported (Ford 1996). In this context, potential effects of global change on animal health status appear as a major concern for a sustainable mollusc aquaculture.

REFERENCES

Anderson, J., Hedgecock, D., Berrigan, M., Criddle, K. R., Dewey, W., Ford, S. Gouletquer, P., Hildreth, R., Paolisso, M., Targett, N;, Whitlach, R. (2004). Non-native Oysters in the Chesapeake Bay. National Research Council, National Academy Press, 325 p;

Arzul, I., Renault, T. Lipart, C., Davison A. J. (2001). Evidence for interspecies transmission of oyster herpesvirus in marine bivalves. *J. Gen. Virol.*, 82: 865-870.

Azevedo, C. (1989). Fine structure of *Perkinsus altanticus* n. sp. (Apicomplexa, Perkinsea) parasite of the clam *Ruditapes decussatus* from Portugal . J. Parasitol. 75: 627-635.

Bachali, S., Jager, M., Hassanin, A., Schoentgen, F., Jollès, P., Fiala-Medioni, A., Deutsch, J. S. (2002). Phylogenetic analysis of invertebrate lysozymes and the evolution of lysozyme function. *J. Mol. Evol.*, 54: 652-654.

Banerjee, B. D., Seth, V., Ahmed, R. S. (2001). Pesticide-indudec ozidative stress: perspectives and trends. *Rev. Environ. Health*, 16(1): 1-40.

Barber, B. J., Davis, C (1994). Prevalence of *Bonamia ostreae* in *Ostrea edulis* populations in *Maine. J. Shellfish Res.*, 13: 298.

Barber, B. J., Mann, R. (1991). Sterile triploid *Crassostrea virginica* (Gmelin 1791) grow faster than diploids but are equally susceptible to *Perkinsus marinus*. *J. Shellfish Res.*, 10: 445-450.

Boettcher, K. J., Geaghan, K. K., Maloy, A. P., Barber, B. J. (2005). *Roseovarius crassostreae* sp. nov., a member of the Roseobacter clade and the apparent cause of juvenile oyster diease (JOD) in cultured Eastern oysters. *Int. J. Syst; Evol. Microbiol.*, 55(4):1531-1537.

Bower, S.M. (1988). Circumvention of mortalities caused by Denman Island oyster disease during mariculure of Pacific oyster. *Am. Fish. Soc. Spec. Publi.*, 18: 246-248.

Bower, S.M. (2001). Synopsis of infectious diseases and parasites of commercially exploited shellfish: Assorted viruses detected in oysters and of unknown significance. ULR: http://www-sci.pac.dfo-mpo.gc.ca/shelldis/assortvirusoy_e.htm.

Bower, S. M., McGladdery, S. E., Price, I. M. (1994). Synopsis of infectious diseases and parasites of commercially exploited shellfish. *Ann. Rev. Fish Dis.*, 4: 1-199.

Bower, S. M., Blackbourn, J., Meyer, G. R., Welch, D. W. (1999). Effect of *Perkinsus qugwadi* on various species and strains of scallops. *Dis. Aquat. Org.*, 36: 143-151.

Breittmayer, J.P., Woaye, C. and Zsurger, N. (1984). Effect of cadmium on mercury uptake by mussels (*Mytilus edulis*). Journal de la Recherche Océanographique 9 (1984) 66-68.

Buchanan, J. T., Nickens, A. D., Cooper, R. K., Tiersch, T. R. (2001a). Transfection of eastern oyster *Crassostrea virginica* embryos. *Mar. Biotechnol.,* 3: 322-335.

Buchanan, J. T., Cheng, T. C., La Peyre, J. F., Cooper, R. K., Tiersch, T. R. (2001b). in vitro transfection of adult eastern oysters *Crassostrea virginica. J. World Aquacult.* Soc., 32: 286-299.

Burge C. A., Judah, L. R., Conquest, L. L., Griffin, F. J., Cheney, D. P, Suhrbier, A., Vadopalas, B., Olin, P. G., Renault T., Friedman, C. S. (2007). Summer seed mortalities of the Pacific oyster, *Crassostrea gigas* Thunberg, grown in Tomales Bay, California, USA: the influence of oyster stock, planting time, pathogens and environmental stressors. *J. Shellfish Res.*, 26 (1): 163-172.

Burreson, E. M. (1991). Effects of *Perkinsus marinus* infection in the eastern oyster, *Crassostrea virginica* (Gmelin, 1791): I. Susceptibility of native and MSX-resistant stocks. *J. Shellfish Res.*, 10: 417-423.

Bushek, D., Allen, Jr. S. K., (1996). Host-parasite interactions among broadly distributed populations of the eastern oyster *Crassostrea gigas* and the protozoan *Perkinsus marinus*; *Mar. Ecol. Prog. Ser.*, 139: 127-141.

Cadoret, J.-P., Bachère, E., Roch, P., Miahle, E., Boulo, V. (2000). Genetic transformation of farmed marine bivalve molluscs. In: M. Fingerman, R. Nagabhushanam (Eds.) Recent advances in marine biotechnology, Vol. 4: Aquaculture, Part A Seaweeds and invertebrates. Science Publishers pp. 111-126.

Calvo, G. W., Luckenbach, M. W., Allen, S. K., Burreson, E. M. (2001). A comparatve field study of *Crassostrea ariakensis* (Fujita 1913) and *Crassostrea virginica* (Gmelin1791) in relation to salinity in Virginia. *J. Shellfish Res.*, 20:221-229.

Canesi, L., Gallo, G., Gavioli, M., Pruzzo, N. (2002). Bacteria-hemocyte interactions and phagocytosis in marine bivalves. *Microsc. Res. Tech.*, 57: 469C-476.

Carnegie, R. B., Burreson, E. M., Hine, P. M., Stokes, N. A., Audemard, C., Bishop, M. J., Peterson, C. H. (2006). *Bonamia perspora* n. sp. (Haplosporidia), a parasite of the oyster *Ostreola equestris*, is the first *Bonamia* species known to produce spores. *J. Eukaryot. Microbiol.*, 53(4): 232-245.

Castro, D., Pujalte, M. J., lopez-Cortes, L., Garay, E., Borrego, J. J. (2002). Vibrios isolated from the cultured manila clam (*Ruditapes philippinarum*): numerical taxonomy and antibacterial activities. *J. Appl. Microbiol.*, 93: 438-447.

Chang, P.H., Kuo, S. T., Lai, S. H., Yang, H. S.,.Ting Y. Y., Hsu, C. L., Chen, H. C. (2005). Herpes-like virus infection causing mortality of cultured abalone *Haliotis diversicolor supertexta* in Taiwan. *Dis. Aquat.* Org., 65: 23-27.

Charlet, M., Chernysh, S., Philippe, H., Hetrut, C., Hoffmann, J., Bulet, P. (1986). Innate immunity. Isolation of several cysteine-rich antimicrobial peptides from the blood of a mollusc, *Mytilus edulis*. *J. Biol. Chem.*, 271: 21808-21813.

Cheng, T.C. (1981). Bivalves. In: Ratcliffe, N.A. and Rowley, A. F. (Eds.), Invertebrate Blood Cells I. Academic Press, London, pp. 233-299.

Cheng, T. C. (1986). Specificity and the role of lysosomal hydrolases in molluscan inflammation. *Int. J. Tissue React.*, 8: 439-45.

Cheng T. C., Rodrick, G. E. (1975) Lysosomal and other enzymes in the hemolymph of *Crassostrea gigas* and *Mercenaria mercenaria*. *Comp. Biochem. Physiol.*, 52 B: 443-447.

Chou, H.-Y., Chang, S.-J., Lee, H.-Y., Chiou, Y.-C. (1998). Preliminary evidence for the effect of heavy metal cations on the susceptibility of Hard Clam (*Meretrix lusoria*) to clam birnavirus infection. *Fish Pathology* 33: 213-219.

Chu, F.L., Volety, A.K., Hale, R.C., Huang, Y. (2002). Cellular responses and disease expression in oysters (*Crassostrea virginica*) exposed to suspended field contaminated sediments. *Marine Environmental Research* 53: 17-35.

Cochennec-Laureau, N., Reece, K. S., Berthe, F. C. J., Hine, P. M. (2003). *Mikrocytos roughleyi* taxonomic affiliation leads to the genus *Bonamia* (*Haplosporidia*). *Dis. Aquat. Org.,* 54: 209-217.

Coles, J. A., Pipe, R. K. (1994). Phenoloxidase activity in the hemolymph and hemocytes of the marine mussel *Mytilus edulis*. Fish Shellfish Immunol., 4: 337-352.

Comps, M. (1970). Observations sur les causes d'une mortalité anormale des huîtres plates (*Ostrea edulis* L.) dans le bassin de Marennes. *Rev. Trav. Inst. Pêches Marit.* 34: 317-326.

Comps, M., Bonami, J. R. (1977). Infection virale associée à des mortalités chez l'huître *Crassostrea angulata Th. C. R. Acad. Sc. D.*, 285: 1139-1140.

Comps, M., Bonami, J. R., Vago, C., Campillo, A. (1976). Une virose de l'huître portugaise (*Crassostrea angulata* Lmk). C. R. *Hebd. Séanc. Acad. Sc. D.*, 282: 1991-1993.

Comps, M., Park, M. S., Desportes, N. (1986). Etude ultrastructurale de *Martelioides chungmuensis* n. g., n. sp. parasite des ovocytes de l'huître *Crassostrea gigas Th. Protistologica,* 22: 279-285.

Culloty, S.C., Cronin, M.A., Mucahy, M.F. (2001). An investigation into the relative resistance of Irish flat oyster Ostrea edulis, L. to the parasite *Bonamia ostreae* (Pichot et al, 1980). *Aquaculture* 19: 229-244.

Culloty, S.C., Cronin, M.A., Mucahy, M.F. (2004). Potential resistance of a number of populations of the oyster *Ostrea edulis* to the paraste *Bonamia ostreae*. *Aquaculture* 237: 41-58

Dinamani, P., Hine, P.M., Jones, J.B. (1987). The occurrence and characteristics of the hemocyte parasite *Bonamia* sp. In New Zealand dredge oyster, *Tiostrea lutaria*. *Dis. Aquat. Org.* 3: 37-43.

Douillet, P. A., Langdon, C. J. (1993). Effects of marine bacteria on the culture of axenic oyster *Crassostrea gigas* (Thunberg) larvae. *The Biological Bulletin,* 184: 36-51.

Elston, R. A., Drum, A. S., Allen, S. K. Jr. (1990). Progressive development of circulating polyploid cells in *Mytilus* with hemic neoplasia. *Dis. Aquat.* Org. 8, 51-59.

Elston, R.A., Farley C.A., Kent, M.L. (1986). Occurrence and significance of bonamiasis in European flat oysters *Ostrea edulis* in North America. *Dis. Aquat. Org.*, 2: 49-54.

Elston, R. A., Kent, M. L., Drum, A. S. (1988). Transmission of hemic neoplasia in the bay musel, *Mytilus edulis*, using whole cells and cell homogenate. *Dev. Comp. Immunol.*, 12: 719-727.

Elston, R., Humphrey, K., Gee, A., cheney, D., Davis, J. (2004). Progress in the development of effective probiotic bacteria for bivalve shellfish hatcheries and nurseries. *J. Shellish Res.*, (23(2): 654.

FAO (2006). State of world aquaculture. FAO Fisheries Technical Paper 500, 129 pp.

Farley, C. A. (1976a). Ultrastructural observations on epizootic neoplasia and lytic virus infection in bivalve molluscs. *Prog. Exp. Tumor Res.*, 20: 283-294.

Farley, C.A. (1976b). Proliferative disorders in bivalve molluscs. Mar. Fish. Rev., 38:30-33.

Farley, C.A. (1978). Viruses and virus-like lesions in marine molluscs. *Mar. Fish. Rev.*, 40: 18-20.

Farley, C.A., Banfield, W. G., Kasnic, J. R. G., Foster, W. S.(1972). Oyster herpes-type virus. *Science*, 178: 759-760.

Farley, C. A., Lewis, E. J., Relyea, D., Zahtila, J., Rivara, G. (1997). Resistance studies for juvenile oyster disease (JOD) 1994-1996. *J. Shellfish Res.*, 16: 331.

Farley, C. A., Lewis, E. J., Relyea, D., Zahtila, J., Rivara, G. (1998). Resistance studies for juvenile oyster disease (JOD) 1997: Some early insights into genetic aspects. *J. Shellfish Res.*, 17: 352-353

Faisal, M., McIntyre, E. A., Adham, K. G., Tall, B. D., Kothary, M. H., La Peyre J. F. (1998). Evidence for the presence of protease inhibitors in eastern (*Crassostrea virginica*) and pacific (*Crassostrea gigas*) oysters. *Comp. Biochem. Physiol.* Part B, 121: 161-168.

Faust, M., Altenburger, R., Backhaus, T., Blanck, H., Boedeker, W., Gramatica, P., Hamer, V., Scholze, M., Vighi, M., Grimme, L.H. (2001). Predicting the joint algal toxicity of multi-component s-triazine mixtures at low-effect concentrations of individual toxicants. *Aquat. Tox*. 56: 13-32.

Ford, S. E. (1996). Range extension by the oyster parasite *Perkinsus marinus* into the northeastern US: Response to climate change? *J. Shellfish Res.*, 15:45-56.

Ford, S. E., Borrero, F. J. (2001). Epizootiology and pathology of Juvenile Oyster Disease in the Eastern Oyster, *Crassostrea virginica*. *J. Invertebr. Pathol.*, 78:141-154.

Ford, S.E., Haskin H. H. (1987). Infection and mortality patterns in strains of oysters *Crassostrea virginica* selected for resistance to the parasite *Haplosporidium nelsoni* (MSX). J. Parasitol., 73:368-376.

Friedman, F C. S., Perkins, F. O. (1994). Range extension of *Bonamia ostreae* to Maine, *U.S.A. J. Invertebr. Pathol.*, 64, 179-181.

Friedman, C. S., McDowell T., Groff, J. M., Hollibaugh, J. T., Manzer, D., Hedrick, R. P. (1989). Presence of *Bonamia ostreae* among populations of the European flat oyster, *Ostrea edulis* Linne, in California, *USA. J. Shellfish Res.*, 8, 133-137.

Friedman, C. S., Estes, R. S., Stokes, N. A., Burge, C. A., Hargove, J. S., Barber, B. J., Elston, R. A., Burreson, E. M., Reece, K. S. (205). Herpes virus in juvenile Pacific oysters *Crassostrea gigas* from Tomales Bay, California, coincides with summer mortality episodes. *Dis. Aquat. Org.*, 63(1): 33-41.

Friedman, C. S., Andree, K. B;, Beauchamp, K. A., Moore, J D., Robbins, T. T., Shields, J. D., Hedrick, R. P. (2000). 'Candidatus Xenohaliotis californiensis', a newly described pathogen of abalone, Haliotis spp., along the west coast of North America. *Int. J. Syst. Evol. Microbiol.* 50(2): 847-855.

Gagnaire, B., Thomas-Guyon, H., Burgeot, T., Renault, T. (2006a). Pollutant effects on Pacific oyster, *Crassostrea gigas* (Thunberg), hemocytes: screening of 23 molecules using flow cytometry. *Cell Biology and Toxicology*, 22: 1-14.

Gagnaire, B., Soletchnik, P. Madec, P., Geairon, P., Le Moine, O., Renault, T. (2006b). Diploid and triploid Pacific oysters, *Crassostrea gigas* (Thunberg), reared at two hieghts bove the sediment in Marennes-Oleron Basin, France: Difference in mortality, sexual maturation and hemocyte parameters. *Aquaculture*, 254: 606-616.

Gagnaire, B, Gay, M., Huvet, A., Daniel' J.-Y., Saulnier' D., Renault, T. (2007). Combination of a pesticide exposure and a bacterial challenge: In vivo effects on immune response of Pacific oyster, *Crassostrea gigas* (Thunberg). *Aquat. Toxicol.*, 84(1):92-102.

Gay, M., Renault, T., Pons, A. M., Le Roux, F. (2004). Two *Vibrio splendidus* related strains collaborate to kill *Crassostrea gigas*: taxonomy and host alterations. *Dis. Aquat. Org.*, 62(1-2): 65-74.

Gibson, L. F., Woodworth, J., George, A. M. (1998). Probiotic activity of *Aeromonas media* on the Paciic oyster, *Crassostrea gigas*, when challenged with *Vibrio tubiashii. Aquaculture,* 169: 111-120.

Gonzalez, M., Gueguen, Y., Desserre, G., de Lorgeril, J., Romestand, B., Bachère. (2007). Molecular characterization of two isoforms of defensin from hemocytes of the oyster, *Crassostrea gigas*. *Dev. Comp. Immunol.*, 31: 332-339.

Goulletquer P., Heral, M. (1997). Marine Molluscan production trends in France: from fisheries to aquaculture, NOAA Technical Report NMFS 129, p137-164.

Grizel, H., Heral, M. (1991). Introduction into France of the Japanese oyster (Crassostrea gigas). ICES J., 47: 399-403.

Gueguen, Y., Cadoret, J. P., Flament, D., Barreau-Roumiguière, C., Girardot, A. L., Garnier, J., Hoareau, A., Bachère, E., Escoubas. J.-M. (2003). Immune gene discovery by expressed sequence tags generated from hemocytes of the bacteria-challenged oyster, *Crassostrea gigas*. *Gene*, 303: 139-145.

Gueguen, Y., Herpin, A., Aumelas, J. Garnier, J. Fievet, Escoubas, J.-M., Bulet, P. Gonzalez, M., Lelong, C., Favrel, P., Bachère, E. (2006). Characterization of a defensin from the oyster *Crassostrea gigas*: recombinant production, folding, solution structure, antimicrobial activities and gene expression. *J. Biol. Chem.*, 281: 313-323 (2006).

Guo X, Ford S, De Brosse, G, Smolowitz, R. (2003). Breeding and evaluation of eastern oyster strains selected for MSX, Dermo and JOD resistance. *J. Shellfish Res.*, 22: 333-334.

Hand, R. E., Nell, J. A., Smith, I. R., Maguire, G. B., (1998). Studies on triploid oystersin australia. XI. Survival of diploid and triploid Sydney rock oysters (*Saccostrea commercialis* (Iredale and Roughley)) throurgh outbreaks of winter mortality caused by *Mikrocytos roughleyi* infestation. *J. Shellfish Res.*, 17: 334-336.

Harvell, C. D., Kim, K., Burkholder, J. M., Colwell, R. R., Epstein, P. R., Grimes, D. J., Hofmann, E. E., Lipp, E. K., Osterhaus, A. D. M. E., Overstreet, R. M., Porter, J., Smith, G., Sutherland, K., Ward, J. (1999). Emerging marine diseases – climate links and anthropogenic factors. *Science,* 285: 1505-1510.

Harvell, C. D., Aronson, R., Baron, N., Connell, J., Dobson, A., Ellner, S., Gerber, L., Kim, K, Kuris, A., McCallum, H., Lafferty, K., McKay, B., Porter, J., Pascual, M., Smith, G., Sutherland, K., Ward, J. (2004). The rising tide of ocean diseases: unsolved problems and research priorities. *Frontiers in Ecology and Environment*, 2: 375-382.

Haskin, H. H., Ford, S. E. (1979). Development of resistance to *Minchinia nelsoni* (MSX) mortality in laboratory-reared and native oyster stocks in Delaware Bay. *Mar. Fish. Rev.*, 54-63.

Hellio, C, Bado-Nilles, A., Gagnaire, B., Renault, T., Thomas-Guyon, H. (2007). Demonstration of a true phenoloxidase activity and activation of a ProPO cascade in Pacific oyster, *Crassostrea gigas* (Thunberg) in vitro. *Fish Shellfish Immunol.*, 22(4): 433-40.

Hine, P.M., Thorne; T. (1997). Replication of herpes-like viruses in haemocytes of adult flat oysters *Ostrea angasi*: an ultrastructural study. *Dis. Aquat. Org.*, 29: 197-204.

Hine, P. M.; Wesney, B., Hay, B. E. (1992). Herpesviruses associated with mortalities among hatchery-reared larval Pacific oysters *Crassostrea gigas*. *Dis Aquat. Org.*, 12(2): 135-142.

Hine, P. M., Cochennec-Laureau, N. Berthe, F. C. J. (2001). *Bonamia exitiosus* n. sp. (*Haplosporidia*) infecting flat oysters *Ostrea chilensis* (Philippi) in New Zealand. Dis. *Aquat. Org.*, 47: 63-72.

Hooper, C, Hardy-Smith, P, Handlinger, J (2007). Ganglioneuritis causing high mortalities in farmed Australian abalone (*Haliotis laevigata* and *Haliotis rubra). Australian Vet. J.*, 85(5): 188-193.

House, M. L., Kim, C. H., Reno, P. W. (1998). Soft shell clams *Mya arenaria* with disseminated neoplasia demonstrate reverse transcriptase activity. *Dis. Aquat. Org.*, 34: 187-192.

Hubert, F., van der Knaap, W., Noël, T., Roch, P. (1996). Cytotoxic and antibacterial properties of *Mytilus galloprovincialis*, *Ostrea edulis* and *Crassostrea gigas* (bivalve molluscs) hemolymph. *Aquat. Liv. Resour.*, 9: 115-124.

Hugh Jones, D. (1999). Breeding ponds as a basis for flat oyster (*Ostrea edulis*) culture and their use to develop resistance to the disease *Bonamia ostreae*. *J. Shellfish Res.*, 18: 718.

Huvet, A, Herpin, A., Degremont, L., Labreuche, Y., Samain, J.-F., Cunningham, C. (2004). The identification of genes from the oyster *Crassostrea gigas* that are differentially expressed in progeny exhibiting opposed susceptibility to summer mortality. *Gene,* 343(1): 211-220.

ICES (1988). Codes of Practice and Manula of Procedures for consideration of the Introduction and Transfers of Marine and Freshwater Organisms. Cooperative Research Report 159. G. E. Turner (ed) (Prepared jointly with the EIFAC Working Party on Introductions) EIFAC publication as EIFAC Occasional Paper No. 23: 44 pp.

ICES (1995). ICES Code of Practice on the Introduction and Transfers of Marine Organisms – 1994. ICES Co-operative Research Report No. 204.

Ito, Y., Yoshikawa, A., Hotani, T., Fukada, S., Sugimura, K., Imoto, T. (1999). Amino acid sequences of lysozymes newly purified from invertebrates imply wide distribution of a novel class in the lysozyme family. *Eur. J. Biochem.*, 259: 456-461.

Jenny, M. J., Ringwood, A. H., Lacy, E. R., Lewitus, A. J., Kempton, J. W., Gross, P. S., Warr, G. W., Chapman, R. W. (2002). Potential indicators of stress response identified by expressed sequence tag analysis of hemocytes and embryos from the American oyster, *Crassostrea virginica*. *Mar. Biotechnol.*, 4: 81-93.

Jepson, P.D., Bennett, P.M., Deaville, R., Allchin, C.R., Baker, J.R., Law, R.J. (2005). Relationships between polychlorinated biphenyls and health status in harbor porpoises (Phocoena phocoena) stranded in the United Kingdom. *Environ. Toxicol. Chem.*, 24: 238-48.

Kent, M. L., Wilkinson, M. T., Drum, A. S., Elston, R. A. (1991). Failure of transmission of hemic neoplasia of bay mussels, *Mytilus trossolus*, to other bivalve species. *J. Invertebr. Pathol.*, 57: 435-436

Kuby, J. (1997). Immunology.WH Freeman Co, NewYork.

Lacoste, A, Jalabert, F., Malham, S., Cueff, A., Gelebart, F., Cordevant, C.,Lange, M., Poulet, S. A., (2001). A *Vibrio splendidus* strain is associated with summer mortality of juvenile oysters *Crassostrea gigas* in the Bay of Morlaix (North Brittany, France). *Dis. Aquat. Org.*, 46(2): 139-145.

Lafferty, K. D., Porter, J. W., Ford, S. E. (2004). Are diseases increasing in the ocean? Annual Review of Ecology and Systematics, 35: 31-54.

Lambert, C., Nicolas, J.-L., Cilia, V. (1998). *Vibrio pectinicida* sp. Nov., a pathogen scallop (*Pecten maximus*) larvae. *Int. J. System. Bacteriol.,* 48: 481-487.

Le Gall, G., Chagot, D., Mialhe, E., Grizel, H. (1988). Branchial Rickettsiales-like infection associated with a mass mortality of sea scallop *Pecten maximus*. *Dis. Aquat. Org.*, 4: 229-232.

Le Roux, F., Goubet, A., Thompson, F. L., Faury, N., Gay, M., Swings, J., Saulnier, D. (2005). *Vibrio gigantis* sp. nov., isolated from the haemolymph of cultured oysters (*Crassostrea gigas*). *Int. J. Syst. Evol. Microbiol.*, 55: 2251-2255.

Lester R. J. G., Davis, G. H. G. (1981). A new *Perkinsus* species (Apicomplexa, Perkinsea) from the abalone *Haliotis ruber*. *J. Invertebr. Pathol.*, 37: 181-187.

Maloy, A. P., Barber, B. J., Boettcher, K. J. (2005). A PCR-based diagnostic assay for the detection of *Roseovarius crassostreae* in *Crassostrea virginica* affected by juvenile oyster disease (JOD). *Dis. Aquat. Org.*, 67(1-2): 155-162.

Marchalonis, J., Schluter, S. (1994). Development of an immune system. In: Primordial Immunity, G Beck, G Habicht, E Cooper and J Marchalonis, Annals NY Acad Sci, 712: 1-12.

McDade, J.E. , Tripp, M. R. (1967). Lysozyme in oyster mantle mucus. *J. Invertebr. Pathol.* 9: 581-582.

Medzhitov, R., Preston-Hurlburt, P., Janeway, C. A. (1997). A human homologue of the *Drosophila* Toll protein signals activation of adaptative immunity. Nature, 338: 394-397.

Means, TK, Golenbock, T., Fenton, M. J. (2000). The biology of Toll-like receptors. *Cytokins Growth Factor Reviews* 11: 219-232.

Meyers, J. A., Burreson, E. M., Barber, B. J., Mann, R. (1991). Susceptibility of diploid and triploid pacific oysters *Crassostrea gigas* (Thunberg, 1793) and eastern oysters, *Crassostrea virginica* (Gmelin, 1791), to *Perkinsus marinus. J. Shellfish Res.*, 10: 433-437.

Mitta, G., Hubert, F., Dyrynda, E. A., Boudry, P., Roch, P. (2000). Mytilin B and MGD2, two antimicrobial peptides of marine mussels: gene structure and expression analysis. *Dev. Comp. Immunol.*, 24: 381-393.

Mortensen, S., Arzul, I., Miossec, L., Paillard, C., Feist, S. W., Stentiford, G. D. (2007) Molluscs and Crustaceans. In "Scientific review of wild-farmed fish and shellfish disease interactions" Ed. By R. Raynard, T. Wahli, I. Vatsos and S. Mortensen, 452 pages, available on-line www.dipnet.info/.

Moss, J. A., Burreson,' E. M., Cordes, J. F., Dungan, C. F., Brown,' G. D., Wang,' A., Wu,' X., Reece, K. S. (2007). Pathogens in *Crassostrea ariakensis* and other Asian oyster species: implications for non-native oyster introduction to Chesapeake Bay *Dis. Aquat.* Org., 77: 207-223

Murell, A., Kleeman, S. N., Barker, S. C., Lester, R. J. G (2002). Synonyny of *Perkinsus olseni* Lester and Davis, 1981 and *Perkinsus atlanticus* Azevedo, 1989 and an update on the phylogenetic position of the genus *Perkinsus*. Bull. *Eur. Assoc. Fish Pathol.*, 22: 258-265.

Naciri-Graven, Y., Martin, A. G., Baud, J. P., Renault, T., Gérard, A. (1998). Selecting the flat oyster *Ostrea edulis* (L.) for survival when infected with the parasite *Bonamia ostreae*. *J. Exp. Mar. Biol. Ecol.* 224: 91-107.

Nell, J. A. (2002). Farming tripolid oysters; *Aquaculture,* 210: 69-88.

Nell, J. A., Perkins, B. (2006). Evaluation of the progeny of the third-generation The Sydney rock oyster *Saccostrea glomerata* (Gould 1850) breeding lines for resistance to QX disease *Marteilia sydneyi* and winter mortality *Bonamia roughleyi*. *Aquacult. Res.*, 37(7): 693-700.

Nell, J. A., Smith, I. R., McPhee, C. C. (2000). The Sydney rock oyster *Saccostrea glomerata* (Gould 1850) breeding programme: progress and goals.*Aquacult. Res.*, 31: 45-4ç.

Nicolas, J.-L., Comps, M., Cochennec, N. (1992). Herpes-like virus infecting Pacific oyster larvae, *Crassostrea gigas*. *Bull. Eur. Assoc. Fish Pathol.,* 12: 11-13.

OIE (2006). Manual of Diagnostic Tests for Aquatic Animals, OIE, Paris, 5th edition.

OIE (2007). International Aquatic Animal Health Code. OIE, Paris, 10th edition.

Oliver, L.M., Fisher, W. S., Ford, S. E., Calvo, L. M., Burreson, E. M., Sutton, E. B., Gandy, J. (1998). *Perkinsus marinus* tissue distribution and seasonal variation in oysters *Crassostrea virginica* from Florida, Virginia and New York. *Dis. Aquat. Org.,* 34: 51-61.

Oprandy, J. J., Chang, P. W., (1983). 5-Bromodeoxyuridine induction of hematopoietic neoplasia and retrovirus activation in the soft-shell clam, *Mya arenaria*. *J. Invertebr. Path*ol., 42: 196-206.

Oprandry, J. J., Chang, P. W., Pronovost, A. D., Cooper, K. R., Brown, C. W.. Yates. V. J. (1981). Isolation of a viral agent causing hematopoietic neoplasia in the soft-shell clam, *Mya arenaria*. *J. Invertebr. Pathol.*, 38: 45-51.

Oubella R., Paillard, C., Maes, P., Auffret, M. (1994). Changes in hemolymph parameters in the manila clam *Ruditapes philippinarum* (Mollusca, Bivalvia) following bacterial challenge. *J. Invertebr. Pathol.*, 64: 33-38.

Paillard, C., Maes, P. (1989). Origine pathogène de l'«anneau brun» chez *Tapes philippinarum* (Mollusque, bivalve). *C. R. Acad. Sci.* Paris, Ser III, 309: 235-241.

Paillard, C., Maes, P. (1990). Etiologie de la maladie de l'anneau brun chez *Tapes philippinarum* : Pathogénicité d'un *Vibrio* sp. C. R. Acad. Sci. Paris, Ser III, 310: 15-20.

Peatman, E.J., Wei, X., Feng, J., Liu, L.,Kucuktas, H., Li, P., He, C., Rouse, D., Wallace, R., Dunham, R., Liu, Z. (2004). Development of expressed sequence tags from Eastern oyster (*Crassostrea virginica*): lessons learned from previous efforts. *Mar. Biotechnol.*, 6: S491-496.

Pichot, Y., Comps, M., Tigé, G., Grizel, H., Rabouin, M. A. (1979). Recherches sur *Bonamia ostreae* gen. n, sp. n., parasite nouveau de I'huître plate *Ostrea edulis* L. *Rev. Trav. Inst.* Pêches Marit. 43: 131-140.

Pipe R. K. (1990). Hydrolytic enzymes associated with the granular haemocytes of the marine mussel *Mytilus edulis*. *Histochem. J.*, 22: 595-603.

Pipe, R. K. (1992). Generation of reactive oxygen metabolites by the hemocytes of the mussel *Mytilus edulis*. *Dev. Comp. Immunol.*, 16: 111-122.

Pipe, R. K., Coles, J. A. (1995). Environmental contaminants influencing immune function in marine bivalve molluscs. Fish and Shellfish Immunology 5: 581-595.

Pipe R. K., Coles, J. A., Thomas, M. E., Fossato, V. U., Apulsford, A. L. (1995). Evidence for environmentally derived immunomodulation in mussels from the Venice Lagoon. *Aquat. Toxicol.*, 32: 59-73.

Ragone-Calvo, L. M., Calvo, G. W., Burreson, E. M., (2003). Dual resistance in a selectively bred eastern oyster, *Crassostrea virginica*, strain tested in Chesapeake Bay. *Aquaculture,* 220: 69-87.

Ramussen, L.P.D. (1986). Occurrence, prevalence and seasonality of neoplasia in the marine mussel *Mytilus edulis* from three sites in Denmark. *Mar. Biol.*, 92: 59-64.

Ray, S. M. (1996). Historical perspective on *Perkinsus marinus* disease of oysters in the Gulf of Mexico. *J. Shellfish Res.*, 15: 9-11.

Riquelme, C., Araya, R., Escribano, R. (2000). Selective incorporation of bacteria by *Argopecten purpuratus* larvae: implicaton for the use of probiotics in culturing systems of the Chilean scallop. *Aquaculture*, 181: 25-36.

Riquelme, C., Araya, R., Escribano, R. (2000). Selective incorporation of bacteria by *Argopecten purpuratus* larvae: implicaton for the use of probiotics in culturing systems of the Chilean scallop. *Aquaculture,* 181: 25-36.

Riquelme, C. E., Jorquera, M. A., Rojas, A. I., Avendano, R. E., Reyes, N. (2001). Addition of inhibitor-producing bacteria to mass cultures of *Argopecten purpuratus* larvae (Lamarck, 1819); *Aquaculture,* 192: 111-119.

Renault, T., Le Deuff, R.-M., Lipart, C., Delsert, C. (2000). Development of a PCR procedure for the detection of a herpes-like virus infecting oysters in France. *J. Virol. Meth.*, 88: 41-50.

Renault, T., Arzul, I. (2001). Herpes-like virus infections in hatchery-reared bivalve larvae in Europe: specific viral DNA detection by PCR. *J. Fish Dis.* 24: 161-167.

Renault, T., Lipart, C., Arzul, I. (2001). A herpes-like virus infects a non-ostreid bivalve species: virus replication in *Ruditapes philippinarum* larvae. *Dis. Aquat. Org.*, 45: 1-7.

Riquelme, C., Toranzo, A. E., Barja, J. L., Vergara, N., Araya, R. (1996). Association of *Aeromonas hydrophila* and *Vibrio alginolyticus* with larval mortalities of scallop (*Argopecten purpuratus*). *J. Invertebr. Pathol.*, 67: 213-218.

Roch, P. (1999). Defense mechanisms and disease prevention in farmed marine invertebrates. *Aquaculture*, 172: 125-145.

Ruiz-Ponte, C., Samain, J.-F., Sanchez, J.J., Nicolas, J.-L. (1999). The benefit of a *Roseobacter* species on the survival of scallop larvae. *Marine Biotechnology*, 1: 52-59.

Seo, J. K., Crawford, J. M., Stone, K. L., Noga, E. J. (2005). Purification of a novel arthropod defensin from the American oyster, *Crassostrea virginica*. Biochem. *Biophys. Res. Commun.*, 338: 1998-2004.

Smolarz, K., Renault, T., Soletchnik, P., Wolowicz, M. (2005). Neoplasia detection in *Macoma balthica* from the Gulf of Gdansk : comparison of flow cytometry, histology and chromosomes analysis. *Dis. Aquat.* Org., 65: 187-195.

Snieszko, S. F. (1974). The effects of environmental stress on outbreaks of infectious diseases of fishes. *J. Fish Biol.*, 6:197-208.

Song, L., Wu, L. Ni, D.., Chang, Y., Xu, W., Xing, K. (2006a). The cDNA cloning and mRNA expression of heat shock protein 70 gene in the haemocytes of bay scallop (*Argopecten irradians,* Lamarck 1819) responding to bacteria challenge and naphthalin stress. *Fish Shellfish Immunol.*, 21: 335-345.

Song, L., Xu, W., Li, C., Li, H., Wu, L., Xiang, J., Guo, X. (2006b) Development of expressed sequence tags from the Bay scallop *Argopecten irradians irradians*. *Mar. Biotechnol.*, 8: 161-169.

Sutton, D.C., Garrick, R. (1993). Bacterial disease of cultured giant clam *Tridacna gigas* larvae. *Dis. Aquat. Org.*, 16:47-53.

Tanguy, A ,Guo, X., Ford, S. E. (2004). Discovery of genes expressed in response to *Perkinsus marinus* challenge in Eastern (*Crassostrea virginica*) and Pacific (*C. gigas*) oysters. *Gene,* 338: 121-131.

United states National Marine fisheries Service (2003). Annual Commercial Landings Statistics: http://www.st.nmfs.gov/st1/commercial/landings/annual_landings.html.

Vásquez-Yeomans, R., Cáceres-Martínez, J., Huerta, A. F. (2004). Herpes-like virus associated with eroded gills of the Pacific oyster *Crassostrea gigas* in Mexico. *J. Shellfish Res.*, 23: 417-419.

Venier, P., Pallavicini, A., De Nardi, B., Lanfranchi, G. (2003). Towards a catalogue of genes transcribed in multiple tissues of *Mytilus galloprovincialis*. *Gene,* 314: 29-40 (2003).

Villaba, A., Carballal, M. J., Lopez, C. (2001). Disseminated neoplasia and large foci indicating heavy haemocytc infiltration in cockles *Cerastoderma edule* from Galicia (NS Spain). *Dis. Aquat. Org.*, 46: 213-216.

Waechter, M., Le Roux, F., Nicolas, J. L., Marissal, E., Berthe, F. (2002). Characterisation of *Crassostrea gigas* spat pathogenic bacteria. C. R. *Acad. Sci. Paris,* 325: 231-238.

Walker, P., Subasinghe, R. P. (2000). DNA-based molecular diagnostic techniques : research needs for standardization and validation of the detection of aquatic animal pathogens and diseases. FAO, Roma (Italia). Fisheries technical paper (FAO). no. 395, 93 p.

Weinberg, J. R., Leavitt, D. F., Lancaster, B. A., McDowell, Capuzzo, J. (1997). Experimental field studies with *Mya arenaria* (Bivalvia) on the induction and effect of hematopoietic neoplasia. *J. Invertebr. Pathol.*, 69: 183-194.

Wood, J.L., Andrews, J. D. (1962). *Haplosporidium costale* (Sporozoa) Associated with a Disease of Virginia Oysters. *Science,* 136(3517): 710-711.

Xue, Q. G., Itoh, N., Schey, K. L., Li, Y. L., Cooper, R. K., La Peyre, J. F. (2007). A new lysozyme from the eastern oyster (*Crassostrea virginica*) indicates adaptive evolution of i-type lysozymes. *Cell. Mol. Life Sci.*, 64: 82-95.

Yu, Z., Guo, X. (2006). Identification and mapping of disease-resistance QTLs in the eastern oyster, *Crassostrea virginica* Gmelin. *Aquaclture*, 254: 160-170.

In: Aquaculture Research Progress
Editor: Takumi K. Nakamura

ISBN 978-1-60456-247-7

Chapter 3

FATTY ACIDS: COMPOSITION AND FUNCTIONS FOR REPRODUCTION [1]

E Hu, Ruo-Jing Wang, Chen-Yi Pan, and Wan-Xi Yang [2]
The Sperm Laboratory, Institute of Cell Biology and Genetics,
College of Life Sciences, Zhejiang University, Hangzhou 310058, China

ABSTRACT

Fatty acids are one of the most important components in the organisms. According to the number of double bonds, fatty acids can be divided into subtypes such as saturated fatty acids, monounsaturated fatty acids (MUFA) and polyunsaturated fatty acids (PUFA). The fatty acid composition in many animals has been identified, and it is found to vary dramatically among tissues. This suggests specific roles for different fatty acids in life systems although their detailed function is far from being revealed. Highly unsaturated fatty acids (HUFAs) are essential for normal physiological actions. The representative fatty acids of HUFAs are DHA, EPA, and AA, which have been intensively investigated in many studies. Conversion processes between fatty acids take place in organisms, which require desaturases and chain elongation and shortening enzymes. Since fatty acids are acquired from food, dietary fatty acids have great effect on the fatty acid composition of tissues and organs, and thus their normal functioning. In this review, we summarized the fatty acid pattern of gonads, and the recent progress on their function during reproduction. We suppose to emphasize the importance of fatty acids and hope to provide insights into further investigation in-depth.

Keywords: Fatty acids; lipids; gonad; reproduction; fatty acid classes; dietary

---

[1] This work was supported by two grants from National Nature Science Fundation of China (No. 30671606 and 40776079) (to Wan-Xi Yang).

1. Introduction

Lipids play multiple important roles in animals, known as efficient energy reserves that can store more energy per unit weight than proteins and carbohydrates (Montero-Torreiro et al. 1998; Castell et al. 2004), participating in physiological processes (e.g. vitamins, hormones and pigments), acting the Protein Kinase during the cellular signal transfer (Roy et al. 1999b), being precursors to essential substances as eicosanoids (Castell et al. 2004), and serving as major components of phospholipids which are building blocks of biomembranes (Takagi et al., 1980). With different composition of fatty acids, lipids have related biological function. In the biomembranes, the fatty acids dominate most of the biophysical characteristics like fluidity and flexibility, which determines the further physiological metabolism such as sperm motility and fertilizing capacity (Ladha 1998). Other researches also suggest that any fluctuation of the lipid storage in vivo or the outside environment such as temperature can make corresponding changes of the membrane fatty acid composition (Cerolini et al. 2000).

In bodies, fatty acids are stored in the form of lipids, organs such as liver, muscle and glands serve as major reservoirs for their storage. Lipids in digestive glands are the most important fraction in terms of metabolic participation, which are hydrolyzed and transferred to other organs (Soudant et al. 1996). Studies have shown in red drum (Craig et al. 2000; Chatzifotis et al. 2004), sand smelt (Chatzifotis et al. 2004) and burbot (Mustonen et al. 2002), liver is a lipid supplier. In *Arctic charr*, muscle is a major lipid depot organ, but limited in dentex (Josrgensen et al. 1997; Chatzifotis et al. 2004). Walker et al.(Walker et al. 1998) found that lipid storage occurs in gonads of green sea urchin (*Strngylocentrotus droebachiensis*) during non-reproduction period. In the mussel bodies, 20 and 22 carbons of polyunsaturated fatty acids and lipids in gonad increase significantly at the end of the maturation of the gonad (De Moreno et al. 1980). And lipid droplets are one of the three most important storage methods in *Aurdia aurita* (Lucas 1994). There is evidence that sexually inactive mussels' fatty acid composition of the gonad varied in the same way (De Moreno et al. 1980). There is also evidence that the organs for lipid storage change during the life history of the aquatic animals, sometimes induced by the environment. For *Argopecten purpuratus*, lipid accumulation in gonads decreases at higher temperatures, even when ample food is available (Martinez et al. 2000). As we known, the composition of the fatty acid in vivo has some direct or indirect connection between organs, whereas it also exist the opposite opinions in studying the *Eledone cirrhosa* and *Eledone moschata* (R. Rosa 2004).

Not only there is great difference between organs, but also the abundance of different fatty acids varies dramatically in types. For example, in *Octopus vulgaris* and many cephalopod species, palmitic acid (16:0), EPA (20:5n-3) and DHA (22:6n-3) are the most abundant fatty acids (Miliou et al. 2006); while Oleic, stearic, and palmitic acid are major constituents of protasome fatty acids (Arienti et al. 1998). To some extent, this variation reflects the composition pattern of phospholipids and their particular roles in cells (Miliou et al. 2006).

[2] Corresponding author and proofs should be addressed to: Dr. Wan-Xi Yang Associate Professor Tel.: ++86 132 1613 0666; fax: 86-571-88206485. E-Mail: wxyang@spermlab.org.

To date, studies mainly focus on the identification and composition of fatty acids in animal tissues, lacking systematic investigation of their specific function in the construction of cells and life processes due to limited approaches for the tagging of fatty acids. In this review, we want to summarize the fatty acid pattern of gonads, and the recent progress on their function during reproduction. We suppose to emphasize the importance of fatty acids and hope to provide insights into further investigation in-depth. And provide information about fatty acids, especially in feeding, for aquaculture.

2 The Fatty Acid Pattern in Gonads Related with Reproduction

2.1. The Importance of Gonad the Fatty Acid

The gonad takes the responsibility for gametogenesis, a great energy- and nutrient-consuming process. Therefore, the nutrient supply of gonad is crucial for normal generation of gametes, which directly affects the reproduction. Lipids provide the structural and energetic materials for the development of gametes (Mathieu et al. 1993). Studies have revealed in female fish, such as sea bass (Fernandez et al. 1989; Cerda et al. 1995; Chatzifotis et al. 2004), striped bass (Lund et al. 2000), mummichog (Jensen et al. 2001) and trout (Wallaert et al. 1994; Bon et al. 1997), vitellogenesis, pre-spawning and spawning occur synchronically with the high lipid content in gonads, and in serum (Chatzifotis et al. 2004), with the highest value in the period prior to spawning. It is also the case during the beginning of the gametogenic process in *Argopecten irradians* that energy metabolism is presumably lipid-based (Ruiz-Verdugo et al. 2001). And the spermatozoa fatty acid composition affects the fertility of some poultry (Surai et al. 2000a). In common dentex, the involvement of visceral fat is an energy source for the development of ovaries and testis (Chatzifotis et al. 2004).

An increase of lipids in gonad is associated with a decline in muscle weight for *catarina scallop* (Ansell 1974). It is supposed that lipids are hydrolysed in the digestive gland and transferred to the ovary where the triglycerides are incorporated into the yolk sac of oocytes (Napolitano et al. 1992; Soudant et al. 1996). Besides, the mobilization and elevated synthesis of lipid storage in liver is accompanied with the development of gonad.

Spermatogenesis is different from the vitellogenesis in many aspects, their structure, size, function and the character in fertilization. For example, the synthesis and transfer of DHA in spermatozoa and egg yolk are sex-specific in poultry, supporting by date (Surai et al. 2000a). Spermatozoa do not accumulate lipid reserves and lipids are mainly incorporated into their membranes (Soudant et al. 1996). Because sperms have a limited metabolic capacity, changes of membrane composition rely on the extracellular milieu, which will affect the ability of sperms to carry out acrosome reaction (Arienti et al. 1998). The fatty acid pattern in sperm is important for the function and stability of sperm membranes (Arienti et al. 1998) and changes of them will bring the sterility in animals like boar (Cerolini et al. 2000). During the protective process, αT with its lipophilic character can enter the plasma membrane (Cerolini et al. 2000) and play a key role. In spermatozoa, inter-regional diffusion is very limited, preventing membrane components from mixing and randomizing, and thus assuring proper and specific functioning (Arienti et al. 1998).

3. HIGHTLY UNSATURATED FATTY ACIDS (HUFAS)

Highly unsaturated fatty acids (HUFAs) are important fatty acids in life systems, essential for the health of all animals (Castell et al. 2004). Study on cuttlefish and squid showed that HUFAs increase fecundity, fertilization and egg quality (Izquierdo et al. 2001), participate in the metabolism of cellular signal transfer (Roy et al. 1999b). HUFAs can be divided into monounsaturated fatty acids (MUFAs) and polyunsaturated fatty acids (PUFAs) according to the number of double bonds. By the position of the first double bond counted from the CH_3 terminal, HUFAs can be divided into many subtypes such n-3, n-6, and n-9 fatty acids. The HUFAs have two types in the cell, one is dissociation; the other is combined with polar lipid like phospholipid (PL) and neutral lipid triglyceride (TG). The proportions of saturated fatty acids, monounsaturated fatty acids (MUFA) and polyunsaturated fatty acids (PUFA) in the polar and neutral lipids always keep steady equilibrium to satisfy the demand in the body (Soudant et al. 1996). And the isotope labeled experiment shows that the TG and PL is the destination of most of the dissociative HUFAs (Retterstol et al. 1998).

The n-3 and n-6 HUFA are considered as the physiologically active factors in many species. The fatty acid composition of eggs (Izquierdo et al. 2001) and their fecundity (Fernandez-Palacios et al. 1995; Izquierdo et al. 2001) in *gilthead seabream* are indirectly affected by the n-3 HUFA such as 18:3n-3, 18:4n-3 and 20:5n-3 (EPA) (Fernandez-Palacios et al. 1995). Some studies show that the tissues like foot, testis and ovary have related signature such as 20:4n-6 (arachidonic acid-AA), 20:5n-3 (eicosapentaenoic acid-EPA) and 18:2,n-6 (linoleic acid-LA) (Grubert et al. 2004). And in marine mollusks, the fatty acids 16:0, 16:1n-7, 18:0, 18:1n-9, 20:5n-3 and 22:6n-3 are at significant level (De Moreno et al. 1980).

The highly unsaturated lipid level in the semen and spermatozoa can cause proxidation spontaneously in mammalian like man (Alvarez JG 1989) and poultry like chicken (Wishart 1984)) and turkey (Cecil 1993). The proxidation process in vivo plays a key role in the sperm function according to date from many scientists' researches. It damages the mobility and fertility of the spermatozoa in many species (Alvarez JG 1984; Wishart 1984; Aitken et al. 1989; Gagnon C 1991; de Lamirande et al. 1992; Aitken et al. 1993; Hammerstedt 1993; Lenzi A 1996), especially the sub-fertile ejaculates (Aitken RJ 1993) and fowl stored semen (Wishart 1989; Cecil 1993) of human being (Cerolini et al. 2000). The mechanism that we know is about a PUFA-ROS (reactive oxygenspecies) interaction. It is proved that the ROS in the bodies increase the DNA fractions (Lopes et al. 1998), change the structure of the cytoskeleton (Hinshaw et al. 1986), have influence on the development of the axoneme (de Lamirande et al. 1992) and sperm-egg fusion process (Aitken RJ 1993). And further research suggests that the PUFAs breakdown by proxidation causes ROS generation and significant decrease the physiological fatty acid index like DHA level, total PUFA level, total n-3 fatty acid level and the percentage of double bond in bodies (Aitken et al. 1989; Aitken 1994; Zalata et al. 1995; Adel A. Zalata 1998), incidentally, interrupt the sperm function and due to infertility (Aitken et al. 1993). To prevent the sperm from oxidant, there is antioxidate system found in mammalian (Jones et al. 1977; Jones et al. 1978; Jones et al. 1979; Smutna et al. 1979; Dawra et al. 1983) and avian semen (Surai et al. 2000a). The antioxidants like ascorbic acid in poultry (Surai et al. 1998), thermolabile antioxidant enzymes SOD and GSH-Px (Surai et al. 2000a) and αT (Maldjian et al. 1998) well protect the unsaturated lipid in vivo, but in vitro, these antioxidants cannot work efficiently (Hammerstedt 1993). When under the oxidation press, the deficiency of αT can be compensated by superoxide dismutase and glutathione peroxidase activety and increased antioxidant activity (Surai et al. 2000a), so the whole system maintains consistent.

3.1. Polyunsaturated Fatty Acids (PUFAs)

Polyunsaturated fatty acids (PUFAs) constitute the biggest group of HUFAs, and their significance in nutrition and metabolism has caught the scientists' eyes. The major varieties of PUFAs are n-3 PUFAs such as DHA (22:6n-3) and EPA (20:5n-3), n-6 PUFAs such as AA (20:4n-6) and n-9 PUFAs. There is a difference between the PUFAs amount (mainly DHA and EPA) in TG and PL fraction in organs. Gonad has the PUFAs mostly from TG fraction, while PUFAs in muscle always come from PL (Chatzifotis et al. 2004).

PUFAs actively participate in gonad maturation, egg quality (Izquierdo et al. 2001) and dentex larval growth (Tulli et al. 1997). And the PL comprised of PUFAs can enhance membrane fluidity and flexibility of the spermatozoa membrane, and are believed to be actively involved in the regulation of cellular movement, gonadal metabolism of lipids and fusion capacity (Miliou et al. 2006). PUFAs also regulate the production of eicosanoid (prostaglandins) and steroid hormones, and gonadal development (ovulation) (Izquierdo et al. 2001). The low environmental temperatures also require increased unsaturated fatty acids in bodies (De Moreno et al. 1980). But some marine fish without the PUFAs synthesizing system, can transfer the saturated fatty acids and MUFAs directly to the maturing ovaries and over the PUFAs during spawning (Zohar et al. 1995). This phenomenon can be considered as a pathway of preserving the PUFAs when lack of the resource outside. The marine species which has the ability to synthesize the PUFAs like mollusks, can be observed a distinct accumulation of 20 and 22 carbons of PUFAs in gonads by the end of the maturation process (De Moreno et al. 1980). In male European sea bass, essential PUFAs significantly modulate testicular prostaglandins series E production in vitro, and therefore, may have important impacts on steroidogenesis and spermiation (Asturiano et al. 2000). Further more, in invertebrates especially crustaceans and mollusks, the PUFAs can not only determine hatching and growth (Xu et al. 1994), but also play a important role in metabolism processes like produce prostaglandins and hormones and regulate ionic fluxes (Stanley-Samuelson 1987).

3.1.1. n-3 PUFAs

The n-3 PUFAs, especially DHA and EPA, are considered as essential nutrients for marine animal in the past years (Sargent et al. 1999). The copepods require n-3 PUFAs during egg production (Støttrup et al. 1990). The n-3 PUFAs also are substrates of metabolic activities in seminal plasma (Miliou et al. 2006). The 18:3n-3 or n-3 PUFAs can be the impetus of 18:1n-9 to 20:1n-9 transformation (Castell et al. 2004). There is evidence that 24 carbon n-3 PUFA like 24:5n-3 and 24:6n-3 which are very long chain intermediates increased in rat and may play a important role of keeping the balance of all the elongation, desaturation and shortening processes (Retterstol et al. 1998). Importantly, docosahexaenoic acid (DHA) and eicosapentaenoic acid (EPA), which are essential in all activities, belong to this group. In European sea bass, they are concerned with gonadal hormone levels, fecundity, egg fatty acid content, egg quality, hatching and survival ratios (Navas et al. 1997; Navas et al. 1998; Bruce et al. 1999).

DHA are essential for tissue development (Chatzifotis et al. 2004). Selective retention of DHA appears during embryogenesis (Izquierdo et al. 2001), and starvation (Tandler et al. 1989), which suggest its importance for the developing embryo and larvae. In sperm storage of the mammalian, the DHA is found to be oxidized preferentially (Jones et al. 1976; Griveau et al. 1995), which suggests its role in male gonad. DHA even can affect the viability of fast growing phospholipids-rich *cephalopos paralarvae* (Navarro et al. 2000; Navarro et al. 2003). Studies have shed light on the role of DHA in gonads: DHA reduces PGE production (Asturiano et al. 2000); it is related to the proportion of motile spermatozoa in human, boar and chicken (Miliou et al. 2006), and has a major role to play at the structural and functional levels of cellular membranes during oogenesis and embryogenesis (Soudant et al. 1996). But some studies speculate on that the DHA may have some other function more than the structural function, such as in the brain and retina (Adel A. Zalata 1998). The ovaries have greater amount of DHA than the testis (Chatzifotis et al. 2004). Even in the testis, sperm with different mobility contain different amount of DHA. To date, the sperm with low mobility has low DHA (Adel A. Zalata 1998).

DHA metabolism in the bodies is always from n-3 precursors in insect (Blomquist et al. 1991). Studies show that the DHA conversion from linolenic acid (18:3n-3) by desaturation and elongation (Surai et al. 2000a) in species like crustaceans and copepods (Brenner 1974; Jones et al. 1979; Moreno et al. 1979). It also explains the phenomenon that linolenic acid and the first intermediates 18:4n-3, 20:4n-3 and 20:5n-3 exist for a long time. Further more, it is proved that all these n-3 fatty acids transform to DHA in the end (Retterstol et al. 1998). However, the source of the DHA requires further investigation, because, in fish cell membranes, no DHA metabolites from the cyclooxygenase pathway are known (German et al. 1983). In turkey sperm, the increased proportion of DHA does not related to decrease in n-6 or n-9 long chain PUFA (LC-PUFA) (Miliou et al. 2006), which suggests the transform dose not exist. The same phenomenon observed in rat research: the fatty acids like DHA are from an in vivo pool, which affected by dietary or fatty acid transfer is still unknown (Adel A. Zalata 1998).

EPA (20:5n-3), a precursor of prostaglandins, can perturb the production of prostaglandins in turbot brain astroglial cells (Bell et al. 1994). EPA can be synthesized by animals, in sea urchin, *P. miliaris*, $\triangle$6-desaturase activity is responsible for the conversion of 18:3n-3 to 20:5n-3 (Bell et al. 2001).

The fact that decrease of EPA and DHA in faster going octopuses and several poikilotherms at the low temperature reflects a higher demand for EPA and DHA, for membrane synthesis rather than a response for maintaining membrane permeability and plasticity (Miliou et al. 2006). EPA and DHA have been reported to attenuate gonadotropin-stimulated steroid production in both goldfish testis and trout and goldfish ovaried (Mercure et al. 1995). The DHA/EPA (22:6n-3/20:5n-3) ratio is an important level for the biological activities, because a higher dietary level of 22:6n-3 and higher dietary DHA/EPA ratio lead to higher hatching rate (Soudant et al. 1996), and increase in dietary DHA/EPA ratio was significantly correlated with an amelioration of growth parameters in the gilthead sea-bream lavae (Mourente et al. 1993). In dentex ovary, the DHA/EPA ration of TG and PL fractionis about 4.6 and 3.5, relatively, the PUFA and DHA in ovary is more than testis (Chatzifotis et al. 2004). Thus it can be expected that dietary EPA and DHA may significantly affect growth and reproduction of animals.

3.1.2. n-6 PUFAs

The n-6 PUFAs are also important for reproduction. Birds' spermatozoa such as chicken (Cerolini et al. 1997; Surai et al. 1998) are rich in n-6 long chain PUFAs (LC-PUFAs), especially C22:4n-6 and C20:4n-6 that play a structural function (Miliou et al. 2006). The isotopes labeled experiment suggests that when the 18:3n-6 and 20:3n-6 decreased, the 20:4n-6, 22:4n-6, 24:4n-6 and 24:5n-6 increased (Retterstol et al. 1998) during the n-6 family metabolism.

Among them arachidonic acid (AA) (20:4n-6) is particularly important in tissues (Miliou et al. 2006). Unlike the other PUFAs, the double bond in AA is not at the $\triangle^{12}$ or $\triangle^{15}$ (Nelson et al. 2004). A number of AA is bound in the sn-2 position of the PL and can be released by the phospholipase A2 as unesterified form (van den Bosch 1980). As the DHA in mammalian spermatozoa, the AA oxidized preferentially during sperm storage (Jones et al. 1976; Griveau et al. 1995). It functions in fish mineral ion exchange across membranes such as gill and kidney (Castell et al. 2004) and in membrane lipid metabolism during oogenesis in *Pecten maximus* (Soudant et al. 1996) and participates in osmoregulation in juvenile green sea urchins (Castell et al. 2004). AA level in octopuses is associated with the maintenance of cell membrane structure and function, egg quality, and an improved growth of *Opuntia vulgaris* (Miliou et al. 2006), as well as survival of fish at early life stages (Bessonart et al. 1999; Koven et al. 2001). Especially, it is the primary precursor of eicosanoid in mammals and fish, which plays an important role in ionic transport across membranes (Soudant et al. 1996; Castell et al. 2004). Through its conversion to prostaglandin (Piomelli et al. 1990), AA induces testosterone production in goldfish *Carassius auratus L.* ovaries and testis (Mercure et al. 1995; Asturiano et al. 2000).

The AA works as the second messenger has been well studied. AA and other unsaturated relative fatty acid are released from membrane phospholipids by phospholipase A2-induced unsaturated fatty acid liberation in response to extracellular stimuli, initiating the "arachidonic cascade" in signaling pathways (Tocher 2003). The Protein Kinase C (PKC) can be activated by the PL and Ca2+ with the second messenger – *cis*-unsaturated fatty acids (Murakami et al. 1986) like AA, oleic, linoleic, linolenic and docosahexaenoic acids. When the cell is lack of the phosphatidylserine, the second messengers can activate PKC efficiently (Shearman et al. 1989), even the Ca2+ at a low level (Nishizuka 1992) . Regulating by the same second messengers, there are other two protein kinase – PKN and PKx. The former exist in the rat testis (Kitagawa et al. 1995). PKx is reported that found in human platelets (Khan et al. 1994) and goat cauda epididymis (Roy et al. 1999a). Unlike PKC and PKN to some extent (Khan et al. 1994; Roy et al. 1999a), PKx is the physiological target of AA and other relative fatty acid (Khan et al. 1994) and its metabolism is regulated by the fertility regulation and signal transduction (Roy et al. 1999a). AA is considered as the best activator of PKx (Roy et al. 1999a). Besides the second messenger of the PK, AA can activate N-methyl-D-aspartate receptor-oriented Ca2+ channels (Miller et al. 1992), GTP binding protein in neutrophil membrane (Abramson et al. 1991) and activate the RhoA GTPase (Amano et al. 1996). Moreover, AA inhibit Ras-GTPase activating protein (Tsai et al. 1989), relate to stress-induced translocation and interact with cytoskeleton proteins (Mukai et al. 1997).

The same as EPA, AA has its own transformation way inside the body, in abalones *Haliotic fulgens* (Durazo-Beltran et al. 2003), *H. laevigata* (Dunstan et al. 1996), and *H. discus hannai* (Miliou et al. 2006), AA is preferably desaturated and elongated from 18:2n-6 (Miliou et al. 2006). Itself also can be chain elongated to 22:4n-6, then by △6 desaturation to 24:5n-6, and by chainshortening reaction in peroxisosomes to 22:5n-6 (Miliou et al. 2006).

3.1.3. Combined view of n-3 and n-6 PUFAs

Because of the lack of enzymes $\triangle^{12}$ and $\triangle^{15}$ desaturase for conversion of 18:1n-9 to 18:2n-6 and 18:3n-3, n-3 and n-6 PUFA families are essential fatty acids (EFA) in all animals (Castell et al. 2004). The EFA in *Eledone* life cycle are from the lipid in the bodies (R. Rosa 2004). For warm blooded animals, 18:2n-6 and 18:3n-3 fatty acids are starting materials for these two families of EFA (Castell et al. 2004). Fresh water fish are capable of elongating and desaturating 18:2n-6 to longer chain HUFAs as 20:4n-6, 18:3n-3, 20:5n-3 and 22:6n-3 (Castell et al. 2004). In this process, the first step is to convert 18:2n-6 to 18:3n-6 and 18:3n-3 to 18:4n-3 by the action of △6-desaturase, except for n-6 family, where first step intermediate was almost absent (Castell et al. 2004). It should be noticed that the PUFAs do not participate in all the metabolism processes.

n-3 family and n-6 family cooperate to improve the reproduction. Diets depleted in either DHA, EPA or AA negatively affect the female gametogenesis and fertilization performance (Soudant et al. 1997). It has been shown that the testis cells have the higher ability to convert the shorter n-6 and n-3 fatty acid to 22:5n-6 and 22:6n-3 than isolated hepatocytes and fibroblasts (Retterstol et al. 1998). Although Sertoli cells and germ cells work together to generate LC-PUFA is still lack of date (Beckman et al. 1978), the conclusion exist shows that there is 22:5n-6 and 22:6n-3 synthesis under the hormonal control (Adel A. Zalata 1998) and the Sertoli cells metabolize actively (Beckman et al. 1980). On the other hand, the n-3 HUFA decrease in octopus will cause the accumulation of 22:4n-6 and 22:5n-6 (Miliou et al. 2006), which suggests the n-3 FA to n-6 FA conversion. Further more, the favoritism of elongation from 20:5n-3 to 22:5n-3 compare with 20:4n-6 to 22:4n-6 in rat can explain the importance of the n-3 FA (Retterstol et al. 1998). Free EPA and DHA probably has the contrast roles with AA, because *in vitro* AA stimulates testicular testosterone in gold fish through the conversion to PGE_2, but EPA or DHA blocked the steroidogenesis from both of AA and PGE_2 (Izquierdo et al. 2001).

In spermium, n-6 FA is higher in unhealthy sample but n-3 FA is lower. The higher n-6/n-3 ratio also shows the unhealthy index (Adel A. Zalata 1998). Large quantities of AA and EPA have been found in macroalgal species, such as *Porphyridinm*, *Ascophyllum* and *Laminaria* (Cook et al. 2000). Dietary EPA and AA levels are correlated with fertilization rates in gilthead seabream broodstock (Fernandez-Palacios et al. 1995; Izquierdo et al. 2001), in the octopuses (Miliou et al. 2006) and the steroidogenesis in the gold fish testis (Wade et al. 1994). Both of EPA and AA are involved in cell-mediated functions and are precursors of eicosanoids (EPA is precursor of PG series Ⅲ, AA is precursor of PG series Ⅱ (Izquierdo et al. 2001). But they seem to be in competition in some cases. EPA competes with AA as a substrate for cyclooxygenase resulting in a reduction in synthesis of PGE_2 and $PGF_{2\alpha}$and an increase in PGE_3 level (Zhang et al. 1992). Also competition between EPA and AA during phosholipid esterification has been found to occur in fish (Bell et al. 1995). The high 22:5n-6/22:6n-3 ratio does not involve in the synthesis favoritism, and there should be a systematic view to explain it (Retterstol et al. 1998).

3.1.4. n-9 PUFAs and other PUFAs

The n-9 PUFAs seem to be less attractive to scientists and only a few reports refer to them. In rat sperm, 22:3n-9, 20:3n-9 and 24:4n-9 are more than in epididymis (Aveldano et al. 1992). But the studies on 22:4n-9 and 20:3n-9 both indicate that there is hindrance in the synthesis process (Retterstol et al. 1998). Until now, no desaturation and elongation products are found to be obtained from n-9 PUFAs such as 18:1n-9, 18:2n-9, 20:2n-9 and 20:3n-9, for some mollusks (Durazo-Beltran et al. 2003). But n-9 LC-PUFA is still significant in rodents' germ plasma (Miliou et al. 2006), and their predominance could represent an adaptation to body temperature (Miliou et al. 2006). The pattern and function of n-9 PUFAs during reproduction remain to be investigated.

Beside the HUFAs mentioned above, there are other members in the big PUFAs group, such as C18:2 and C18:3. They serve as precursors for the synthesis of important structural and functional lipids including sphingolipids and eicosanoids. We still know little about these fatty acids.

3.2. Monounsaturated Fatty Acids (MUFA)

Mostly, the double bond is between C-9 and C-10 in MUFAs. And it can be transformed into PUFAs and other fatty acids if needed. In capelin, MUFAs as 20:1(n-9) and 22:1(n-11) are mainly catabolized to provide the necessary energy while PUFAs of n-3 series are transported to eggs (Henderson et al. 1995). Longer chain MUFAs as 20:1n-11, 20:1n-9 and 22:1n-9 are produced by the sea urchins or specifically retained by them (Castell et al. 2004). Probably, they are involved in some unknown processes. 16:1 and 16:0 have no significantly correlated in both polar and non-polar lipids (De Moreno et al. 1980). In marine finfish exhibit an increased 18:1n-9/n-3 MUFA ratio may be because an increase in 18:1n-9 due to essential fatty acid deficiency (Ibeas et al. 1996). In addition, the ratio of 18:1n-9/18:1n-7 fatty acid has been found to give a useful indication of the contribution of bacteria to the nutrition of marine organisms (Cook et al. 2000).

3.3. Saturated Fatty Acids

The Saturated fatty acids have different physical character compare to the PUFAs: 12 to 18 carbons SFA can resolve slightly in water (Nelson et al. 2004). The SFA in the TG fractions are more than PUFA in organs like muscle, liver and gonad (Chatzifotis et al. 2004). The stearic acid (18:0) among SFA is much higher in the unhealthy sperm than the normal sample (Adel A. Zalata 1998). The amount of SFA, especially 18:0 will increase with the decreased DHA level. This may be the only information we known about the relation between

SFA and DHA (Cerolini et al. 2000). The absence of double bonds of SFA affects the biochemical characters of it, for example, the SFA is not able to interact with the kinase (Roy et al. 1999a).

4. PHOSPHOLIPID

The simplest form of phospholipid is composed of glycerol bonded to two fat-soluble *fatty acids* and a charged water-soluble phosphate group. Thus, the phospholipid has bipolar character, which is essential to biological function in cell membranes. To be noticed, the two fat-soluble fatty acids consist of a saturated fatty acid in sn-1 position and a *cis*-unsaturated fatty acid in sn-2 position. Phospholipids play an important role in the membrane fluidity, permeability and plasticity, enzyme activation (Farkas 1984; Chatzifotis et al. 2004). All these attributes are important in gametogenesis and germs fusion. In addition, some phospholipids have a chemical group bonded to the phosphate, for example, phosphatidylcholine, phosphatidylglycerol, phosphatidylinositol, phosphatidylserine and phosphatidylethanolamine. These compounds also play important roles in cell. To take the phospholipid as a whole rather than as fractions is to know them.

Different species have different phospholipids fractions of fatty acids in the germ cell. The pathway of C_{20-22} PUFAs synthesis and the incorporation into the sperm phospholipids may be different among species (Surai et al. 2000b). In the phospholipids of avian spermatozoa, the dominant unsaturated fatty acids are n-6 PUFAs (Kelso et al. 1997a), mainly C20-22 n-6 PUFAs, such as docosatetraenoic acids (22:4,n-6) and arachidonic acids (20:4,n-6) (Nissen et al. 1983; Blesbois et al. 1997; Cerolini et al. 1997; Kelso et al. 1997b). In phospholipids of mammalian spermatozoa, the main fatty acids are n-3 polyunsaturated fatty acids, especially docosahexaenoic acid (22:6,n-3) (Jones et al. 1977; Nissen et al. 1983; Kelso et al. 1997c; Connor et al. 1998), such like in human normozoospermia DHA contribute >60% of total PUFA (Adel A. Zalata 1998). The proportion and fatty acids composition of phospholipids are tissue-specific in higher vertebrates (Soudant et al. 1997), fishes (Bell et al. 1985; Soudant et al. 1997) and mollusk (De Moreno et al. 1980; Watson et al. 2003). In the germ cell, the fatty acid composition of phospholipids is different in different location, phospholipids which contain 22:6n-3 mainly locate in the tail of mammalian spermatozoa (Connor et al. 1998).

Polar lipids play important role in gametes (Pollero et al. 1988), to maintain reproductive functions associated with membranes (Soudant et al. 1997), or to be involved in the synthesis of essential compounds of germ cell. Lack of total n-3 HUFAs in the lipid fraction of phospholipid occurs in the spermatozoa of idiopathic oligo/asthenozoospermia patients (Adel A. Zalata 1998).

5. THE IMPACT OF DIET OF THE FATTY ACID COMPOSITION OF GANAD

The diets have a significant influence on the fatty acids of tissues in many species. A correlation between the increase of C20 and C22 PUFAs in the tissues of *Mesodesma mactroides* and the food abundance has been revealed (De Moreno et al. 1980), as well as *Mytilus platensis* d'Orbigny (De Moreno et al. 1980). An increase of 16:0 and 16:1 acids, which can be easily synthesized de nove by the mollusc, occur when food is less abundant (De Moreno et al. 1980).

Marine fish lack the desaturase enzymes and require pre-formed n-3 HUFA in diet (Castell et al. 2004), the clam showed little or no ability to synthesize long chain n-3 and n-6 fatty acids from dietary fatty acids (De Moreno et al. 1980), and the content of EPA in egg is more easily influenced by dietary n-3 HUFA than DHA (Izquierdo et al. 2001). There is increasing evidence that high dietary n-3 HUFA levels could affect the brain-pituitary-gonad endocrine axis, since both EPA and DHA have been found to reduce in vitro the steroidogenic process of gonadotropin in the ovary of teleost fish (Mercure et al. 1995). A decrease of DHA and EPA in diets could cause atresia of copepod, thus, reducing fecundity (Lacoste et al. 2001). In zooplankton, fecundity is found to be related to the DHA and EPA in diets (Pond et al. 1996; Payne et al. 2000)

The matters in gonads and eggs are affected by different kinds of feeding microalgae in *Pect en maximus* (Soudant et al. 1996). A major amount of lipids are gained via diet, and therefore, the feeding behavior tremendously alters the fatty acid composition of certain herbivorous copepods, like *Calanus spp.* (Cook et al. 2000). Variations in the contents of 20:4n-6, 20:5n-3 and 22:6n-3 fatty acids in the gonads of 12 different species of sea urchin collected from various locations in Japan have been attributed to differences in their respective habitats, particularly the type of food available (Tagaki et al. 1986; Cook et al. 2000). In rat testis, 22:5n-6 is in larger proportion than 22:6n-3, but in rats fed a diet rich n-3 fatty acids 22:6n-3 is the dominant C22 fatty acid in testis (Ayala et al. 1977; Retterstol et al. 1998).

The broodstock is the most frequent method used in the fishery culture, and good nutrition can improve the quality and fertilization rate of sperms and eggs (Izquierdo et al. 2001). The dietary fatty acids are one of the main nutritients, which significantly affect fish reproduction (especially the fatty acid composition and the quality of eggs) (Soudant et al. 1997), the development of the gonad and body in *P. miliaris* (Cook et al. 2000), and the somatic growth in juvenile green sea urchins (Takagi et al. 1980), and in some species of teleosts (Luquet et al. 1986). The spawning of *Nile tilapia* increases while fed with soybean oil rich in n-6 fatty acids (Soudant et al. 1997).

Some animals developed a mechanism to retain the proportions of DHA, EPA and AA in the lipids of reproductive organs, even under deficient conditions (Soudant et al. 1996). The contents of EPA and DHA in diets have no correlation with fecundity and egg viability in marine copepods (Kleppel et al. 1995). Sperm fatty acid composition determines the membrane integrity after freeze-thawing, but no dietary n-3 and n-6 PUFA are found to be related to sperm freeze-thaw fertilizing ability (Labbe et al. 1996).

6. Conclusion and Propects

Fatty acids are considered as the crucial component of the organisms, because of the insolubility of water and molecular structure like double bonds. The unsaturated fatty acids in vivo seldom present as the *trans*-fatty acid type. Only a few *trans*-fatty acids can be found in the bacteria of the rumen of the dairy (Nelson et al. 2004) and most of *trans*-fatty acids we can see is artificial. The naturally occurring *cis*-fatty acids with the different three-dimensional structure from *trans*- can help the researches of specific binding domain of fatty acid-stimulate protein like PKx. Moreover, all the *cis*-unsaturated fatty acids in the sn-2 position of the PL must be the determinant of the fluidity of the membrane, since the weak interaction produced by them.

From the phospholipid related review mentioned above, the sn-1 and sn-2 are binding with different fatty acid. The structure analysis shows that the sn-1 position is thermodynamic steady, which means that the unsaturated fatty acids in sn-2 position are able to move to sn-1 after oxidation. In the other hand, the quantity of saturated fatty acid "stored" in the sn-1 position could be another source of the unsaturated fatty acid in the sn-2 position and in cytoplast. At last, it forms a compensatory pathway of the fatty acid metabolism from endoplasmic reticule to the biomembrane.

The physiological function of fatty acids at the molecular level is mysterious and attractive, but the methods are limited. Most of experiments are just feeding and chemical analysis and few are isotope labeled injection (Chatzifotis et al. 2004) which maybe the last choice of many laboratories. The newly methods are focus on the lipid-protein interaction, for example, locating the fatty acid desaturases to locating the related fatty acids (Dyer et al. 2001). In this way, more lipid-specific proteins are required for distribution studies of fatty acid.

The fatty acid pattern in tissues and organs is very complicated, since there exist many different kinds of fatty acids. The fatty acid composition in many animals has been identified by intensive studies, and it is found to vary dramatically among tissues. This may suggest different roles of various fatty acids in membrane structure and cellular metabolism. However, we know little about the specific function of these fatty acids. It is supposed that in the future the concentration of studies will focus on the revelation of individual fatty acids or groups, which have similar characters, and their interaction with other components. However more effective approaches for research of fatty acids should be invented. Also the metabolism of fatty acids, especially the key enzymes such as Fad2 and Fad3 involved in this process, is expected to gain more attention.

Reviewed by : D. Sc. Dr. Hans-U. Dahms Visiting Professor Institute of Marine Biology National Taiwan Ocean University Tel: +886-2-24698084, Fax: +886-2-24629464 E-Mails: hansd@ntou.edu.tw or hansd@mail.ntou.edu.tw

References

Abramson, S.B., Leszczynska-Piziak, J., Weissmann, G. (1991). Arachidonic acid as a second messenger. Interactions with a GTP-binding protein of human neutrophils. *J. Immunol.* 147, 231-236.

Adel A. Zalata, A.B.C., Christophe E. Depuydt, Frank Schoonjans and Frank H. Comhaire, (1998). The fatty acid composition of phospholipids of spermatozoa from infertile patients. *Molecular Human Reproduction* 4, 111-118.

Aitken, R. (1994). Pathophysiology of human spermotozoa. Curr Opin Obstet Gynecol 6, 128-135.

Aitken, R.J., Buckingham, D., Harkiss, D. (1993). Use of a xanthine oxidase free radical generating system to investigate the cytotoxic effects of reactive oxygen species on human spermatozoa. *J.Reprod. Fertil.* 97, 441-450.

Aitken RJ, C.J., Fishel S, (1993). Relationship between iron-catalysed lipid peroxidation potential and human sperm function. *Journal of Reproduction and Fertility* 98, 257-265.

Aitken, R.J., Clarkson, J.S., Hargreave, T.B., Irvine, D.S., Wu, F.C. (1989). Analysis of the relationship between defective sperm function and the generation of reactive oxygen species in cases of oligozoospermia. *J. Androl.* 10, 214-220.

Alvarez JG, S.B. (1984). Assessment of cell damage caused by spontaneous lipid peroxidation in rabbit spermatozoa. *Biology of Reproduction* 30, 323-332.

Alvarez JG, S.B. (1989). Role of glutathione peroxidase in protecting mammalian spermatozoa from loss of motility caused by spontaneous lipid peroxidation. *Gamete Research* 23, 77-89.

Amano, M., Mukai, H., Ono, Y., Chihara, K., Matsui, T., Hamajima, Y., Okawa, K., Iwamatsu, A., Kaibuchi, K., (1996). Identification of a putative target for Rho as the serine-threonine kinase protein kinase N. *Science* 271, 648-650.

Ansell, G.B. (1974). Pharmacological control of lipid metabolism. : Edited by W. L. Holmes R. Paoletti and D. Kritchevsky. Advances in experimental medicine and biology Vol. 26 (1972), Plenum Press, New York, 359pp, $26.00. *Neuropharmacology* 13, 1095.

Arienti, G., Carlini, E., Polci, A., Cosmi, E.V., Palmerini, C.A. (1998). Fatty Acid Pattern of Human Prostasome Lipid. Archives of Biochemistry and Biophysics 358, 391-395.

Asturiano, J.F., Sorbera, L.A., Zanuy, S., Carrillo, M. (2000). Effects of polyunsaturated fatty acids and gonadotropin on prostaglandin series E production in a primary testis cell culture system for the European sea bass. *Journal of Fish Biology* 57, 1563-1574.

Aveldano, M.I., Rotstein, N.P., Vermouth, N.T. (1992). Occurrence of long and very long polyenoic fatty acids of the n-9 series in rat spermatozoa. *Lipids* 27, 676-680.

Ayala, S., Brenner, R.R., Dumm, C.G. (1977). Effect of polyunsaturated fatty acids of the alpha-linolenic series on the development of rat testicles. *Lipids* 12, 1017-1024.

Beckman, J.K., Coniglio, J.G. (1980). The metabolism of polyunsaturated fatty acids in rat Sertoli and germinal cells. *Lipids* 15, 389-394.

Beckman, J.K., Gray, M.E., Coniglio, J.G. (1978). The lipid composition of isolated rat spermatids and spermatocytes. *Biochim. Biophys. Acta* 530, 367-374.

Bell, J.G., John, D.C., Douglas, R.T., Fiona, M.M., John, R.S. (1995). Effects of different dietary arachidonic acid : docosahexaenoic acid ratios on phospholipid fatty acid compositions and prostaglandin production in juvenile turbot (Scophthalmus maximus). *Fish Physiology and Biochemistry* 14, 139-151.

Bell, J.G., Tocher, D.R., Sargent, J.R. (1994). Effect of supplementation with 20:3(n-6), 20:4(n-6) and 20:5(n-3) on the production of prostaglandins E and F of the 1-, 2- and 3-series in turbot (Scophthalmus maximus) brain astroglial cells in primary culture. *Biochim. Biophys. Acta* 1211, 335-342.

Bell, M.V., Dick, J.R., Kelly, M.S. (2001). Biosynthesis of eicosapentaenoic acid in the sea urchin Psammechinus miliaris. Lipids 36, 79-82.

Bell, M.V., Henderson, R.J., Sargent, J.R. (1985). Changes in the fatty acid composition of phospholipids from turbot (Scophthalmus maximus) in relation to dietary polyunsaturated fatty acid deficiencies. *Comp. Biochem. Physiol.* B, 81, 193-198.

Bessonart, M., Izquierdo, M.S., Salhi, M., Hernandez-Cruz, C.M., Gonzalez, M.M., Fernandez-Palacios, H. (1999). Effect of dietary arachidonic acid levels on growth and survival of gilthead sea bream (Sparus aurata L.) larvae. *Aquaculture* 179, 265-275.

Blesbois, E., Lessire, M., Grasseau, I., Hallouis, J.M., Hermier, D. (1997). Effect of dietary fat on the fatty acid composition and fertilizing ability of fowl semen. *Biol. Reprod.* 56, 1216-1220.

Blomquist, G.J., Borgeson, C.E., Vundla, M. (1991). Polyunsaturated fatty acids and eicosanoids in insects. *Insect. Biochem.* 21, 99–106.

Bon, E., Corraze, G., Kaushik, S., Le Menn, F. (1997). Effects of Accelerated Photoperiod Regimes on the Reproductive Cycle of the Female Rainbow Trout: I-Seasonal Variations of Plasma Lipids Correlated with Vitellogenesis - aspects of absorption, transportation, deposition and mobilization. Comparative Biochemistry and Physiology - Part A: *Physiology* 118, 183-190.

Brenner, R.R. (1974). The oxidative desaturation of unsaturated fatty acids in animals. Mol. *Cell. Biochem.* 3, 41–52.

Bruce, M., Oyen, F., Bell, G., Asturiano, J.F., Farndale, B., Carrillo, M., Zanuy, S., Ramos, J., Bromagc, N. (1999). Development of broodstock diets for the European Sea Bass (Dicentrarchus labrax) with special emphasis on the importance of n-3 and n-6 highly unsaturated fatty acid to reproductive performance. *Aquaculture* 177, 85-97.

Castell, J.D., Kennedy, E.J., Robinson, S.M.C., Parsons, G.J., Blair, T.J., Gonzalez-Duran, E. (2004). Effect of dietary lipids on fatty acid composition and metabolism in juvenile green sea urchins (Strongylocentrotus droebachiensis). *Aquaculture* 242, 417-435.

Cecil, H.C., Bakst, M. T. (1993). In vitro lipid peroxidation of turkey spermatozoa. poultry science 9. Cerda, J., Zanuy, S., Carrillo, M., Ramos, J., Serrano, R., 1995. Short- and long-term dietary effects on female sea bass (Dicentrarchus labrax): seasonal changes in plasma profiles of lipids and sex steroids in relation to reproduction. Comparative Biochemistry and Physiology Part C: *Comparative Pharmacology and Toxicology* 111, 83-91.

Cerolini, S., Kelso, K.A., Noble, R.C., Speake, B.K., Pizzi, F., Cavalchini, L.G. (1997). Relationship between spermatozoan lipid composition and fertility during aging of chickens. *Biol. Reprod.* 57, 976-980.

Cerolini, S., Maldjian, A., Surai, P., Noble, R. (2000). Viability, susceptibility to peroxidation and fatty acid composition of boar semen during liquid storage. Animal Reproduction Science 58, 99-111.Chatzifotis, S., Muje, P., Pavlidis, M., Agren, J., Paalavuo, M., Molsa, H. (2004). Evolution of tissue composition and serum metabolites during gonadal development in the common dentex (Dentex dentex). *Aquaculture* 236, 557-573.

Connor, W.E., Lin, D.S., Wolf, D.P., Alexander, M. (1998). Uneven distribution of desmosterol and docosahexaenoic acid in the heads and tails of monkey sperm. *J. Lipid. Res.* 39, 1404-1411.

Cook, E.J., Bell, M.V., Black, K.D., Kelly, M.S. (2000). Fatty acid compositions of gonadal material and diets of the sea urchin, Psammechinus miliaris: trophic and nutritional implications. *Journal of Experimental Marine Biology and Ecology* 255, 261-274.

Craig, S.R., MacKenzie, D.S., Jones, G., Gatlin, D.M. (2000). Seasonal changes in the reproductive condition and body composition of free-ranging red drum, Sciaenops ocellatus. *Aquaculture* 190, 89-102.

Dawra, R.K., Sharma, O.P., Makkar, H.P. (1983). Lipid peroxidation in bovine semen. *Int. J. Fertil.* 28, 231-234.

de Lamirande, E., Gagnon, C. (1992). Reactive oxygen species and human spermatozoa. I. Effects on the motility of intact spermatozoa and on sperm axonemes. *J. Andro*l. 13, 368-378.

De Moreno, J.E.A., Pollero, R.J., Moreno, V.J., Brenner, R.R. (1980). Lipids and fatty acids of the mussel (Mytilus Platensis d'Orbigny) from South Atlantic waters. *Journal of Experimental Marine Biology and Ecology* 48, 263-276.

Dunstan, G.A., Baillie, H.J., Barrett, S.M., Volkman, J.K. (1996). Effect of diet on the lipid composition of wild and cultured abalone. *Aquaculture* 140, 115-127.

Durazo-Beltran, E., D'Abramo, L.R., Toro-Vazquez, J.F., Vasquez-Pelaez, C., Viana, M.T. (2003). Effect of triacylglycerols in formulated diets on growth and fatty acid composition in tissue of green abalone (Haliotis fulgens). *Aquaculture* 224, 257-270.

Dyer, J.M., Mullen, R.T. (2001). Immunocytological localization of two plant fatty acid desaturases in the endoplasmic reticulum. *FEBS Letters* 494, 44-47.

Farkas, T. (1984). Adaptation of fatty acid composition to temperature--a study on carp (Cyprinus carpio L.) liver slices. *Comp. Biochem. Physiol.* B 79, 531-535.

Fernandez-Palacios, H., Izquierdo, M.S., Robaina, L., Valencia, A., Salhi, M., Vergara, J. (1995). Effect of n - 3 HUFA level in broodstock diets on egg quality of gilthead sea bream (Sparus aurata L.). *Aquaculture* 132, 325-337.

Fernandez, J., Gutierrez, J., Carrillo, M., Zanuy, S., Planas, J. (1989). Annual cycle of plasma lipids in sea bass, Dicentrarchus labrax L.: Effects of environmental conditions and reproductive cycle. Comparative Biochemistry and Physiology Part A: *Physiology* 93, 407-412.

Gagnon C, I.A., De Lamirande E, Kovalski N. (1991). Reactive oxygen species and human spermatozoa. annals of the New York Academy of Science 637, 436-445.

German, B., Bruckner, G., Kinsella, J. (1983). Evidence against a PGF4a prostaglandin structure in trout tissue--a correction. *Prostaglandins* 26, 207-210.

Griveau, J.F., Dumont, E., Renard, P., Callegari, J.P., Le Lannou, D. (1995). Reactive oxygen species, lipid peroxidation and enzymatic defence systems in human spermatozoa. *J. Reprod. Fertil* .103, 17-26.

Grubert, M.A., Dunstan, G.A., Ritar, A.J. (2004). Lipid and fatty acid composition of pre- and post-spawning blacklip (Haliotis rubra) and greenlip (Haliotis laevigata) abalone conditioned at two temperatures on a formulated feed. *Aquaculture* 242, 297-311.

Hammerstedt, R.H. (1993). Maintenance of bioenergetic balance in sperm and prevention of lipid peroxidation: a review of the effect on design of storage preservation systems. *Reprod. Fertil. Dev.* 5, 675-690.

Henderson, R.J., Moira, T.P., John, R.S. (1995). The desaturation and elongation of 14C-labelled polyunsaturated fatty acids by pike (Esox lucius L.) in vivo. *Fish Physiology and Biochemistry* 14, 223-235.

Hinshaw, D.B., Sklar, L.A., Bohl, B., Schraufstatter, I.U., Hyslop, P.A., Rossi, M.W., Spragg, R.G., Cochrane, C.G. (1986). Cytoskeletal and morphologic impact of cellular oxidant injury. *Am. J. Pathol* .123, 454-464.

Ibeas, C., Cejas, J., Gomez, T., Jerez, S., Lorenzo, A. (1996). Influence of dietary n - 3 highly unsaturated fatty acids levels on juvenile gilthead seabream (Sparus aurata) growth and tissue fatty acid composition. *Aquaculture* 142, 221-235.

Izquierdo, M.S., Fernandez-Palacios, H., Tacon, A.G.J. (2001). Effect of broodstock nutrition on reproductive performance of fish. *Aquaculture* 197, 25-42.

Jensen, B.H., Taylor, M.H. (2001). Lipid transport in female Fundulus heteroclitus during the reproductive season. *Fish Physiology and Biochemistry* 25, 141-151.

Jones, R., Mann, T. (1976). Lipid peroxides in spermatozoa; formation, role of plasmalogen, and physiological significance. *Proc. R. Soc. Lond. B. Biol. Sci.* 193, 317-333.

Jones, R., Mann, T. (1977). Damage to ram spermatozoa by peroxidation of endogenous phospholipids. *J. Reprod. Fertil.* 50, 261-268.

Jones, R., Mann, T., Sherins, R. (1979). Peroxidative breakdown of phospholipids in human spermatozoa, spermicidal properties of fatty acid peroxides, and protective action of seminal plasma. *Fertil. Steril* .31, 531-537.

Jones, R., Mann, T., Sherins, R.J. (1978). Adverse effects of peroxidized lipid on human spermatozoa. *Proc. R. Soc. Lond B. Biol .Sci.* 201, 413-417.

Josrgensen, E.H., Johansen, S.J.S., Jobling, M. (1997). Seasonal patterns of growth, lipid deposition and lipid depletion in anadromous Arctic charr. *Journal of Fish Biology.* 51. 2. 312-326

Kclso, K.A., Cerolini, S., Noble, R.C., Sparks, N.H., Speake, B.K. (1997a). The effects of dietary supplementation with docosahexaenoic acid on the phospholipid fatty acid composition of avian spermatozoa. *Comp. Biochem. Physiol. B. Biochem. Mol. Biol.* 118, 65-69.

Kelso, K.A., Cerolini, S., Speake, B.K., Cavalchini, L.G., Noble, R.C. (1997b). Effects of dietary supplementation with alpha-linolenic acid on the phospholipid fatty acid composition and quality of spermatozoa in cockerel from 24 to 72 weeks of age. *J. Reprod. Fertil.* 110, 53-59.

Kelso, K.A., Redpath, A., Noble, R.C., Speake, B.K. (1997c). Lipid and antioxidant changes in spermatozoa and seminal plasma throughout the reproductive period of bulls. *J. Reprod. Fertil* .109, 1-6.

Khan, W.A., Blobe, G.C., Richards, A.L., Hannun, Y.A. (1994). Identification, partial purification, and characterization of a novel phospholipid-dependent and fatty acid-activated protein kinase from human platelets. *J. Biol. Chem.* 269, 9729-9735.

Kitagawa, M., Mukai, H., Shibata, H., Ono, Y. (1995). Purification and characterization of a fatty acid-activated protein kinase (PKN) from rat testis. Biochem J 310 (Pt 2), 657-664.

Kleppel, G.S., Burkart, C.A. (1995). Egg production and the nutritional environment of Acartia tonsa: the role of food quality in copepod nutrition. I*CES Journal of Marine Science* 52, 297-304.

Koven, W., Barr, Y., Lutzky, S., Ben-Atia, I., Weiss, R., Harel, M., Behrens, P., Tandler, A. (2001). The effect of dietary arachidonic acid (20:4n-6) on growth, survival and resistance to handling stress in gilthead seabream (Sparus aurata) larvae. *Aquaculture* 193, 107-122.

Labbe, C., Maisse, G. (1996). Influence of rainbow trout thermal acclimation on sperm cryopreservation: relation to change in the lipid composition of the plasma membrane. *Aquaculture* 145, 281-294.

Lacoste, A., Poulet, S.A., Cueff, A., Kattner, G., Ianora, A., Laabir, M. (2001). New evidence of the copepod maternal food effects on reproduction. *Journal of Experimental Marine Biology and Ecology* 259, 85-107.

Ladha, S. (1998). Lipid Heterogeneity and Membrane Fluidity in a Highly Polarized Cell, the Mammalian Spermatozoon. *Journal of Membrane Biology* 165, 1-10.

Lenzi A, P.M., Gandini L, Dondero F. (1996). Lipids of the sperm plasma membrane: from polyunsaturated fatty acids considered as markers of sperm function to possible scavenger therapy. *Human Reproduction Update* 2, 246-257.

Lopes, S., Jurisicova, A., Sun, J.G., Casper, R.F. (1998). Reactive oxygen species: potential cause for DNA fragmentation in human spermatozoa. *Hum Reprod* 13, 896-900.

Lucas, C.H. (1994). Biochemical composition of Aurelia aurita in relation to age and sexual maturity. *Journal of Experimental Marine Biology and Ecology* 183, 179-193.

Lund, E.D., Sullivan, C.V., Place, A.R. (2000). Annual cycle of plasma lipids in captive reared striped bass: effects of environmental conditions and reproductive cycle. *Fish Physiology and Biochemistry* 22, 263-275.

Luquet, P., Watanabe, T. (1986). Interaction "nutrition-reproduction" in fish. *Fish Physiology and Biochemistry* 2, 121-129.

Maldjian, A., Cerolini, S., Surai, P., Speake, B.K. (1998). The effect of vitamin E, green tea extracts and catechin on the in vitro storage of turkey spermatozoa at room temperature. *Poultry and Avian Biology Reviews* 9, 143-151.

Martinez, G., Brokordt, K., Aguilera, C., Soto, V., Guderley, H. (2000). Effect of diet and temperature upon muscle metabolic capacities and biochemical composition of gonad and muscle in Argopecten purpuratus Lamarck 1819. Journal of Experimental Marine *Biology and Ecology* 247, 29-49.

Mathieu, M., Lubet P. (1993). Storage tissue metabolism and reproduction in marine bivalves - a brief review. *Invert. Reprod. Develop.* 23, 123-129.

Mercure, F., Van Der Kraak, G. (1995). Inhibition of gonadotropin-stimulated ovarian steroid production by polyunsaturated fatty acids in teleost fish. *Lipids* 30, 547-554.

Miliou, H., Fintikaki, M., Tzitzinakis, M., Kountouris, T., Verriopoulos G.. (2006). Fatty acid composition of the common octopus, Octopus vulgaris, in relation to rearing temperature and body weight. *Aquaculture* 256, 311-322.

Miller, B., Sarantis, M., Traynelis, S.F., Attwell, D. (1992). Potentiation of NMDA receptor currents by arachidonic acid. *Nature* 355, 722-725.

Montero-Torreiro, M.F., Garcia-Martinez, P., Catoira, J.L., Mosquera, G. (1998). Seasonal variation in biochemical composition in gonads of the sea urchin, Paracentrotus lividus Lmk. In: Mooi, R., Telford, M. (Eds.), echioderms. Proceedings of the 9th International Echinoderm Meeting, San Francisco 1998. A.A. Balkema, Rotterdam 753-758.

Moreno, V.J., Moreno, D., J.E.A., Brenner, R.R. (1979). Fatty acid metabolism of the calanoid copepod Paracalanus parvus: 1. Polyunsaturated fatty acids. *Lipids* 14, 313–317.

Mourente, G., Rodriguez, A., Tocher, D.R., Sargent, J.R. (1993). Effects of dietary docosahexaenoic acid (DHA; 22:6n-3) on lipid and fatty acid compositions and growth in gilthead sea bream (Sparus aurata L.) larvae during first feeding. *Aquaculture* 112, 79-98.

Mukai, H., Toshimori, M., Shibata, H., Takanaga, H., Kitagawa, M., Miyahara, M., Shimakawa, M., Ono, Y. (1997). Interaction of PKN with alpha-actinin. *J. Biol. Chem.* 272, 4740-4746.

Murakami, K., Chan, S.Y., Routtenberg, A. (1986). Protein kinase C activation by cis-fatty acid in the absence of Ca2+ and phospholipids. *J. Biol. Chem.* 261, 15424-15429.

Mustonen, A.M., Nieminen, P., Hyv, auml, rinen, H. (2002). Liver and plasma lipids of spawning burbot. *Journal of Fish Biology* 61, 1318-1322.

Napolitano, G.E., Ackman, R.G. (1992). Anatomical distributions and temporal variations of lipid classes in sea scallops Placopecten magellanicus (Gmelin) from Georges Bank (Nova Scotia). Comp. *Biochem. Physiol.* 103B, 645-650.

Navarro, J.C., Villanueva, R. (2000). lipid and fatty acid composition of early stages of cephalopods: and approach to their lipid requirements. *Aquaculture* 183, 167-177.

Navarro, J.C., Villanueva, R. (2003). The fatty acid composition of Octopus vulgaris paralarvae reared with live and inert food: deviation from their natural fatty acid profile. *Aquaculture* 219, 613-631.

Navas, J.M., Bruce, M., Thrush, M., Farndale, B.M., Bromage, N., Zanuy, S., Carrillo, M., Bell, J.G., Ramos, J. (1997). The impact of seasonal alteration in the lipid composition of · broodstock diets on egg quality in the European sea bass. *JFish Biol.,* 51(4), 760-773

Navas, J.M., Mananos, E., Thrush, M., Ramos, J., Zanuy, S., Carrillo, M., Zohar, Y., Bromage, N. (1998). Effect of dietary lipid composition on vitellogenin, 17beta-estradiol and gonadotropin plasma levels and spawning performance in captive sea bass (Dicentrarchus labrax L.). *Aquaculture* 165, 65-79.

Nelson, D., Cox, M.M. (2004). Lehninger Principle of Biochemistry. W.H. Freeman. Nishizuka, Y., 1992. Intracellular signaling by hydrolysis of phospholipids and activation of protein kinase C. *Science* 258, 607-614.

Nissen, H.P., Kreysel, H.W. (1983). Polyunsaturated fatty acids in relation to sperm motility. *Andrologia* 15, 264-269.

Payne, M.F., Rippingale, R.J. (2000). Evaluation of diets for culture of the calanoid copepod Gladioferens imparipes. *Aquaculture* 187, 85-96.

Piomelli, D., Greengard, P. (1990). Lipoxygenase metabolites of arachidonic acid in neuronal transmembrane signalling. *Trends Pharmacol .Sci.* 11, 367-373.

Pollero, R.J., Iribarne, O.O. (1988). Biochemical changes during the reproductive cycle of the small patagonian octopus, Octopus tehuelchus, D'Orb. Comparative Biochemistry and Physiology Part B: *Biochemistry and Molecular Biology* 90, 317-320.

Pond, S.M., Gordon, R.A., Simi, A.L., Winzor, D.J. (1996). Further Observations on the Measurement of Fatty Acid Incorporation by Erythrocyte Ghosts to Quantify Unbound Palmitate Concentration in Albumin-Palmitate Mixtures. *Analytical Biochemistry* 237, 232-238.

R. Rosa, P.R.C., J. Pereira, M. L. Nunes (2004). Biochemical dynamics of spermatogenesis and oogenesis in Eledone cirrhosa and Eledone moschata (Cephalopoda: Octopoda). *Comparative Biochemistry and Physiology* 139, 299-311.

Retterstol, K., Haugen, T.B., Woldseth, B., Christophersen, B.O. (1998). A comparative study of the metabolism of n-9, n-6 and n-3 fatty acids in testicular cells from immature rat. Biochimica et Biophysica Acta (BBA) - *Lipids and Lipid Metabolism* 1392, 59-72.

Roy, K., Mandal, A.K., Sikdar, R., Majumdar, S., Ono, Y., Sen, P.C. (1999a). Unsaturated fatty acid-activated protein kinase (PKx) from goat testis cytosol. Biochimica Et Biophysica Acta-Protein Structure and Molecular Enzymology 1434, 161-169.

Roy, K., Mandal, A.K., Sikdar, R., Majumdar, S., Ono, Y., Sen, P.C. (1999b). Unsaturated fatty acid-activated protein kinase (PKx) from goat testis cytosol. Biochimica et Biophysica Acta (BBA) - Protein Structure and Molecular Enzymology 1434, 161-169.

Ruiz-Verdugo, C.A., Racotta, I.S., Ibarra, A.M. (2001). Comparative biochemical composition in gonad and adductor muscle of triploid and diploid catarina scallop (Argopecten ventricosus Sowerby II, 1842). *Journal of Experimental Marine Biology and Ecology* 259, 155-170.

Sargent, J.R., McEvoy, L.A., Estevez, A., Bell, G., Bell, M., Henderson, J., Tocher, D. (1999). Lipid nutrition of marine fish during early development: current status and future directions. *Aquaculture* 179, 217-229.

Shearman, M.S., Naor, Z., Sekiguchi, K., Kishimoto, A., Nishizuka, Y. (1989). Selective activation of the gamma-subspecies of protein kinase C from bovine cerebellum by arachidonic acid and its lipoxygenase metabolites. *FEBS Lett* 243, 177-182.

Smutna, M., Synek, O. (1979). Lipid peroxidation in semen of the boar. Acta Vet Brno 48, 35-43.Soudant, P., Marty, Y., Moal, J., Robert, R., Quere, C., Le Coz, J.R., Samain, J.F., 1996. Effect of food fatty acid and sterol quality on Pecten maximus gonad composition and reproduction process. *Aquaculture* 143, 361-378.

Soudant, P., Moal, J., Marty, Y., Samain, J.F. (1997). Composition of polar lipid classes in male gonads of Pecten maximus (L.): Effect of nutrition. *Journal of Experimental Marine Biology and Ecology* 215, 103-114.

Støttrup, J.G., Jensen, J. (1990). Influence of algal diet on feeding and egg-production of the calanoid copepod Acartia tonsa Dana. *J. Exp. Mar. Biol. Ecol.* 141, 87–105.

Stanley-Samuelson, D.W. (1987). Physiological roles of prostaglandins and other eicosanoids in invertebrates. *Bio. Bull.* 173, 92-109.

Surai, P.F., Brillard, J.P., Speake, B.K., Blesbois, E., Seigneurin, F., Sparks, N.H. (2000a). Phospholipid fatty acid composition, vitamin E content and susceptibility to lipid peroxidation of duck spermatozoa. *Theriogenology* 53, 1025-1039.

Surai, P.F., Brillard, J.P., Speake, B.K., Blesbois, E., Seigneurin, F., Sparks, N.H.C. (2000b). Phospholipid fatty acid composition, vitamin e content and susceptibility to lipid peroxidation of duck spermatozoa. *Theriogenology* 53, 1025-1039.

Surai, P.F., Cerolini, S., Wishart, G.J., Speake, B.K., Noble, R.C., Sparks, N.H.C. (1998). Lipid and antioxidant composition of chicken semen and its susceptibility to peroxidation. *Poultry and Avian Biology Reviews* 9, 11-23.

Tagaki, T., Kaneniwa, M., Itabashi, Y., Ackman, R.G., (1986). Fatty acids in Echinoidea: unusual cis-5-Olefinic acids as distinctive lipid components in sea urchins. *Lipids* 21, 558-565.

Takagi, T., Eaton, C.A., Ackman, R.G. (1980). Distribution of fatty acids in lipids of the common Atlantic sea urchin Strongylocentrotus droebachiensis. Can, *J. Fish. Aquat. Sci.* 37, 195-202.

Tandler, A., Watanabe, T., Satoh, S., Fukusho, K. (1989). The effect of food deprivation on the fatty acid and lipid profile of red seabream (Pagrus major) larvae. *British Journal of Nutrition* 62, 349-361.

Tocher, D.R. (2003). Metabolism and Functions of Lipids and Fatty Acids in Teleost Fish. *Reviews in Fisheries Science* 11, 107-184.

Tsai, M.H., Yu, C.L., Wei, F.S., Stacey, D.W. (1989). The effect of GTPase activating protein upon ras is inhibited by mitogenically responsive lipids. Science 243, 522-526.

Tulli, F., Tibaldi, E. (1997). Changes in amino acids and essential fatty acids during early larval rearing of dentex. *Aquaculture International* 5, 229-236.

van den Bosch, H. (1980). Intracellular phospholipases A. *Biochim. Biophys. Acta* 604, 191-246.

Wade, M.G., Van der Kraak, G., Gerrits, M.F., Ballantyne, J.S. (1994). Release and steroidogenicactions of polyunsaturated fatty acids in the goldfish testis. *Biology of Reproduction* . 51. 131-139

Walker, C., McGinn, N.A., Harrington, L.M., Lesser, M.P., (1998). New perspectives on sea urchin gametogenesis and thei relevance to aquaculture. *J. Shell. Res* 17, 1507-1514.

Wallaert, C., Babin, P.J. (1994). Age-related, sex-related, and seasonal changes of plasma lipoprotein concentrations in trout. *J. Lipid Res.,* 35, 1619-1633

Watson, G.J., Bentley, M.G., Gaudron, S.M., Hardege, J.D. (2003). The role of chemical signals in the spawning induction of polychaete worms and other marine invertebrates. *Journal of Experimental Marine Biology and Ecology* 294, 169-187.

Wishart, G.J. (1984). Effects of lipid peroxide formation in fowl semen on sperm motility, ATP content and fertilizing ability. J Reprod Fertil 71, 113-118.

Wishart, G.J. (1989). Physiological changes in fowl and turkey spermatozoa during in vitro storage. *Br. Poult. Sci.* 30, 443-454.

Xu, Y., Cleary, L.J., Byrne, J.H. (1994). Identification and characterization of pleural neurons that inhibit tail sensory neurons and motor neurons in Aplysia: correlation with FMRFamide immunoreactivity. *J. Neurosci.* 14, 3565-3577.

Zalata, A., Hafez, T., Comhaire, F. (1995). Evaluation of the role of reactive oxygen species in male infertility. *Hum .Reprod.* 10, 1444-1451.

Zhang, Z., Benson, B., Logan, J.L. (1992). Dietary fish oil delays puberty in female rats. *Biology of Reproduction.* 47. 998-1003

Zohar, Y., Harel, M., Hassin, S., Tandler, A. (1995). Broodstock management and manipulation of spawning in the gilthead seabream, Sparus aurata. In: "Broodstock Management and Egg and Larval Quality" (N. Bromage and R.J. Roberts, eds.). Blackwell Sci. Press, London.

In: Aquaculture Research Progress
Editor: Takumi K. Nakamura

ISBN 978-1-60456-247-7

Chapter 4

PHOTOPERIOD MANIPULATION OF GROWTH PERFORMANCE AND STRESS RESPONSE IN RED SEA BREAM *PAGRUS MAJOR*

Amal Kumar Biswas[1]
Fisheries Laboratory, Kinki University, Uragami 468-3, Nachikatsuura, Wakayama 649-5145, Japan

ABSTRACT

A series of studies have been carried out to establish a light regime giving optimal fish growth for a complete production cycle of red sea bream *Pagrus major*, which is one of the most important fish in Japan because of its multipurpose uses. In different trials, red sea bream were reared under four photoperiods (6L:6D, 12L:12D, 16L:8D and 24L:0D) with constant light intensity 1500 lx on the water surface to investigate their growth performance, apparent digestibility coefficient (ADC) of nutrients and energy, and stress response. Fish were reared for eight weeks in each of three trials. The growth performance and stress response were investigated in trial 1 in which when fish were reared from 1 to 30 g. In trials 2 and 3, growth performance and ADC, and stress response, respectively, were investigated in fish reared from 20 to 100 g.

In trial 1, the results demonstrated that the growth performance, in terms of weight gain, specific growth rate (SGR), feed intake and feed conversion efficiency (FCE) of red sea bream reared from 1 to 30 g can be stimulated significantly by using a continuous (24L:0D) photoperiod without any measurable significant stress response in fish. Similarly, trial 2 demonstrated that the higher growth performance in fish under continuous photoperiod was attributed to improved appetite, greater food intake and higher FCE as well as higher ADC of lipid and energy in red sea bream reared from 20 to 100 g. In addition, trial 3 demonstrated that photoperiod manipulation did not cause significant stress response in red sea bream reared from 20 to 100 g.

These results together help to establish a light regime giving optimal fish growth for a complete production cycle of red sea bream. It is assumed that more attention will be given to on-land and outdoor aquaculture in the near future. Photoperiod manipulation

---

[1] Tel: +81 735 58 0116, Fax: +81 735 58 1246, E-mail: ns_akb@nara.kindai.ac.jp.

will definitely be the option of choice to yield higher output from on-land and outdoor farms.

INTRODUCTION

Photoperiod is the recurring cycle of light and dark periods. It is also the term applied to the length of time that light shines. The natural photoperiod is approximately 24 hours, and the ratio of light to dark hours slowly changes over the course of a year. However, artificial photoperiod is a technology whereby the length of light and dark periods is modified according to interest. In artificial photoperiod, the basic methodology implies application of light pulses for a defined period of time, aiming to modify the physiological performances of animals.

During the last few decades, a great amount of work has been carried out to study the influence of PPD manipulation in fish (Lundqvist, 1980; Boehlert, 1981; Eriksson and Lundqvist, 1982; Saunders et al., 1985; Clarke and Shelbourn, 1986; Saunders and Harmon, 1988; Thorarensen and Clarke, 1989; Stefansson et al., 1991; Berg et al., 1992; Hansen et al., 1992; Mäkinen and Ruhonen, 1992; Mason et al., 1992; Imsland et al., 1995; Imsland et al., 1997; Porter et al., 1998; Simensen et al., 2000; Biswas and Takeuchi, 2003; Biswas et al., 2005), birds (Prabakaran et al., 1991; Charles et al., 1992; Basco et al., 1996; Boon et al., 2000), and mammals (Saito and Noma, 1980; Schanbacher and Crouse, 1981; Bartness and Wade, 1984; Briynklow and Forbes, 1984; Shambacher, 1984; Sorensen, 1984; Wade and Bartness, 1984; Negus et al., 1986; Spears and Clarke, 1986; Canguilhem et al., 1988; Reynolds and Lavigne, 1989; Larkin et al., 1991; Nagy and Negus, 1993; Bartness et al., 1995; Bronson and Kerbeshian, 1995; Vilaplana et al., 1995; Boon et al., 1997; Campuzano et al., 1999). In fish, some contradictory findings with regard to growth performance have been demonstrated in some species. Even within a species, the growth response of different types—which have different life history patterns—to different photoperiods can vary widely (Clarke et al., 1989). The mechanisms underlying these differences are not known; they may be genetic or environmental.

The concept of stress, as applied to biological systems, has a long and controversial history, primarily because of difficulties in formulating a universally acceptable definition of stress (Pickering, 1981) and many have advocated abandoning its use altogether. However, most fish farmers are conscious of when their fish are failing to perform to their full potential and the term 'stress' is still widely used when referring to such situations. It is generally regarded that any environmental disturbance is a potential source of stress, and theoretically can be detected by changes of hormone or substrate concentrations in plasma, or by changes in erythrocyte parameters (Donaldson, 1981). Fish are sensitive to acute and chronic environmental changes and demonstrate a classic stress response (Barton and Iwama, 1991; Wendelaar Bonga, 1997). Although the changes in photoperiod are reported to influence hormone levels and diel activity, stress is not an apparent consequence (Pickering and Pottinger, 1983; Audet et al., 1986).

The red sea bream *Pagrus major* (family: Sparidae and order: Perciformes) is a subtropical (34°N - 15°N) species distributed in the northwest Pacific in the northeastern part of the South China Sea (excluding the Philippines) northward to Japan (Eggleston, 1974). It occupies depths ranging from 10 to 50 m, often on rough grounds, but also on softer bottoms. Adults migrate into shallower parts of their depth range to spawn in late spring and summer;

juveniles live mainly in the shallower areas (Eggleston, 1974). This species feeds on benthic invertebrates, including echinoderms, worms, mollusks and crustaceans, and also on fishes. It is a popular food fish throughout its range. This fish is used in Chinese medicine (Tang, 1987). Red sea bream holds great importance in Japan because of its multipurpose use. It is known as 'madai' and is usually eaten raw as 'sashimi' or 'sushi', and presented in ceremonies such as weddings, where it is seasoned with salt and grilled. It therefore comes as no surprise that its production has steadily increased and is only second to that of yellowtail (Watanabe and Vassallo-Aqius, 2003). Consequently, researchers are continuously trying to seek technology to improve the production system in this species. In Japan, the growth performance of this species has been controlled by the manipulation of food ingredients or by the addition of feeding stimulants and digestibility enhancers (Foscarini, 1988; Fushimi, 2001; Watanabe and Vassallo-Agius, 2003). However, these ingredients are sometimes more costly, which may result in a greater production cost. Although photoperiod manipulation has been practiced in a number of fish species for decades as an economic way to stimulate growth performance, there had been no report on the effect of photoperiod manipulation on the growth performance and stress response in red sea bream until 2005.

This chapter summarizes data from a series of experiments that were devoted to establishing a light regime giving optimal growth for a complete production cycle of red sea bream. Apart from growth performance, effects on the apparent digestibility of nutrients and energy in fish under different photoperiods have also been discussed. It would be premature to propose an artificial photoperiod for rearing fish without careful analysis of how it affects the stress response in fish. Therefore, physiological response in fish under different photoperiods has also been discussed.

PHOTOPERIOD DESIGN

Four different light regimes were established: i) 6 h light:6 h dark (6L:6D), ii) 12 h light:12 h dark (12L:12D), iii) 16 h light:8 h dark (16L:8D), and iv) continuous light (24L:0D). A 12-h cycle (6L:6D) was selected because higher growth performance was observed in other species (Biswas and Takeuchi, 2003). A programmed time controller (Matsushita Electric Works Ltd., Osaka, Japan) was used to maintain the periods of light and dark, including dimming over a 30-minute period. The three tanks in each treatment were illuminated with one 40 W fluorescent tube suspended 45 cm above the water surface. Light intensity was maintained at 1500 lx on the water surface during light phase and 0 lx during dark phase throughout the experiment. Each set of three replicates was isolated from the other set and from stray light by shielding with an opaque partition. Feeding times were manipulated according to the design of photoperiods; feeding schedules are given in Figure 1.

Many farmers from Norway and Scotland nowadays use continuous lighting during the autumn or winter to improve growth. The main concern is how the different photoperiods could be controlled in outdoor farms. Some possible ways are discussed by Bromage et al. (2001). Generally, this involved the installation of light-proof covers over farms and the provision of artificial lighting controlled by automatic time clocks.

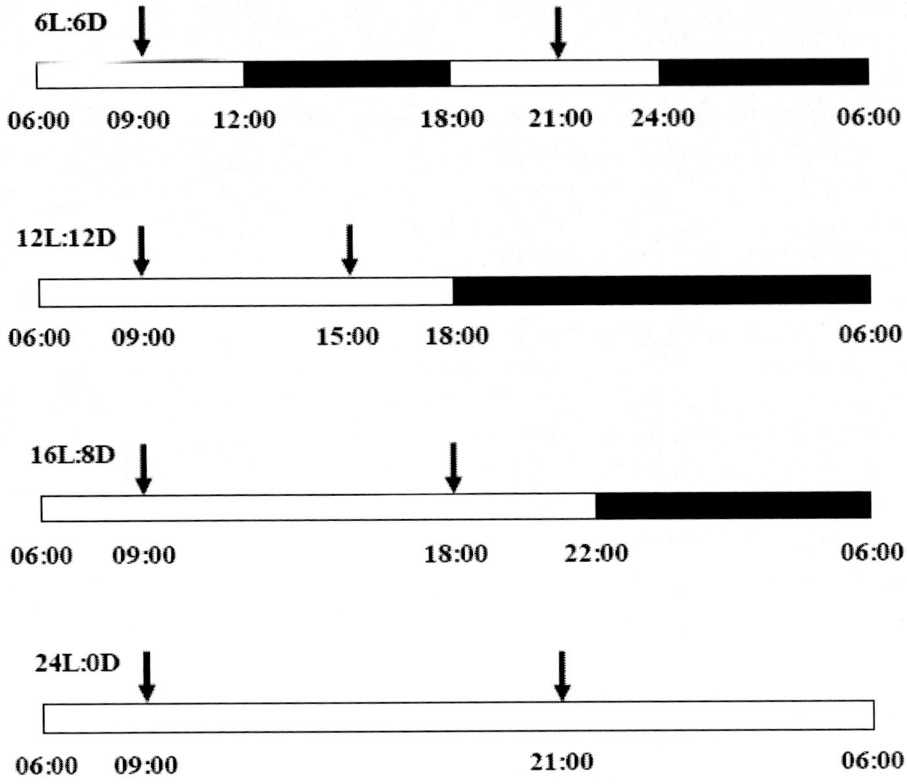

Figure 1. Feeding schedule for fish exposed to different photoperiods.

The heavy-duty polythene or butyl linings over a simple metal, plastic pipe or wooden framework provide a cheap and effective method of blacking out the farm areas. Total blackout is recommended, inasmuch as artificial light is different from ambient. Tungsten or fluorescent sources of illumination can be used, preferably with a spectrum as close as possible to that of natural light. They should provide intensities of at least 100 lux at the water's surface in all areas of the enclosures. Intensities less than 20 lux should be avoided, as they may lead to inconsistent results. To reduce stress due to abrupt changes in light during the switch-on and switch-off periods, fluorescent lights may require installation of a second system of less bright lights, which are switched on shortly before and after the main system and, hence, provide the necessary twilight periods.

EXPERIMENTAL DESIGN

For all purposes, red sea bream of the selected strain (Murata et al., 1996) were obtained from the Fish Nursery Center of Kinki University, Uragami, Japan. Growth experiments were conducted at different developmental stages: i) 1 to 30 g and ii) 20 to 100 g. Fish were reared for eight weeks in each of two experiments under the photoperiods mentioned earlier. The tanks were supplied with filtered seawater and the rearing tank was aerated to maintain the oxygen level near 100% saturation. The water flow was 4 L min^{-1} and the temperature was maintained at 21±1°C throughout all experiments. Fish were fed to apparent satiation in all experiments with a commercial diet (protein 47.7%, lipid 10.7%, Marubeni Nisshin Feed Co. Ltd., Tokyo, Japan) twice daily during the light phase according to the feeding schedules given in Figure 1. Individual body length and weight were taken to investigate the growth performances in terms of specific growth rate (SGR), feed conversion efficiency (FCE), retention efficiencies of protein and energy, and apparent digestibility of nutrients and energy. For apparent digestibility coefficient (ADC) of nutrients and energy, fecal samples were collected when the fish were reared from 20 to 100g under different photoperiods. Chromic oxide (Cr_2O_3) was used at 0.5% as an inert marker to investigate the digestibility.

Blood samples were taken during all experiments to analyze different stress indicators. In addition, an acute handling and confinement stress was carried out according to Biswas et al. (2004) to investigate the stress-induced levels of different stress indicators in fish reared from 20 to 100g. Briefly, this consisted of netting the fish from the rearing tanks and transferring them to a small confinement tank (30 L) which was half filled with water and supplied with constant flow (2 L/min) of water for 1 h. Fish were then sampled at 0.5, 1, 4, 24, 48 and 72 h post-stress. Initial samples were also taken to measure the base levels of different stress indicators.

Samples (diets, fish and feces) were analyzed for dry matter, protein and ash using a standard method (AOAC, 1995). Crude lipid was determined using a chloroform-methanol extraction procedure (Folch et al., 1957). Plasma cortisol levels were determined by ELISA according to the manufacturer's instructions (Oxford Biomedical Research, Inc., U.S.A.). Plasma levels of glucose and total cholesterol were measured using analytical kits (Wako Pure Chemical Ind. Ltd., Osaka, Japan). In all cases, data were statistically analyzed by SPSS program for Windows (v. 10.0) using Tukey's test of multiple comparison with a 95% significance level.

Growth and Digestibility

When fish reared from 1 to 30 g, fish exposed to 24L:0D photoperiod showed significantly higher total weight gain and SGR than those exposed to 16L:8D, 6L:6D and 12L:12D photoperiods (Table 1). Food intake differed significantly among the treatments and significantly higher value was in fish in the 24L:0D group than that of 6L:6D, 16L:8D and 12L:12D photoperiods (Table 1, $P<0.05$). There was no significant difference in the FCE between fish exposed to 24L:0D and 16L:8D photoperiods, but the FCE in both photoperiods was significantly higher than that of 6L:6D and 12L:12D photoperiods (Table 1, $P<0.05$).

Fish exposed to 24L:0D photoperiod also showed the highest total weight gain and SGR followed by 16L:8D, 6L:6D and 12L:12D photoperiods when reared from 20 to 100 g (Table 1, P<0.05). Food intake differed significantly among the treatments and the highest value was in fish in the 24L:0D group followed by 16L:8D, 6L:6D and 12L:12D photoperiods (Table 1). There was no significant difference in the FCE between fish exposed to 24L:0D (97.7±0.5%) and 16L:8D (98.1±0.7%) photoperiods, but the FCE in both treatments was significantly higher than in the 12L:12D photoperiod (Table 1, P<0.05). Fish reared under 24L:0D photoperiod showed significantly higher condition factors than those reared under other photoperiods (Table 1, P<0.05). The apparent digestibility coefficients (ADC) of protein, lipid and energy of fish exposed to different photoperiods are given in Figure 2.

Table 1. Growth performance in red sea bream exposed to different photoperiods. Each value is a mean±SD derived from 3 samples each of 40 fish for body size 1-30g and 32 fish body size 20-100g for each treatment

Parameters	Photoperiods			
	6L:6D	12L:12D	16L:8D	24L:0D
Body size: 1-30g				
Initial body weight (g)	1.3±0.2	1.4±0.2	1.4±0.2	1.4±0.2
Final body weight (g)	22.9±0.6[c]	22.5±1.0[c]	26.1±0.2[b]	29.6±0.4[a]
Weight gain (g)	898.7±44.1[c]	844.0±42.3[c]	986.7±6.1[b]	1,129.3±15.1[a]
SGR (%)[1]	5.1±0.1[b]	5.0±0.1[c]	5.2±0.0[b]	5.5±0.1[a]
Feed intake (g)	771.3±8.4[b]	664.5±21.9[d]	706.61±5.6[c]	849.7±11.5[a]
FCE (%)[2]	120.1±4.0[c]	127.0±2.3[bc]	139.7±2.5[a]	132.9±1.0[ab]
CF[3]	3.6±0.1	3.7±0.1	3.7±0.1	3.6±0.1
Survival rate (%)	97.5±0.0	96.7±1.4	97.5±0.0	98.3±1.4
Body size: 20-100g				
Initial body weight (g)	19.6±0.4	19.5±0.1	19.0±0.6	19.4±0.2
Final body weight (g)	74.1±1.7[c]	68.6±1.2[d]	79.8±0.8[b]	90.3±2.0[a]
Weight gain (g)	1692.4±38.5[c]	1571.9±31.1[d]	1947.5±36.9[b]	2266.8±29.2[a]
SGR (%)[1]	2.34±0.02[c]	2.25±0.01[d]	2.57±0.05[b]	2.74±0.04[a]
Feed intake (g)	1745.8±26.2[c]	1645.6±32.0[c]	1984.6±25.7[b]	2320.6±35.7[a]
FCE (%)[2]	97.0±0.6[a,b]	95.5±0.6[b]	98.1±0.7[a]	97.7±0.5[a]
CF[3]	4.09±0.02[b]	4.07±0.05[b]	4.12±0.03[b]	4.27±0.03[a]
Survival rate (%)	100	100	100	100

Values in a row with different letters are significantly different (P <0.05).
[1] SGR (%) = 100 × (ln final weight - ln initial weight) / rearing period (days)
[2] FCE (%) = 100 × (wet weight gain / dry fed intake)
[3] CF = 100 × (W L-3), where W = body weight (g) and L = standard body length (cm)
(References: Biswas et al., 2005, 2006b)

There was no significant difference in the ADC of lipid and energy between 16L:8D and 24L:0D photoperiods (P>0.05), but the ADC of lipid and energy in both treatments was significantly higher than those in 12L:12D photoperiod (P<0.05).

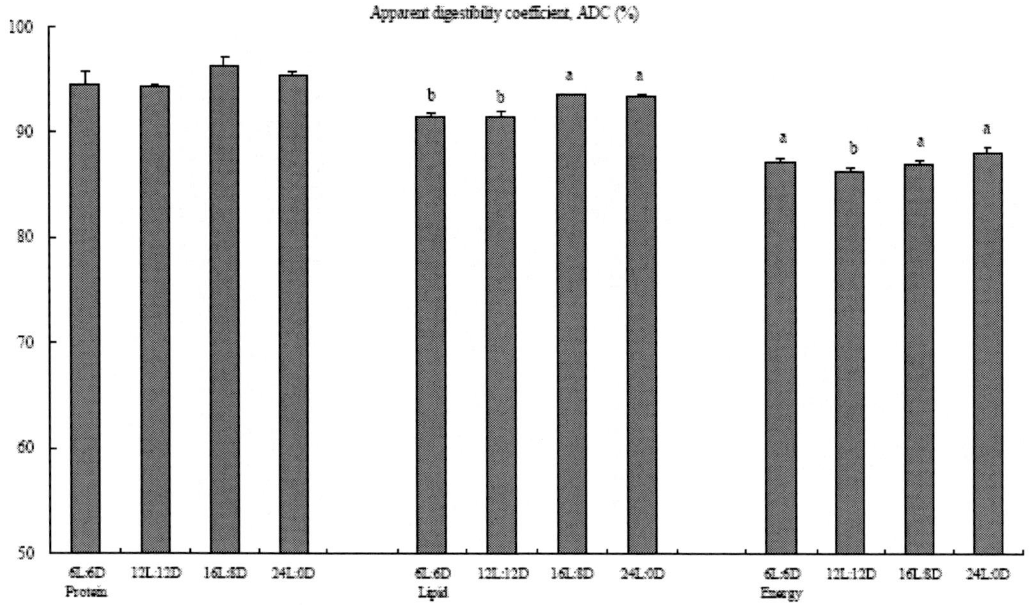

Figure 2. Apparent digestibility coefficient (ADC) of protein, lipid and energy in fish exposed to different photoperiods. Bars with different letters are significantly different (n=2, P<0.05). ADC of nutrients and energy (%) = 100 × [1 – {(dietary Cr_2O_3 / fecal Cr_2O_3) × (fecal nutrient or energy / dietary nutrient or energy)}]. (adapted from Biswas et al., 2005).

Although fish exposed to 16L:8D and 24L:0D photoperiods showed significantly higher whole body crude protein content than those exposed to 12L:12D photoperiod when reared from 20 to 100 g, there were no major variations in other parameters either in fish reared from 1 to 30 g or from 20 to 100 g (Table 2).

The early efforts by Brown (1946) using static daylengths suggested an inverse relation of growth rate with daylength, whereas those of Anderson (1959) and Bjorklund (1958) appeared to show no relationship at all. Most studies indicated that, whatever the response, it took at least six to eight weeks to become manifest (Eisler, 1957). However, in the subsequent years, the high correlation of increasing daylength with growth in the spring was confirmed as the stimulating agent for endocrine activity (Swift 1959, 1960, 1961). One of the most in-depth enquiries was that by Gross et al. (1965) on the green sunfish, *Lepomis cyanellus*, using both fixed and variable photoperiods. From four 6-week experiments conducted during different seasons of the year, a comparison was made on each occasion among different photoperiods and concluded that the greatest weight gain accompanied from longer light period. After that, Hogman (1968) also demonstrated that seasonal changes in growth rate were more closely related to daylength than to changes in partially-controlled water temperature. The application of photoperiod manipulation on growth is greatly practiced to stimulate the growth performance in larval (Tandler and Helps, 1985; Duray and Kohno, 1988; Person-Le Ruyet et al., 1991; Ronzani Cerqueira et al., 1991; Hallaråker et al., 1995a; Hart et al., 1996), juvenile and adult stages (Lundqvist, 1980; Boehlert, 1981; Kadmon et al., 1985; Saunders et al., 1985; Hoar, 1988; Saunders and Harmon, 1988; Thorarensen and Clarke, 1989; Kråkenes et al., 1991; Hansen et al., 1992, 1995; Mäkinen and Ruhonen, 1992; Mason et al., 1992; Boeuf, 1993; Folkvord and Otterå, 1993; Mortensen and Damsgård, 1993; Saunders et al., 1994; Barlow et al., 1995; Forsberg, 1995; Imsland et al., 1995; Sigholt et al., 1995;

Table 2. Proximate composition and retention efficiency in fish exposed to different photoperiods. Each value is a mean±SD derived from 3 samples each of 40 fish for body size 1-30g and 32 fish body size 20-100g for each treatment.

Parameters	Photoperiods			
	6L:6D	12L:12D	16L:8D	24L:0D
Body size: 1-30g				
Moisture (%)	68.7±0.8	67.5±1.7	67.6±1.5	67.4±0.6
Crude protein (%)	16.1±0.2	16.3±0.4	16.4±0.4	16.7±0.5
Crude lipid (%)	10.6±1.7	11.6±1.1	11.1±1.3	10.5±0.4
Crude Ash (%)	3.9±0.1	4.0±0.5	4.1±0.3	4.1±0.6
PRE (%)[1]	46.9±0.6	46.6±1.0	47.2±0.7	47.5±0.5
ERE (%)[2]	45.7±1.8	46.9±2.9	46.6±2.3	45.8±2.5
Body size: 20-100g				
Moisture (%)	68.10.8	66.91.7	67.61.5	67.30.6
Crude protein (%)	16.00.1^{a,b}	15.80.2^{b}	16.40.3^{a}	16.30.3^{a}
Crude lipid (%)	11.31.7	12.31.1	11.51.3	11.70.4
Crude Ash (%)	4.20.1	4.10.5	4.30.3	4.20.6
PRE (%)[1]	45.60.3	45.60.9	45.70.4	45.40.3
ERE (%)[2]	46.52.3	47.03.3	49.53.0	49.12.8

Values in a row with different letters are significantly different ($P < 0.05$).

[1] Protein retention efficiency (PRE, %) = 100 × (final whole body protein - initial whole body protein)/total protein intake

[2] Energy retention efficiency (ERE, %) = 100 × ($GE_1W_1 - GE_0W_0$)/(GE_f × total feed intake), where GE_0 and GE_1 are the initial and final whole body energy concentration (J/g), W_0 and W_1 are the initial and final fish weights, and GE_f is the gross energy of feed (J/g) (references: Biswas et al., 2005, 2006b)

Silva-García, 1996; Karlsen et al., 2000; Simensen et al., 2000; Biswas and Takeuchi, 2003; Petit et al., 2003; Biswas et al., 2005) in different fish species.

From the results from this series of experiments, the positive effect of long (16L:8D) and continuous (24L:0D) photoperiods on the growth performance of red sea bream in both size groups is comparable with a closely-related species gilthead sea bream, *Sparus aurata*, where long and constant photoperiods of both 16:00 hours of light (16L:8D) and continuous light (24L:0D) have improved growth (Kissil et al., 2001; Ginés et al., 2004). The finding that food intake was greater under continuous (24L:0D) photoperiods in both size groups parallels the findings in other species such as gilthead sea bream *Sparus aurata* L. (Kissil et al., 2001; Ginés et al., 2004), largemouth bass *Micropterus salmoides* (Petit et al., 2003), Atlantic salmon *Salmo salar* (Thorpe et al., 1988) and halibut *Hippoglossus hippoglossus* (Jonassen et al., 2000). The higher food intake may be due to diurnal fishes being more active under continuous photoperiods, and having greater foraging activity when food is delivered, or may be related to the hormonal stimulation of appetite under continuous photoperiod; for example, growth hormone, known to have a positive effect on appetite, increased with increasing day length during smoltification in salmon (Johnsson and Björnsson, 1994; McCormick et al., 1995). The difference in food intake and FCE among the treatments may also be due to the differences in the time of day when the food was delivered. The higher food intake and FCE in fish exposed to 16L:8D and 24L:0D photoperiods may be because the feeding strategy in fish exposed to these photoperiods reflected most closely the times of maximum appetite. The time of feeding has also been reported to affect food intake and growth performance in Indian catfish (Sundararaj et al., 1982), goldfish (Noeske-Hallin and Spieler, 1984), channel catfish (Noeske-Hallin et al., 1985), sea bream (Divanach et al., 1986; Ginés et al., 2004), rainbow trout (Boujard et al., 1995) and tilapia (Biswas and Takeuchi, 2003). Azzaydi et al. (1999) also demonstrated that improved growth and feed efficiency are achievable if the quantity of food supplied is modulated in accordance with the time of maximum appetite. The finding that higher growth accompanied with both higher food intake and FCE under long and continuous photoperiods parallels the findings in other species, such as largemouth bass *Micropterus salmoides* (Petit et al., 2003), haddock *Melanogrammus aeglefinus* (Trippel and Neil, 2003) and gilthead seabream *Sparus aurata* L. (Kissil et al., 2001; Ginés et al., 2004). As noted in Boeuf and LeBail's (1999) review, Gross et al. (1965) were the first to specify that growth might be influenced by light through better food conversion efficiency and not just stimulated food intake. The longer time interval between feeding times in fish exposed to 16L:8D and 24L:0D photoperiods may allow a slower and more efficient digestive process, which might improve digestion and retention efficiency in both treatments. This is supported by the ADC value of nutrients and energy in fish reared from 20 to 100 g. The significantly higher ADC of lipid and energy, and superior retention efficiency of energy under 16L:8D and 24L:0D photoperiods, implied that the meals offered to fish exposed to these photoperiods might be ingested, digested and assimilated more efficiently than meals offered to fish exposed to 12L:12D photoperiod. Therefore, it has been assumed that the feeding strategy and intervals between feeding times might cause a significantly higher FCE in fish exposed to 16L:8D and 24L:0D photoperiods. However, the lower FCE in the 6L:6D photoperiod, in spite of higher food intake and a longer time interval between feeding times, may be attributed to the existence of a dark period which reduced the digestive and absorptive performance because of the reduced activity of fish during this period. However, this area needs to be clarified.

When fish are reared from 20 to 100 g, the significantly higher ADC of lipid and inferior whole body lipid content in fish exposed to 16L:8D and 24L:0D photoperiods than those exposed to 12L:12D photoperiod indicated that fish exposed to the former photoperiods may have used lipids to mobilize energy to compensate for a greater energy demand for growth and an elevated metabolic rate. Cook et al. (2000) reported similar results in transgenic salmon. Although there was no significant difference in protein digestibility among the treatments, fish exposed to 16L:8D and 24L:0D photoperiods showed significantly higher whole body protein content than those exposed to 12L:12D photoperiod. This may be attributed to the protein being deposited in the body to enhance growth because the major share of growth in terms of body weight increase consists of protein growth (Sumpter, 1992).

Although the photoperiod manipulation could be used to stimulate the growth performance in both size groups in red sea bream, its effect on growth performance is not as conclusive as for other species. In Atlantic salmon *Salmo salar*, Brännäs (1987) failed to demonstrate an influence of PPD during the yolk sac phase. With the same species, Berg et al. (1992) obtained a good relationship between the duration of lighting and growth after first feeding: growth decreased with reduced daylength. In halibut *Hippoglossus hippoglossus,* Hallaråker et al. (1995b) reared fish from 5 to 20 g exposed to changing light regimes (7–12L and 12–18L) with no effect on growth rates, whereas, Jonassen et al. (2000) and Simensen et al. (2000) observed that continuous light could be used to improve growth in juvenile halibut. In turbot *Scophthalmus maximus* (maintained at natural, constant 16L:8D and 24L:0D), continuous light slightly enhanced the growth rate above that of other regimes, after at least three months of exposure at 10 and 16°C, but not throughout the six-month experiment (Imsland et al., 1995). However, Imsland et al. (1995, 1997) observed a better long-term growth in turbot exposed to extended daylength during the first winter.

Stress Response

In fish reared from 1 to 30 g, there were no significant differences in the hematocrit levels and plasma levels of cortisol and glucose among the treatments at the end of experiment (Table 3).

In the fish reared from 20 to 100 g, fish exposed to 6L:6D, 16L:8D and 24L:0D photoperiods had significantly higher cortisol levels than fish with the 12L:12D photoperiod during the first week ($P<0.05$, Table 3). However, there were no significant differences in other parameters among the treatments during the first week. Although higher plasma levels of cortisol were observed in fish exposed to 6L:6D, 16L:8D and 24L:0D photoperiods than to the 12L:12D photoperiod, the differences were not statistically significant at the end of experiment (Table 3). Similarly, there were no significant differences in other parameters among the treatments at the end of experiment (Table 3).

An acute handling and confinement stress was carried out to identify the stress-induced levels of different stress indicators in fish reared from 20 to 100 g. Plasma cortisol levels in fish exposed to acute handling and confinement stress rose rapidly to maximum value of 190.1 ng ml^{-1} within 30 minutes and returned to levels indistinguishable from those of the control, undisturbed fish within 24 h of the initial disturbance (Figure 3). Similarly, glucose levels also increased following handling, peaking at 109.5 mg dl^{-1} 4 h after acute handling and confinement stress and returned to levels indistinguishable from those of the control, undisturbed fish within 24 h of the initial disturbance (Figure 3). However, plasma levels of cholesterol did not recover within 72 h of the initial disturbance (Figure 3).

Table 4 compares different stress indicators observed in both fish groups exposed to different photoperiods with those of stress-induced levels from acute handling and confinement stress. In both fish groups, the levels of different stress indicators in fish exposed to different photoperiods were far from the stress-induced levels.

Table 3.The mean ± S.D. (n=3) of different stress indicators for fish exposed to different photoperiod regimes. Means with different letters are significantly different (P<0.05).

Parameters	Initial	1st week				8th week			
	levels	6L:6D	12L:12D	16L:8D	24L:0D	6L:6D	12L:12D	16L:8D	24L:0D
Body size: 1-30g									
Cortisol (ng/ml)	11.2±0.7	13.7±2.1	12.6±0.9	12.2±1.2	13.1±1.7	12.2±1.2	11.7±0.5	10.9±1.5	12.5±1.1
Glucose (mg/dl)	75.1±3.2	80.5±5.2	78.6±3.1	81.3±4.4	76.9±3.7	77.2±4.2	75.5±3.9	76.8±3.7	75.9±2.9
Total cholesterol (mg/dl)	250.8±19.2	242.3±20.1	255.8±16.9	261.3±30.8	249.2±19.8	245.1±26.4	261.3±30.9	241.8±29.7	252.9±17.6
Hematocrit (%)	35.0±0.7	36.1±1.2	35.8±0.6	34.9±1.0	35.2±0.9	35.8±0.9	34.9±0.6	34.5±1.0	35.1±1.1
Body size: 20-100g									
Cortisol (ng/ml)	7.7±0.9	10.9±1.4^a	7.4±0.9^b	10.2±0.9^a	10.1±1.1^a	7.7±1.2	7.1±0.5	7.4±0.8	7.7±1.1
Glucose (mg/dl)	74.7±2.8	68.7±4.1	71.1±3.1	72.9±2.8	74.2±3.9	76.4±4.4	73.6±4.1	70.8±3.3	72.9±3.8
Total cholesterol (mg/dl)	241.6±17.9	236.0±25.2	247.2±19.5	240.5±25.4	242.4±18.6	246.2±31.5	244.8±18.7	238.0±24.1	248.5±32.7
Hematocrit (%)	36.9±0.7	36.2±1.8	36.7±1.1	37.2±1.3	36.9±0.6	37.3±0.9	37.0±0.6	36.9±1.0	37.2±1.1

(References: Biswas et al., 2006a,b)

The concept of stress has been with us for a long time. In the English language of the 1600s, stress meant ‘hardship, straits, adversity, and affliction’. In the 1700s and 1800s, stress first came to mean a force exerted on a person, and later the response or resistance of the person to such forces; this was the meaning adopted by physical and engineering sciences (Reite, 1985). As eloquently discussed by Weiner (1992), the underlying idea for the concept of stress may be attributed to Darwin (1859). He pointed out in his elaboration of the concept of natural selection that the environment was in constant change, challenging and stressful to the organism. Only those organisms able to cope with such natural challenges were capable of surviving to reproduce. Perhaps one of the earliest modern attempts to articulate the concept of stress in a physiological context was that by Cannon (1935), who described the normal resting state of the organism as ‘homeostasis’. Threats to homeostasis involved the autonomic nervous system response to a variety of stimuli, or what Cannon referred to as stress. Selye (1936) proposed the first major encompassing theory of stress and its effects on the organism. He noticed a non-specificity and similarity of responses in his test animals regardless of the disturbance, which he considered the basis of a general response syndrome exclusive of specific responses to individual stimuli.

Researchers usually measure plasma cortisol to assess stress in fish because of its responsiveness to acute stressors, its relative ease of measurement, and its functional significance in physiological processes affecting fish health. Although plasma cortisol levels are widely used as a reliable indicator of stress response activation (Barton and Iwama, 1991), this chapter uses a number of primary (plasma levels of cortisol) and secondary (plasma levels of glucose and total cholesterol) features that have previously been used as stress indicators in fish (Deitinger and McCauley, 1990; Goede and Barton, 1990; Wedemeyer et al., 1990) to investigate the physiological response in fish exposed to different photoperiods.

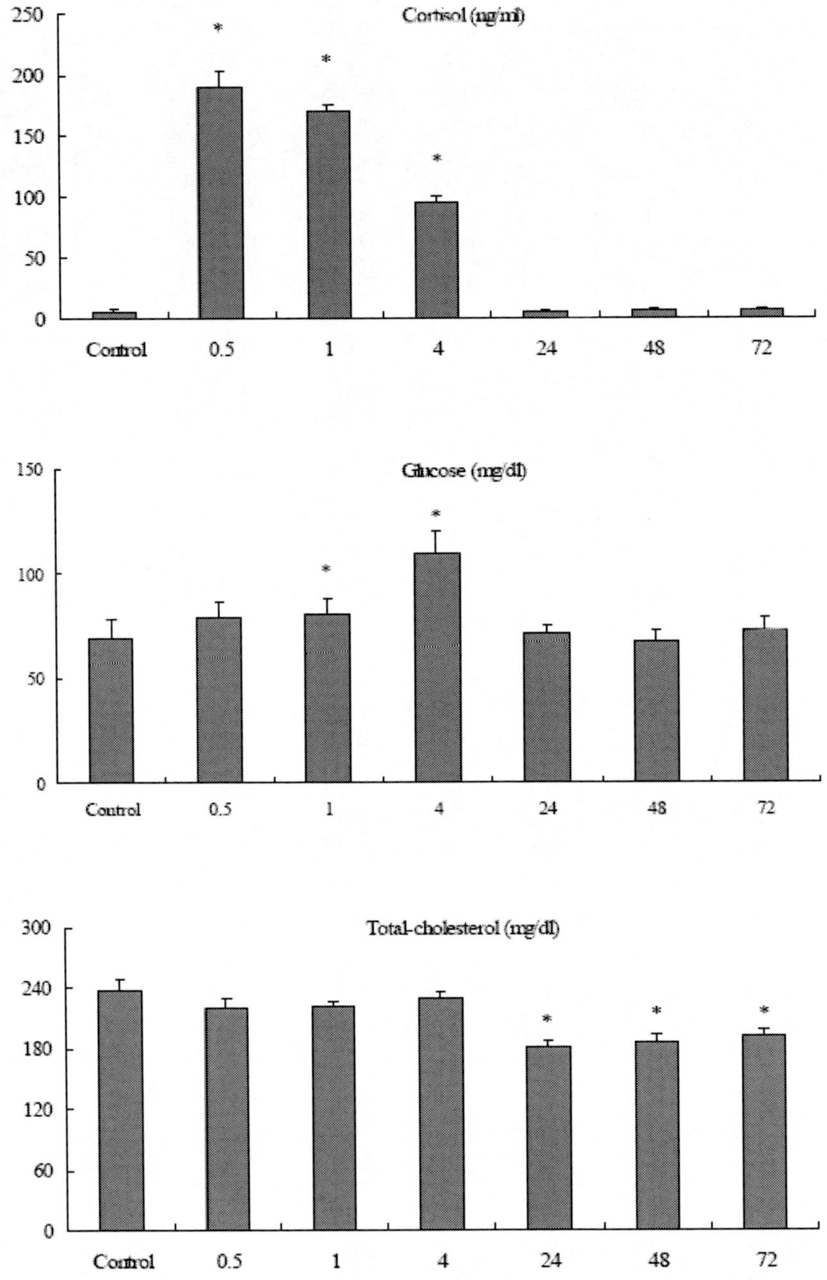

Figure 3. Plasma levels of cortisol, glucose and total cholesterol in *Pagrus major* during and after acute stress. Values given are the means±S.D. (n=5). *P<0.05 compared with control zero time values. (Reference: Biswas et al., 2006a).

The major problem in using the cortisol level alone is the high sensitivity and rapid response of the hypothalamic-pituitary-interrenal (HPI) axis to a wide variety of stressful stimuli (Foo and Lam, 1993b).

Table 4. Comparison of different stress indicators observed in fish exposed to different photoperiods with those stress-induced levels

Photoperiod	Cortisol (ng/ml)	Glucose (mg/dl)	T-cholesterol (mg/dl)
Stress-induced levels	190.1±12.7[b] ↑	109.5±10.7[b] ↑	180.5±6.9[b] ↓
Body size: 1-30g			
6L:6D	12.2±1.2[a]	77.2±4.2[a]	245.1±26.4[a]
12L:12D	11.7±0.5[a]	75.5±3.9[a]	261.3±30.9[a]
16L:8D	10.9±1.5[a]	76.8±3.7[a]	241.8±29.7[a]
24L:0D	12.5±1.1[a]	75.9±2.9[a]	252.9±17.6[a]
Body size: 20-100g			
6L:6D	8.6±1.9[a]	73.4±4.1[a]	240.9±5.1[a]
12L:12D	8.1±0.5[a]	73.6±4.1[a]	244.8±18.7[a]
16L:8D	8.4±1.6[a]	72.0±1.1[a]	241.3±3.8[a]
24L:0D	8.6±1.2[a]	73.6±0.7[a]	244.6±3.4[a]

(References: Biswas et al., 2006a,b).

On the other hand, fish that appear to be most stressed using one indicator may not necessarily be stressed using another (Barton and Iwama, 1991). It is therefore important to use different parameters to investigate the stress responses in fish.

The results demonstrated that photoperiod manipulation did not cause a significant acute stress response in red sea bream—as mentioned earlier, the levels of different parameters used as stress indicators in fish exposed to different photoperiods were far from the acute stress-induced levels. The finding that acute stress increased the plasma levels of cortisol and glucose parallels the findings in other species (Demers and Bayne, 1997; Barcellos et al., 1999; Lohner et al., 2001; Biswas et al., 2004; Geslin and Auperin, 2004).

On the other hand, no apparent signs of chronic stress were observed in fish exposed to different photoperiods. Chronic stress generally results in a higher elevation of cortisol and glucose levels in fish (Barton and Iwama, 1991; Barton, 1997; Van Weerd and Komen, 1998; Barcellos et al., 1999; Pottinger et al., 1999) than the levels observed in fish exposed to different photoperiods in this series of studies. Recently, Leonardi and Klempau (2003) observed that the chronic stress from an artificial photoperiod resulted in a pronounced and progressive increase in cortisol levels to mean value of 120 ng ml^{-1} until after the time point where normal photoperiod was resumed. A steady but slow decrease was observed over the following two months, but fish exposed to the artificial photoperiod regime showed levels of more than 100 ng ml^{-1}, whereas the controls remained at about 20 ng ml^{-1}. However, the mean plasma cortisol levels in fish exposed to different photoperiods during the courses of the studies were within the range of basal or resting levels reported for *Pagrus major* (Chopin et al., 1996). Chronic stress was reported to reduce the growth rate of different fish species because of the reallocation of energy towards activities aimed at restoring homeostasis, such as respiration, locomotion, hydromineral regulation and tissue repair (Barton and Iwama, 1991; Barton, 1997; Wendelaar Bonga, 1997; Van Weerd and Komen, 1998; Barcellos et al., 1999). Chronic stress is also reported to suppress food intake (Bernier and Peter, 2001; Wang et al., 2004); however, fish exposed to different photoperiods showed neither a decreased growth rate nor reduced food intake. In sunfish populations, Lohner et al. (2001) suggested that chronic stress decreased cholesterol levels to lower than the normal range; however, there was no significant change in plasma cholesterol levels among the treatments in this study. Therefore, the photoperiod manipulation did not cause a significant chronic stress response in fish.

It has been argued that cortisol itself is closely linked to many adverse consequences of stress including effects on growth, reproduction and the immune system (Pickering et al., 1987; Pickering and Pottinger, 1989; Barton and Iwama, 1991; Barton, 1997; Van Weerd and Komen, 1998). However, remarkably elevated levels of cortisol are required to cause these adverse consequences in fish (Pickering et al., 1987; Pickering and Pottinger, 1989; Foo and Lam, 1993a,b). Therefore, it can be concluded that cortisol levels observed in fish under different photoperiods were not sufficient to cause adverse consequences, since the levels of various parameters remained near to levels indistinguishable from those of control, undisturbed fish.

CONCLUSION

As in many other fish species, these results demonstrated that the growth performance of juvenile red sea bream reared from 1 to 100 g can be stimulated significantly by using a continuous (24L:0D) photoperiod. Improved appetite, greater food intake and higher feed conversion efficiency, higher digestibility and superior retention efficiency are factors reportedly responsible for the faster growth rate under the continuous photoperiod. In addition, the photoperiod manipulation used in this series of studies did not cause a significant stress response, and can therefore be used to stimulate the growth performance of juvenile red sea bream *P. major*. These results together help to establish a light regime giving optimal fish growth for a complete production cycle of red sea bream *Pagrus major*. It is assumed that more attention will be given to on-land and outdoor aquaculture in the near future. The photoperiod manipulation will definitely be the option of choice to yield higher output from on-land and outdoor farms.

ACKNOWLEDGEMENTS

This study was financially supported by the 21st Century COE program of the Ministry of Education, Culture, Sport, Science and Technology, Japan. The expenses were defrayed in part by a grant from Kansai Electric Power Corporation, Japan.

REFERENCES

Anderson RO. The influence of season and temperature on growth of the bluegill, *Lepomis machrochirus* (Rafinesque). Ph.D. Thesis, Univ. of Michigan, Ann Arbor. 1959.

AOAC, 1995. Official Methods of Analysis of AOAC International. Vol I. Agricultural Chemicals; Contaminants, Drugs, 16th edition. AOAC International, Arlington, VA, USA; 1995.

Audet, C; FitzGerald, GJ; Guderley, H. Photoperiod effects on plasma cortisol levels in *Gasterosteus aculeatus*. *Gen. Comp. Endocrinol.* 1986, 61, 76-81.

Azzaydi, M; Martínez, FJ; Zamora, S; Sánchez-Vázquez, FJ; Madrid, JA. Effect of meal size modulation on growth performance and feeding rhythms in European sea bass (*Dicentrarchus labrax,* L.). *Aquaculture* 1999, 170, 253-266.

Barlow, CG; Pearce, MG; Rodgers, LJ; Clayton, P. Effects of photoperiod on growth, survival, and feeding periodicity of larval and juvenile barramundi, *Lates calcarifer* (Bloch). *Aquaculture* 1995, 138, 159-168.

Bartness TJ; Wade GN. Photoperiod control of body weight and energy metabolism in Syrian hamster (*Mesocricetus auratus*): Role of pineal gland, melatonin, gonads, and diet. *Endocrinol.* 1984, 114, 492-498.

Bartness TJ; Morley JE; Levine AS. Effect of food deprivation and metabolic fuel utilization on the photoperiodic control of food intake in Siberian hamster. *Physiol. Behav.* 1995, 57, 61-68.

Barton, BA 1997. Stress in finfish: past, present and future-a historical perspective. In: Iwama, GK; Pickering, AD; Sumpter, JP; Schreck, CB; editors. *Fish Stress and Health in Aquaculture*. Society for Experimental Biology Seminar Series 62; Cambridge University Press; Cambridge; U. K.; 1997; pp. 1-33.

Barton, BA; Iwama, GK. Physiological changes in fish from stress in aquaculture with emphasis on the response and effects of corticosteroids. *Ann. Rev. Fish Dis.* 1991, 1, 3-26.

Basco, PS; Rashotte, ME; Stephan, FK. Photoperiod duration and energy balance in the pigeon. *Physiol. Behav.* 1996, 60, 151-159.

Barcellos, LJG; Nicolaiewsky, S; de Souza, SMG; Lulhier, F. Plasmatic levels of cortisol in the response to acute stress in Nile tilapia *Oreochromis niloticus* (L.), previously exposed to chronic stress. *Aquacult. Res.* 1999, 30, 437-444.

Berg, A; Hansen, T; Stefansson, S. First feeding of Atlantic salmon (*Salmo salar* L.) under different photoperiods. *J. Appl. Ichthyol.* 1992, 8, 251-256.

Bernier, NJ; Peter, RE. The hypothalamic-pituitary-interrenal axis and the control of food intake in teleost fish. *Comp. Biochem. Physiol. Part B Biochem. Mol. Biol.* 2001, 129, 639-644.

Biswas, AK; Takeuchi, T. Effects of photoperiod and feeding interval on food intake and growth rate of Nile tilapia *Oreochromis niloticus* L. *Fish. Sci.* 2003, 69, 1010-1016.

Biswas, AK; Maita, M; Yoshizaki, G; Takeuchi, T. Physiological responses in Nile tilapia exposed to different photoperiod regimes. *J. Fish Biol.* 2004, 65, 811-821.

Biswas, AK; Seoka, M; Inoue, Y; Takii, K; Kumai, H. Photoperiod influences the growth, food intake, feed efficiency and digestibility of red sea bream (*Pagrus major*). *Aquaculture* 2005, 250, 666-673.

Biswas, AK; Seoka, M; Takii, K; Maita, M; Kumai, H. Stress response of red sea bream *Pagrus major* to acute handling and chronic photoperiod manipulation. *Aquaculture* 2006a, 252, 566-572.

Biswas, AK; Seoka, M; Tanaka, Y; Takii, K; Kumai, H. Effect of photoperiod manipulation on the growth performance and stress response of juvenile red sea bream (*Pagrus major*). *Aquaculture* 2006b, 258, 350-356.

Bjorklund, RG. The biological function of the thyroid and the effect of length of day on growth and maturation of goldfish, *Carassius auratus* Linn. Ph.D. Thesis, Univ. of Michigan, Ann Arbor. 1958.

Boeuf, G. Salmonid smolting: a preadaptation to the oceanic environment. In: Rankin, JC; Jensen, FB; editors. *Fish Ecophysilogy*, Chapman and Hall, London. 1993, pp. 105-135.

Boehlert, GW. The effects of photoperiod and temperature on laboratory growth of juvenile *Sebastes diploproa* and a comparison with growth in the field. *Fish. Bull.* 1981, 79, 789-794.

Boeuf, G; Le Bail, PY. Does light have an influence on fish growth? *Aquaculture* 1999, 177, 129-152.

Boon, P; Visser, H; Daan, S. Effect of photoperiod on body mass, and daily energy intake and energy expenditure in young rats. *Physiol. Behav*. 1997, 62, 913-919.

Boon, P; Visser, GH; Daan, S. Effect of photoperiod on body weight gain, and daily energy intake and energy expenditure in Japanese quail (*Coturnix c. japonica*). *Physiol. Behav*. 2000, 70, 249-260.

Boujard, T; Gelineau, A; Corraze, G. Time of a single daily meal influences growth performance in rainbow trout (*Oncorhynchus mykiss*). *Aquacult. Res*. 1995, 26, 341-349.

Brännäs, E. Influence of photoperiod and temperature on hatching and emergence of Baltic salmon (*Salmo salar* L.). *Can. J. Zool.* 1987, 65, 1503-1508.

Bromage, NR; Porter, M; Randall, CF. The environmental regulation of maturation in farmed finfish with special reference to the role of photoperiod and melatonin. *Aquaculture* 2001, 197; 63-98.

Bronson, FH; Kerbeshian, MC. Reactions of reproductively photoresponsive versus unresponsive meadow voles to simulated winter conditions. *Can. J. Zool*. 1995, 73, 1479-1488.

Brown, ME. The growth of brown trout (*Salmo trutta* Linn.). II. Growth of 2-year old trout at a constant temperature of 11.5°C. *J. Exp. Biol*. 1946, 22, 130-144.

Briynklow, RB; Forbes, JM. Effect of extended photoperiod on the growth of sheep. In: Roche, JF; O'Callaghan, DO; editors. *Manipulation of growth in farm animals*. Boston: Martinus Nijhoff Publishers. 1984, 984: 260-273.

Campuzano, A; Campras, T; Vilaplana, J; Canal, MM; Carulla, M; Díez-Noguera, A. Period length of the light-dark cycle influences the growth rate and food intake in mice. *Physiol. Behav*. 1999, 67, 791-797.

Canguilhem, B; Vaultie,r JP; Pevet, P; Coumaros, G; Masson-Pevet, M; Bentz, I. Photoperiodic regulation of body mass, food intake, hibernation, and reproduction in intact and castrated male European hamsters, *Cricetus cricetus*. *J. Comp. Physiol*. 1988, 163A, 549-557.

Cannon, W. Stresses and strains of homeostasis. *Am. J. Med. Sci*. 1935, 189, 1-14.

Charles, RG; Robinson, FE; Hardin, RT; Yu, MW; Feddes, J; Classen, HL. Growth, body composition, and plasma androgen concentration of male broiler chickens subjected to different regimens of photoperiod and light intensity. *Poult. Sci.* 1992, 71, 1595-1605.

Clarke, WC; Shelbourn, JE. Delayed photoperiod produces more uniform growth and greater seawater adaptability in underyearling coho salmon (*Oncorhynchus kisutch*). *Aquaculture* 1986, 56, 287-299.

Clarke, WC; Shelbourn, JE; Ogasawara, T; Hirano, T. Effect of initial daylength on growth, seawater adaptability and plasma growth hormone levels in underyearling coho, chinook, and chum salmon. *Aquaculture* 1989, 82, 51–62.

Chopin, FS; Arimoto, T; Inoue, Y. A comparison of the stress response and mortality of sea bream *Pagrus major* captured by hook and line and trammel net. *Fish. Res.* 28, 1996, 277-289.

Cook, JT; McNiven, MA; Richardson, GF; Sutterlin, AM. Growth rate, body composition and feed digestibility/conversion of growth-enhanced transgenic Atlantic salmon (*Salmo salar*). *Aquaculture* 2000, 188, 15-32.

Darwin, CR. On the Origin of Species by Means of Natural Selection, or, the Preservation of Favoured Races in the Struggle for Life. Murray, London. 1859.

Deitinger, TL; McCauley, RW. Whole animal physiological processes for the assessment of stress in fishes. *J. Great Lakes Res.* 1990, 16, 542-575.

Demers, NE; Bayne, CJ. The immediate effects of stress on hormones and plasma lysozyme in rainbow trout. *Dev. Comp. Immunol.* 1997, 21, 363-373.

Divanach, P; Kentouri, M; Dewavrin, G. The weaning and the development of biological performance of extensively reared sea bream, *Sparus auratus*, fry after replacing continuous feeders by self-feeding distributors. *Aquaculture* 1986, 52, 21-29.

Donaldson, EM. The pituitary-interrenal axis as an indicator of stress in fish. In: Pickering, AD, editor. *Stress and fish,* London: Academic Press, 1981, pp. 11-41.

Duray, M; Kohno, H. Effects of continuous lighting on growth and survival of first-feeding larval rabbitfish, *Siganus guttatus*. *Aquaculture* 1988, 72, 73-79.

Eggleston, D. Sparidae. In Fischer, W; Whitehead, PJP; editors. *FAO species identification sheets for fishery purposes*. Eastern Indian Ocean (Fishing Area 57) and Western Central `acific (Fishing Area 71), Volume 4. FAO, Rome, 1974.

Eisler, T. The influence of light on the early growth of chinook salmon. *Growth* 1957, 21, 197-203.

Eriksson, LO; Lundqvist, H. Circannual rhythms and photoperiod regulation of growth and smolting in Baltic salmon (*Salmo salar* L.). *Aquaculture* 1982, 28, 113-121.

Folch, J; Lees, M; Sloane, GH. Simple method for isolation and purification of total lipids from animal tissues. *J. Biol. Chem.* 1957, 226, 497-507.

Folkvord, A; Otterå, H. Effects of initial size distribution, day length, and feeding frequency on growth, survival, and cannibalism in juvenile Atlantic cod (*Gadus morhua* L.). *Aquaculture* 1993, 114, 243-260.

Foo, TW; Lam, TJ. Retardation of ovarian growth and depression of serum steroid levels in the tilapia *Oreochromis mossambicus*, by cortisol implantation. *Aquaculture* 1993a, 115, 133-143.

Foo, JTW; Lam, TJ. Serum cortisol response to handling stress and the effect of cortisol implantation on testosterone in the tilapia, *Oreochromis mossambicus*. *Aquaculture* 1993b, 115, 145-158.

Forsberg, OI. Empirical investigations on growth of post-smolt Atlantic salmon (*Salmo salar* L.) in land-based farms. Evidence of a photoperiodic influence. *Aquaculture* 1995, 133, 235-248.

Foscarini, R. A review: intensive farming procedure for red sea bream (*Pagrus major*) in Japan. *Aquaculture* 1988, 72, 191-246.

Fushimi, H. Production of juvenile marine finfish for stock enhancement in Japan. *Aquaculture* 2001, 200, 33-53.

Geslin, M; Auperin, B. Relationship between changes in mRNA of the genes encoding steroidogenic acute regulatory protein and P450 cholesterol side chain cleavage in head kidney and plasma levels of cortisol in response to different kinds of acute stress in rainbow trout (*Oncorhynchus mykiss*). *Gen. Comp. Endocrinol.* 2004, 135, 70-80.

Ginés, R; Afonso, JM; Argüello, A; Zamorano, MJ; López, JL. The effects of long-day photoperiod on growth, body composition and skin colour in immature gilthead sea bream (*Sparus aurata* L.). *Aquacult. Res.* 2004, 35, 1207-1212.

Goede, RW, Barton, BA. Organismic indices and an autopsy-based assessment as indicators of health and condition in fish. *Am. Fish. Soc. Symp.* 1990, 8, 93-108.

Gross, WL, Roelofs, EW, Fromm, PO. Influence of photoperiod on growth of green sunfish, *Lepomis cyanellus*. *J. Fish. Res. Board Canada* 1965, 22, 1379-1386.

Hallaråker, H; Folkvord, A; Stefansson, SO. Growth of juvenile halibut (*Hippoglossus hippoglossus*) related to temperature, day length and feeding regime. *Neth. J. Sea Res.* 1995a, 34, 139-147.

Hallaråker, H; Folkvord, A; Stefansson, SO. Growth of juvenile halibut (*Hippoglossus hippoglossus*) related to temperature, day length and feeding regime. *Netherlands J. Sea Res.* 1995b, 34, 139-147.

Hansen, T; Stefansson, SO; Taranger, GL. Growth and time of ovulation of Atlantic salmon (*Salmo salar*) in sea cages at two different light regimes. *Aquacul. Fish. Man.* 1992, 23, 135-140.

Hansen T, Kjesbu OS, Holm JC, Karlsen Ø. Growth, gonadal development and spawning time of Atlantic cod (*Gadus morhua*) reared under different photoperiods. In: Goetz FW, Thomas P, editors. *The 5th International Symposium on Reproductive Physiology of Fish*. University of Texas, Austin. 1995, pp. 186.

Hart, PR; Hutchinson, WG; Purser, GJ. Effects of photoperiod, temperature and salinity on hatchery-reared larvae of the greenback flounder (*Rhombosolea tapirina* Günther, 1862). *Aquaculture* 1996, 144, 303-311.

Hoar, WS. The physiology of smolting salmonids. In: Hoar, WX; Randall, DJ; editors. *Fish Physiology*, Vol. 11,B. Academic Press, New York. 1988, pp. 275-343.

Hogman, WJ. Annulus formation on scales of four species of coregonids reared under artificial conditions. *J. Fish. Res. Board Can.* 1968, 25, 2111-2112.

Imsland, A; Folkvord, AF; Stefansson, SO. Growth, oxygen consumption and activity of juvenile turbot (*Scophthalmus maximus* L.) reared under different temperatures and photoperiods. *Neth. J. Sea Res.* 1995, 34, 149-159.

Imsland, AK; Folkvord, AF; Jónsdóttir, ODB; Stefansson, SO. Effects of exposure to extended photoperiods during the first winter on long-term growth and age at first maturity in turbot (*Scophthalmus mximus*). *Aquaculture* 1997, 159, 125-141.

Johnsson, JI; Björnsson, BT. Growth hormone increases growth rate, appetite and dominance in juvenile rainbow trout, *Oncorhynchus mykiss*. *Anim. Behav.* 1994, 48, 177-186.

Jonassen, TM; Imsland, AK; Stefansson, SO. Interaction of temperature and light on growth in Atlantic halibut, *Hippoglossus hippoglossus* L. *Aquacult. Res*. 2000, 31, 219-227.

Kadmon, G; Gordin, H; Yaron, Z., 1985. Breeding-related growth of captive *Sparus aurata* (Teleostei, Perciformes). *Aquaculture* 1985, 46, 299-305.

Karlsen, Ø; Taranger, GL; Dahle, R; Norberg, B. Effect of exercise and continuous light on early sexual maturation in farmed Atlantic cod (*Gadus morhua* L.). In: Norberg, B; Kjesbu, OS; Taranger, FL; Andersson, E; Stefansson, SO; editors. *Reproductive Physiology of Fish*. John Grieg Forlag A/S, Bergen. 2000, 328-330.

Kissil, GW; Lupatsch, I; Elizur, A; Zohar, Y. Long photoperiod delayed spawning and increased somatic growth in gilthead seabream (*Sparus aurata*). *Aquaculture* 2001, 200, 363-379.

Kråkenes, R; Hansen, T; Stefansson, SO; Taranger, GL. Continuous light increases growth rate of Atlantic salmon (*Salmo salar* L.) postsmolts in sea cages. *Aquaculture* 1991, 95, 281-287.

Larkin, LM; Moore, BJ; Stern, JS; Horwitz, BA. Effect of photoperiod on body weight and food intake of obese and lean Zucker rats. *Life Sci*. 1991, 49, 735-745.

Leonardi, MO; Klempau, AE. Artificial photoperiod influence on the immune system of juvenile rainbow trout (*Oncorhynchus mykiss*) in the Southern Hemisphere. *Aquaculture* 2003, 221, 581-591.

Lohner, TW; Reash, RJ; Willet, VE; Fletcher, J. Assessment of tolerant sunfish populations (*Lepomis* sp.) inhabiting selenium-laden coal ash effluents. *Ecotoxicol. Environ. Saf.* 2001, 50, 225-232.

Lundqvist, H. Influence of photoperiod on growth in Baltic salmon parr (*Salmo salar* L.) with special reference to the effect of precocious sexual maturation. *Can. J. Zool.* 1980, 58, 940-944.

Mäkinen, T; Ruhonen, K. Effect of delayed photoperiod on the growth of a Finnish rainbow trout (*Oncorhynchus my*kiss Walbaum) stock. *J. Appl. Ichthyol.* 1992, 8, 40-50.

Mason, EG; Gallant, RK; Wood, L. Productivity enhancement of rainbow trout using photoperiod manipulation. *Bull. Aquaculturists Assoc. Can.* 1992, 91, 44-46.

McCormick, SD; Björnsson, BT; Sheridan, M; Eilertson, C; Carey, JB; O'Dea, M. Increased day length stimulates plasma growth hormone and gill Na+, K+-ATpase in Atlantic salmon (*Salmo salar*). *J. Comp. Physiol.* 1995, 165B, 245-254.

Mortensen, A; Damsgård, B. Compensatory growth and weight segregation following light and temperature manipulation of juvenile Atlantic salmon (*Salmo salar* L.) and Arctic charr (*Salvelinus alpinus* L.). *Aquaculture* 1993, 114, 261-272.

Murata, O; the late Harada, T; Miyashita, S; Izumi, K; Maeda, S; Kato, K; Kumai, H. Selective breeding for growth in red sea bream. *Fish. Sci.* 1996, 62: 845-849.

Nagy, TR; Negus, NC. Energy acquisition and allocation in male collared lemmings (*Dicrostonyx groenlandicus*): effects of photoperiod, temperature, and diet quality. *Physiol. Zool.* 1993, 66, 537-560.

Negus, NC; Berger, PJ; Brown, BW. Microtine population dynamics in a predictable environment. *Can. J. Zool.* 1986, 64, 785-792.

Noeske-Hallin, TA; Spieler, RE. Circadian feeding time affects growth of fish. *Trans. Am. Fish. Soc.* 1984, 113, 540-544.

Noeske-Hallin, TA; Spieler, RE; Parker, NC; Suttle, MA. Feeding time differentially affects fattening and growth of channel catfish. *J. Nutr.* 1985, 115, 1228-1232.

Petit, G; Beauchaud, M; Attia, J; Buisson, B. Food intake and growth of largemouth bass (*Micropterus salmoides*) held under alternated light/dark cycle (12L:12D) or exposed to continuous light. *Aquaculture* 2003, 228, 397-401.

Person-Le Ruyet, J; Baudin Laurencin, F; Devauchelle, N; Métailler, R; Nicolas, JL; Robin, J; Guillaume, J. Culture of turbot (Scophthalmus maximus). In: McVey, JP; editor. *Handbook of Mariculture. Finfish aquaculture, Vol. 2*, CRC Press, Boston. 1991, 21-41.

Pickering, AD. Introduction: the xoncept of biological stress. In: Pickering AD, editor. *Stress and Fish*. Academic Press, London, 1981, pp. 1-9.

Pickering, AD; Pottinger, TG. Seasonal and diel changes in plasma cortisol levels of brown trout, *Salmo trutta* L. *Gen. Comp. Endocrinol.* 1983, 49, 232-239.

Pickering, AD; Pottinger, TG. Stress responses and disease resistance in salmonid fish: effects of chronic elevation of plasma cortisol. *Fish Physiol. Biochem*. 1989, 7, 253-258.

Pickering, AD; Pottinger, TG; Carragher, J; Sumpter, JP. The effects of acute and chronic stress on the levels of reproductive hormones in the plasma of mature male brown trout, Salmo trutta L. *Gen. Comp. Endocrinol.* 1987, 68, 249-259.

Porter, MJR; Randall, CF; Bromage, NR; Thorpe, JE. The role of melatonin and the pineal gland on development and smoltification of Atlantic salmon (*Salmo salar*) parr. *Aquaculture* 1998, 168, 139-155.

Pottinger, TG; Yeomans, WE;Carrick, TR. Plasma cortisol and 17β-oestradiol levels in roach exposed to acute and chronic stress. *J. Fish Biol.* 1999, 54, 525-532.

Prabakaran, R; Babu, M; Sundararasu, V. Effect of photoperiod on the growth performance of Japanese quails. *J. Vet. Anim. Sci*. 1991b, 22, 9-11.

Reite, M. Implantable biotelemetry and social separation in monkeys. In: Moberg, GP; editor. *Animal Stress*, American Physiological Society, Bethesda, Maryland. 1985, pp. 141-160.

Reynolds, PS; Lavigne, DM. Photoperiodic effects on post-weaning growth and food consumption in the calloared lemming *Dicrostonyx groenlandicus*. *J. Zool. London* 1989, 218,109-121.

Ronzani Cerqueira, V; Chatain, B; Lavens, P; Jaspers, E; Ollevier, F. Photoperiodic effects on the growth and feeding rhythm of European seabass, *Dicentrarchus labrax*, larvae in intensive rearing. *Larvi'91, Special Publ. Eur. Aquacult. Soc*. 1991, 15, 304-306.

Saito, M; Noma, H. Food intake and growth of rats red with adiurnal periodicity. *Physiol. Behav*. 1980, 24, 87-91.

Saunders, RL; Harmon, PR. Extended daylength increases postsmolt growth of Atlantic salmon. *World Aquacult*. 1988, 19, 72-73.

Saunders, RL; Henderson, EB; Harmon, PR. Effects of photoperiod on juvenile growth and smolting of Atlantic salmon and subsequent survival and growth in sea cages. *Aquaculture* 1985, 45, 55-66.

Saunders, RL; Duston, J; Benfey, TJ. Environmental and biological factors affecting growth dynamics in relation to smolting of Atlantic salmon, *Salmo salar* L. *Aquacult. Fish. Manage.* 1994, 25, 9-20.

Schanbacher, BD; Crouse, JD. Photoperiodic regulation of growth: a photosensitive phase during light-dark cycle. *Am. J. Physiol*. 1981, 241, E1-5.

Selye, H. A syndrome produced by diverse nocuous agents. *Nature* 1936, 138, 32.

Shambacher, BD. Hormonal and photoperiodic control of growth. In: Roche, JF; O'Callaghan, DO; editors. Manipulation of growth in farm animals. Boston: Martinus Nijhoff Publishers 1984, 289-299.

Sigholt, T; Staurnes, M; Jakobsen, HJ; Asgård, T. Effects of continuous light and short-day photoperiod on smolting, seawater survival and growth in Atlantic salmon (*Salmo salar*). *Aquaculture* 1995, 130, 373-388.

Silva-García, AJ. Growth of juvenile gilthead seabream (*Sparus aurata* L.) reared under different photoperiod regimes. *The Is. J. Aquacult-Bamidgeh* 1996, 48, 84-93.

Simensen, LM; Jonassen, TM; Imsland, AK; Stefansson, SO. Photoperiod regulation of growth of juvenile Atlantic halibut (*Hippoglossus hoppoglossus* L.) *Aquaculture* 2000, 190, 119-128.

Sorensen, MT. Photoperiodic effect on growth and feed consumption of young bulls. In Roche, JF; O'Callaghan, DO; editors. *Manipulation of growth in farm animals*. Boston: Martins Nijhoff Publishers 1984, 275-287.

Spears, N; Clarke, JR. Effect of male presence and of photoperiod on the sexual maturation of the field vole (*Microtus agrestis*). *J. Reprod. Fertil.* 1986, 78, 231-238.

Stefansson, SO; Björnsson, BTh; Hansen, T; Haux, C; Taranger, GL; Saunders, RL. Growth, parr-smolt transformation, and changes in growth hormone of Atlantic salmon (*Salmo salar*) reared under different photoperiods. *Can. J. Fish. Aquat. Sci.* 1991, 48, 2100-2108.

Sumpter, JP. Control of growth of rainbow trout (*Oncorhynchus mykiss*). *Aquaculture* 1992, 100, 299-320.

Sundararaj, B; Nath, P; Hallberg, F. Circadian meal timing in relation to lighting schedule optimizes catfish body weight gain. *J. Nutr.* 1982, 112, 1085-1097.

Swift, DR. Seasonal variation in the activity of the thyroid gland of yearling brown trout (*Salmo trutta* Linn.). *J. Exp. Biol.* 1959, 36, 120-125.

Swift, DR. Cyclical activity of the thyroid gland of fish in relation to environment changes. *Symp. Zool. Soc. London No. 1*, 1960, pp. 17-27.

Swift, DR. The annual growth rate cycle in brown trout (*Salmo trutta* Linn.) and its cause. *J. Exp. Biol.* 1961, 38, 595-604.

Tandler, A; Helps, S. The effects of photoperiod and water exchange rate on growth and survival of gilthead sea bream (*Sparus aurata*, Linnaeus; Sparidae) from hatching to metamorphosis in mass rearing systems. *Aquaculture* 1985, 48, 71-82.

Tang, WC. Chinese medicinal materials from the sea. *Abstracts of Chinese Medicine* 1987, 1, 571-600.

Thorarensen, H; Clarke, WC. Smoltification induced by a 'skeleton' photoperiod in underyearling coho salmon (*Oncorhynchus kisutch*). *Fish Physiol. Biochem.* 1989, 6, 11-18.

Thorpe, JE; Morgan, RIG; Pretswell, D; Higgins, PJ. Movements rhythms in juvenile Atlantic salmon, *Salmo salar* L. *J. Fish Biol.* 1988, 33, 931-940.

Trippel, EA; Neil, SRE. Effects of photoperiod and light intensity on growth and activity of juvenile haddock (*Melanogrammus aeglefinus*). *Aquaculture* 2003, 217, 633-645.

Wade, GN; Bartness, TJ. Seasonal obesity in Syrian hamsters: effects of age, diet, photoperiod, and melatonin. *Am. J. Physiol.* 1984, 247, R328-334.

Wang, C; King, W; Woods, C. Physiological indicators of divergent stress responsiveness in male striped bass broodstock. *Aquaculture* 2004, 232, 665-678.

Watanabe, T; Vassallo-Agius, R. Broodstock nutrition research on marine finfish in Japan. *Aquaculture* 2003, 227, 35-61.

Wedemeyer, GA; Barton, BA; McLeay, DJ. Stress and acclimation. In: Schreck, CB; Moyle, PB; editors, *Methods for fish biology*, American Fisheries Society, Bethesda, MD. 1990, pp. 451-489.

Wendelaar Bonga, SE. The stress response in fish. *Physiol. Rev.* 1997, 77, 591-625.

Weiner, H. Perturbing the Organism: the Biology of Stressful Experience. University of Chicago Press, Chicago. 1992.

Van Weerd, JH; Komen, J. The effects of chronic stress on growth in fish: a critical appraisal. *Comp. Biochem. Physiol.* 1998, 120A, 107-112.

Vilaplana, J; Madrid, JA; Sánchez-Vázquez, J; Campuzano, A; Cambras, T; Díez-Noguera, A. Influence of period length of light/dark cycles on the body weight and food intake of young rats. *Physiol. Behav.* 1995, 58, 9-13.

In: Aquaculture Research Progress
Editor: Takumi K. Nakamura
ISBN 978-1-60456-247-7

Chapter 5

MARINE SOURCES OF OMEGA-3: IS THEIR FUTURE IN AQUACULTURE?

Candida Savage[1]
Department of Marine Science,
310 Castle Street University of Otago, Dunedin, New Zealand

ABSTRACT

Omega-3 polyunsaturated fatty acids are essential factors in human health and nutrition that are derived primarily from consuming seafood and marine oil supplements. The commercially important long-chain omega-3 fatty acids, docosahexaenoic acid (DHA; C22:6 n-3) and eicosapentaenoic acid (EPA; C20:5 n-3) are synthesized mainly in phytoplankton and passed onto higher trophic levels in their diets. The proportion of omega-3 fatty acids in bivalves and fishes thus shows strong seasonal fluctuations that correspond with shifts in the quantity and nutritional quality of each dietary component, as well as spawning periodicity, and environmental variability. This chapter reviews the sources of variation in omega-3 content in bivalves and fishes and discusses how farming marine products can overcome some of this variation. Finally, sustainable alternatives for the production of omega-3s are examined to address the growing market for seafood and associated health products.

1. INTRODUCTION

Omega-3 fatty acids have experienced an upsurge of interest in recent years due to their role in promoting human health and nutrition. Omega-3s are essential long-chain polyunsaturated fatty acids that cannot be effectively synthesized in the human body [1]. Therefore, they must be obtained by consuming products rich in omega-3 fatty acids, such as seafood and products incorporating marine oils. The main sources of these marine oils are extracted from pelagic cold-water fish species with oily flesh, such as menhaden, mackerel

[1] Tel +64-3-479 8324, Fax + 64-3-479 8336, Email: candida.savage@stonebow.otago.ac.nz.

and herring. However, with diminishing fish stocks, the industry is being pressurized to look at more sustainable alternatives to capture fisheries for food and fish oils.

This chapter describes the group of fatty acids referred to as omega-3 polyunsaturated fatty acids (PUFA) and briefly mentions their human health benefits. The long-chain highly unsaturated omega-3 fatty acids docosahexaenoic acid (DHA; C22:6 n-3) and eicosapentaenoic acid (EPA; C20:5 n-3) are highlighted here. The chapter reviews their main sources – marine finfish and bivalves – and explains factors that affect variation in PUFA content in seafood. We look at industry trends and how aquaculture production is responding to the worldwide decline in capture fisheries. The use of vegetable oils as a substitute for fish oil in aquaculture feeds is discussed and its effects on omega-3 content of farmed fish. Finally, we explore the potential to farm bivalves as a rich source of omega-3 that may offer a sustainable alternative to meet the growing market for functional foods and nutraceuticals in the future.

2. Omega-3 Fatty Acids

Fatty acids consist of a hydrocarbon chain (CH_2) with an acid or carboxyl group (HO-C=O) at one end and a methyl group (CH_3) at the other. The hydrocarbon chain varies in length from 2 to 20 or more carbons, with the most common fatty acids in food varying from 12 to 22 carbons [2]. Fatty acids are further classified as saturated or unsaturated, depending on the number of hydrogen atoms present. Saturated fats have the maximum number of hydrogen atoms and therefore no double bonds, while polyunsaturated fatty acids contain two or more double bonds. The group of polyunsaturated fatty acids termed omega-3 (abbreviated as n-3 or ω3) fatty acids are distinguished by having the first double bond positioned on the third carbon atom from the methyl or N- terminal end of the fatty acid chain [2]. By convention, fatty acids are written as, for example, C20:5 n-3, eicosapentaenoic acid (EPA), where the numbers from left to right designate the number of carbon atoms in the chain (i.e. 20 in this example), the number of double bonds (i.e. five in EPA), and the position of the first double bond (i.e. third carbon-carbon bond in this example). A further distinguishing characteristic of fatty acids is the orientation of the molecule at the site of the double bond. When both sections of the molecule lie on the same side of the double bond, a *cis* configuration is formed, which creates a kink in the fatty acid chain, making the molecule more flexible. Most naturally occurring fatty acids in seafoods have *cis* configuration [2].

The fatty acids that are important in human nutrition and health are the long-chain (C_{20} and C_{22}) omega-3 polyunsaturated fatty acids (PUFA), including docosahexaenoic acid (DHA; C22:6 n-3) and eicosapentaenoic acid (EPA; C20:5 n-3) (Figure 1). In living cells, EPA and DHA are normally esterified to form complex lipid molecules. They are major constituents of cell membranes where they are stored as phospholipids. DHA and EPA phospholipids are believed to be important in regulating membrane fluidity in finfish and bivalves, respectively [3]. At a cellular level, these phospholipids block or disrupt the formation of islands of gel-phase lipids, a high proportion of which interrupt permeability and restrict movement of electron carriers necessary for energy production, in the membrane [4]. Thus, the net effect of DHA and EPA is hypothesized to increase energy availability to the cell, both via energy conservation and enhanced energy production [4].

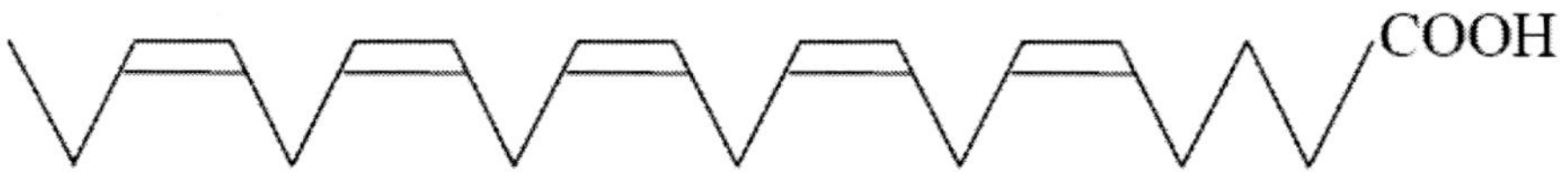

EPA (C20:5 n-3)

DHA (C22:6 n-3)

Figure 1. Schematic of the chemical structure of the long-chain omega-3 polyunsaturated fatty acids, eicosapentaenoic acid (EPA) and docosahexaenoic acid (DHA) that are derived principally from marine fish and shellfish.

3. Polyunsaturated Fatty Acids and Health Benefits

Omega-3 fatty acids have been recognized as effective factors in human health and nutrition. Early work comparing Greenland Eskimos living on their traditional diet to Eskimos living in Denmark noted a significantly reduced prevalence of cardiovascular disease in the Greenland population [5]. This suggested that intake of long-chain omega-3 PUFA from fish and marine mammals may protect against coronary heart disease [5]. More recently, the biological effects of consuming seafood have been intensively explored. Clinical and epidemiological research suggest that long-chain n-3 PUFA such as EPA and DHA are especially beneficial for the prevention of human cardiovascular diseases [6] and for the alleviation of inflammatory conditions such as rheumatoid and osteoarthritis [7] and asthma [8]. These long-chain omega-3 fatty acids are also essential for normal growth and development, especially for retinal and brain development in infants [9, 10]. The highly unsaturated fatty acids like EPA and DHA play a crucial role in higher animals and humans as a precursor of a group of eicosanoids that are critical in regulating developmental and regulatory physiology [11]. The eicosanoids are hormone-like substances including prostaglandins, thromboxanes, and leukotrienes [12]. Consequently, there has been a burgeoning interest in identifying foods rich in omega-3 for human health and nutrition.

The second major group of unsaturated fatty acids involved in human health is the omega-6 fatty acids, which are characterized by containing the first double bond at the sixth carbon atom position from the methyl end. The modern Western diet is rich in omega-6 fatty acids from vegetable oils, cereals and wholegrain bread that contain linoleic acid (C18:2 n-6). Linoleic acid is the precursor for the long-chain product, arachidonic acid (AA; C20:4 n-6) and intermediate γ-linolenic acid (GLA; C18:3 n-6) [13]. Several lines of evidence suggest that human beings evolved eating marine organisms, probably an intertidal shellfish diet with a healthy 1:1 balance of omega-3 and omega-6 fatty acids [14]. In contrast, modern diets are generally depleted in long-chain omega-3 fatty acids and typically have an omega-6:omega-3 ratio of 17:1 [15]. The highly unsaturated long-chain omega-3 fatty acids EPA (C20:5 n 3) and DHA (C22:6 n-3) cannot be synthesized effectively in the human body [1, 16] and are therefore termed 'essential' fatty acids [17]. The precursor for EPA and DHA, α-linolenic acid (ALA, C18:3 n-3), is available from certain plants including the seeds and oils of flax and linseed [18, 19]. Humans can synthesize up to 5% EPA and DHA from dietary ALA through a series of chain elongation and desaturation steps [13]. However, this pathway is considered inefficient to meet our requirements for long-chain fatty acids. Furthermore, with a diet rich in omega-6 fatty acids, these conversion rates may be reduced by as much as 40-50% [20] due to competitive interactions between the synthesis of n-6 and n-3 fatty acids. Therefore, essential fatty acids like EPA and DHA cannot be synthesized in adequate quantities for optimal physiological performance and must be obtained via dietary intakes or supplements derived from marine sources.

The recommended average daily intake of omega-3 fatty acids differs among the various nutritional organizations, but generally ranges between 1.0 – 1.5 g EPA + DHA per day for adults. The British Nutrition Foundation has recommended a diet including two or three portions of fatty fish per week (equivalent to 3-4 g standardized fish oil per day), or more specifically 1.25 g EPA + DHA per day [20]. The adequate intake published by the US NAS Institute of Medicine is 1.1 and 1.6 g per day for adult females and males, respectively [21]. A caveat, however, is that some studies have reported adverse effects on immune function with excessive intakes of fish oils. For example, at exposures 7-15 times greater than typical US intakes of EPA and DHA, the potential of the immune system to attack pathogens was diminished [22, 23]. Nevertheless, the benefits derived from eating moderate quantities of marine products are considered to outweigh the health risks.

To increase the intake of omega-3 PUFA, researchers have proposed the inclusion of a fish or shellfish component in the diet, daily supplements of marine oil, or incorporation of stabilized marine oil in food products [24]. A number of techniques have been developed to concentrate omega-3 PUFA from marine products including molecular distillation, urea complexation, chromatographic separation and lipase-assisted hydrolysis [25]. Consequently, omega-3 concentrated oils, salmon oil and cod-liver oil are readily available as supplements in the liquid or capsule form in pharmacies and health stores. Incorporation of marine oils into food products is also popular and recent advances in the microencapsulation and refinement technologies have allowed production of marine oils that do not cause any flavor reversion [13]. Foods in which omega-3 oils are added include bakery products, pastas, dairy products and a plethora of other food items [13]. However, some research suggests that the bioavailability of n-3 PUFA and thus the health benefits are enhanced when fatty fish are consumed compared with supplementation of the diet with fish oils in the form of capsules [26]. Although when fish is suggested as a means of improving health, both fat content and the PUFA composition must be considered. According to Kinsella [27], an increase in PUFA content in fish is correlated with a decrease in the cholesterol level. Moreover, the omega-3:omega-6 ratio has been suggested to be a better index for comparing the relative nutritional value of different species [28]. However, this index is of limited value without consideration of which fatty acids are present. Generally the C_{20} and C_{22} fatty acids are more nutritionally valuable than the C_{18} fatty acids [17]. Thus the optimal dietary sources are those rich in the highly unsaturated C_{20} and C_{22} omega-3 fatty acids, specifically EPA and DHA.

4. Marine Sources of Omega-3 Fatty Acids

Marine organisms are generally a good source of essential omega-3 polyunsaturated fatty acids, particularly the highly unsaturated eicosapentaenoic (EPA; C20:5 n-3) and docosahexaenoic (DHA; C22:6 n-3) acids that are principally obtained from fish and shellfish [13]. The nutritional value of fish and shellfish is mainly due to their lipid composition with many marine, cold-water species containing elevated levels of PUFA, including EPA and DHA. Many marine animals have no ability or a limited ability to synthesize EPA and DHA from precursor fatty acids such as α-linolenic acid [29]. Therefore they acquire their fatty acid profiles from their diet, with phytoplankton at the base of the food web being the primary source of fatty acids in the marine environment.

4.1. Phytoplankton

The biosynthesis of omega-3 PUFA occurs through a series of reactions that can be divided into two distinct steps. First is the *de novo* synthesis of oleic acid (C18:1 n-9) from acetate. This is followed by the conversion of oleic acid to linoleic acid (LA; C18:2 n-6) and α-linoleic acid (ALA; C18:3 n-3) [11]. Subsequently, a number of stepwise desaturation and elongation steps form the group of omega-3 fatty acids including EPA and DHA (Figure 2) [11].

There is considerable variation in nutritional quality between classes of phytoplankton and even between species within the same class [31, 32]. Microalgae generally have a dominance of one particular fatty acid, either DHA or EPA [13]. For example, diatoms usually have high concentrations of C_{16} PUFA and C20:5n-3 (EPA), and low concentrations of C22:6n-3 (DHA) relative to dinoflagellates [31, 32]. In contrast, green algae usually do not contain high concentrations of the essential polyunsaturated C_{20} and C_{22} fatty acids [31]. Thus, seasonal shifts in plankton community structure alter the lipid concentration and component fatty acid signatures available to bivalves and fish through their diets [33]. In general, a predominance of dinoflagellates causes increased levels of DHA, while diatoms present in large amounts cause increased levels of EPA [34-37].

Laboratory studies have shown that major changes in the fatty acid composition of phytoplankton can result from modification of the culture environment [11].

Specifically, the fatty acid composition of marine phytoplankton can vary in response to variation in temperature [38], nutrient availability [39], irradiance [40] including UV-B irradiation [41], and a combination of these environmental factors [42]. When eight species of microalgae were grown under different temperature regimes, there was a modest ($r^2 = 0.11$, $p<0.05$) by significant inverse relationship between the percentage of PUFA and temperature when all the species data were pooled [43]. Individually, however, none of the species had a significant linear relationship between total PUFA content and temperature [38]. In one study on the effects of temperature on PUFA content of the marine diatom *Phaeodactylum tricornutum,* a widely used food organism in aquaculture due to its ability to accumulate up to 30% of the total fatty acids as EPA, temperature was shown to be a significant factor on EPA content [44]. When temperature was shifted from 25 to 10 °C for 12 hours, EPA concentration increased by 120% compared with the control [44]. Similarly, low temperature improved EPA yields in *Nannochloropsis* sp. [39]. In this latter study, excessive nitrate and low salinity also increased EPA levels in *Nannochloropsis* sp. [39]. It is generally believed that low-temperature stress leads to a relatively high omega-3 PUFA content since the algae attempt to maintain proper fluidity of cell membranes by producing more PUFA [11, 45]. The high level of intracellular oxygen may be another reason for the increased PUFA at low temperature, because the enzymes responsible for desaturation and elongation of PUFA depend on the availability of molecular oxygen [46]. Further, the light environment can affect the fatty acid profile of phytoplankton, with experiments showing an increase in saturated fats when grown at high light and conversely an increase in PUFA content in phytoplankton when grown under low light conditions [40]. Thus variation in environmental conditions can cause significant physiological and biochemical changes in plankton that can, in turn, affect the nutritional value of phytoplankton that is passed onto filter-feeding bivalves and ultimately fish. Moreover, natural fluctuations in phytoplankton community structure cause seasonal patterns in the fatty acids available to other trophic levels.

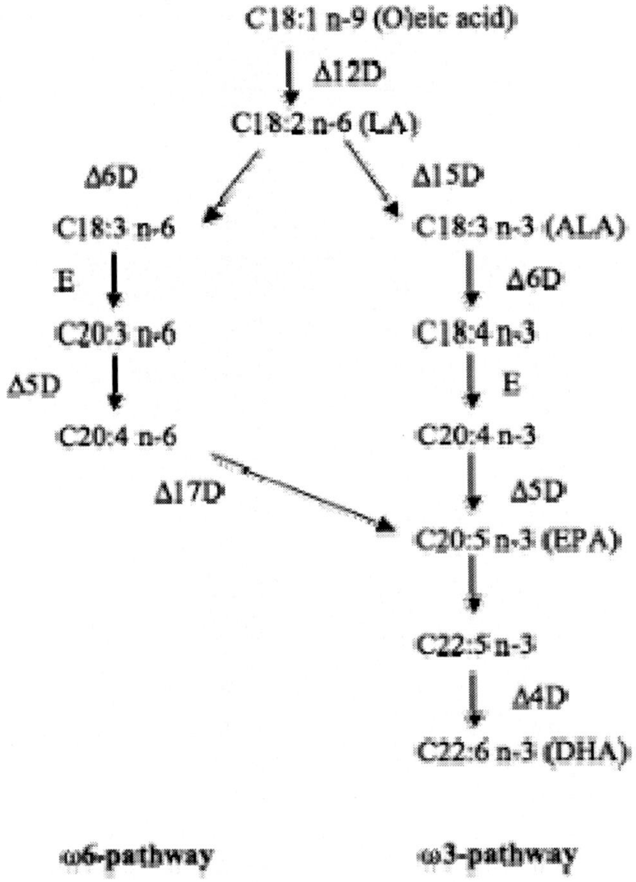

Figure 2. Two pathways for the biosynthesis of EPA and DHA in phytoplankton. D = desaturase and E = elongase. Modified from [11] and [30].

Table 1. Levels of total polyunsaturated fatty acids (ΣPUFA), omega-3 PUFA (ΣPUFA n-3), EPA, DHA, and the ratio of omega-3:omega-6 in selected bivalves. Values are reported as % of total fatty acids and are based on whole organisms unless stated otherwise. [a] Values are averaged over several seasons. [1]NZGLM = New Zealand green-lipped mussel

Scientific name	Common name	ΣPUFA	ΣPUFA n-3	C20:5 n-3 (EPA)	C22:6 n-3 (DHA)	n-3/n-6	Notes	Reference
Argopecten purpuratus	Chilean scallop	53.5	49.4	16.4	25.6	12.1	abductor muscle, n= 3	[57]
Atrina pectinata	Pen shell	40.7	36.6	12.3	23.9	9.9	n = 21 [a]	[53]
Chlamys islandica	Icelandic scallops	63.4	58.5	26.9	25.9	19.0	n = 5	[76]
Crassostrea gigas	Pacific cupped oyster	43.4	38.7	16.5	20.7	8.5	n = 21 [a]	[53]
Crassostrea gigas	Pacific cupped oyster	58.1	42.7	11.3	20.3	2.8	gonad-visceral mass, n = 24 [a]	[52]
Cyclina sinesis	Venus clam	36.3	30.6	14.1	15.3	4.8	n = 21 [a]	[53]
Mactra veneriformis	Surf clam	37.0	28.3	13.4	17.0	5.5	n = 21 [a]	[53]
Meretrix lusoria	Orient hard clam	34.6	28.2	8.9	17.7	5.6	n = 21 [a]	[53]
Mytilus coruscus	Hard-shelled mussel	40.9	36.1	14.6	20.4	7.7	n = 21 [a]	[53]
Mytilus edulis	Blue mussel	57.2	43.6	19.6	13.2	11.0	n = 3	[76]
Mytilus edulis	Tasmanian blue mussel	47.8	41.5	17.9	21.2	8.3	n = 9	[48]
Mytilus edulis	Blue mussel	61.9	47.8	12.0	21.5	8.8	n = 67 [a]	[47]
Mytilus edulis	Blue mussel	60.8	49.5	17.0	20.0	8.0	n = 75 [a]	[47]
Perna canaliculus	NZGLM [1]	44.2	37.1	14.6	13.5	6.9	n = 9	[48]
Perna canaliculus	NZGLM [1]	48.6	42.9	18.5	17.9	7.7	n = 20 [a]	[51]
Perna canaliculus	NZGLM [1]	46.7	42.0	24.7	11.3	9.0	n = 20 [a]	[51]
Perna viridis	Green-lipped mussel	61.1	56.5	17.9	38.1	15.6	n = 8	[49]

Scientific name	Common name	ΣPUFA	ΣPUFA n-3	C20:5 n-3 (EPA)	C22:6 n-3 (DHA)	n-3/n-6	Notes	Reference
Peronidia venulosa	Pink butterfly shell	37.1	31.3	10.9	19.8	5.6	n = 15 [a]	[53]
Ruditapes philippinarium	Manilla clam	32.3	25.7	9.1	16.0	4.2	n = 21 [a]	[53]
Scapharca broughtonii	Ark shell	40.9	34.9	15.9	18.2	6.6	n = 21 [a]	[53]
Serripes groenlandicus	Greenland cockle	56.2	45.2	22.6	16.5	13.8	n = 7	[76]
Sinonoacula constricta	Jack-knife clam	34.3	28.6	11.2	15.7	5.1	n = 21 [a]	[53]
Spisula solidissima	Surf clam	50.2	41.2	22.9	14.3	6.1	n = 1	[76]
Tegillarca granosa	Granulated ark shell	32.6	28.2	14.8	12.3	6.7	n = 18 [a]	[53]

4.2. Bivalves

Marine bivalves offer a potentially important source of long-chain omega-3 polyunsaturated fatty acids, especially EPA (C20:5 n-3) and DHA (C22:6 n-3). Individual studies have shown that bivalves including the blue mussel (*Mytilus edulis*) [47, 48], green-lipped mussel (*Perna viridis*) [49], the New Zealand green-lipped mussel (*Perna canaliculus*) [48, 50, 51], Pacific cupped oyster (*Crassostrea gigas)* [52] and various other commercially important shellfish species [53], are rich sources of DHA and EPA (Table 1). In all these studies, total lipid content was low (<2%) with polyunsaturated fatty acids generally dominating. Moreover, the commercially important omega-3 fatty acids, DHA and EPA, were in the top percentage among all fatty acids present. In most studies, the nutritionally important omega-3 group comprised about 40-50% of fatty acids in mussels [47, 54-56]. For example, in *Perna canaliculus*, an important aquaculture mussel in New Zealand, PUFA dominated (~50%) with EPA and DHA together accounting for 70-80% of PUFA content in spring and summer [51].

Lipid content and fatty acid composition are generally reported for the edible portions of seafood. In the case of bivalves, analyses are generally based on the whole organism. However, lipids and their constituent fatty acids are distributed variously in the different organs of molluscs. A study by Caers et al. [57] investigated the fatty acid composition of six organs in the Chilean scallop, *Argopecten purpuratus* and demonstrated that lipid content was lowest but n-3 PUFA levels (mainly EPA and DHA) were highest in adductor muscle tissue [57]. Irrespective of the organ analyzed, however, omega-3 fatty acids were the most abundant fatty acid type and were dominated by DHA and EPA [57]. This finding bodes well for the consumption of these organisms as part of a healthy diet.

4.2.1. Variation in Omega-3 Content in Bivalves Due to Changes in Dietary Fatty Acids

Bivalves have a very limited or no capacity to synthesize PUFA [58, 59] and therefore derive their fatty acid signatures from their diet. The diet of suspension-feeding bivalves consists mainly of phytoplankton (diatoms and dinoflagellates), as well as bacteria and detritus [36]. However, the diet varies at different stages of the life history of the bivalve, owing in part to ontogenetic shifts in feeding [47]. In mariculture, the nutritional requirement for omega-3 PUFA in diets of marine molluscs during egg and larval development is well established. Researchers have found positive correlations between growth rates of mussel larvae (*Mytilus edulis*) [55] and oyster larvae (*Ostrea edulis*) [60] with total n-3 fatty acids and DHA content, respectively. Even in the adult form, phytoplankton of different sizes or classes and with different fatty acid signatures may be consumed during the various growth stages, causing corresponding changes in the fatty acid composition of the bivalves. Not surprisingly therefore, fatty acid profiles of small (~44 mm) New Zealand green-lipped mussels (*Perna canaliculus*) differed significantly from their larger (~86 mm) counterparts [51]. Small mussels had less total n-3 fatty acids, EPA and DHA content than large mussels from Marlborough Sounds, New Zealand [51]. Equally, cultured blue mussels (*Mytilus edulis*) in Newfoundland, Canada, showed a linear relationship in the total essential fatty acids in mussels from seed to harvest [47]. These studies suggest, therefore, that mussels acquire greater omega-3 levels as they grow, either as a function of increased lipid content or as they consume larger quantities and/or higher quality phytoplankton in terms of omega-3 content.

Several studies have reported pronounced seasonal shifts in lipid and fatty acid composition in bivalves due to temporal variation in the phytoplankton classes available [37, 47, 51, 53, 56, 57, 61-63]. In these studies, the omega-3 content and thus the nutritional value of the bivalves peaked during periods of maximum plankton availability, which was normally coincident with the spring diatom bloom. For example, cultured mussels (*Mytilus edulis*) showed significant seasonal variation in essential fatty acids with peak concentrations of DHA in autumn and maxima for EPA in spring-summer that coincided with periods of high plankton availability [47]. Similarly, in studies of mussels (*Mytilus edulis*) that separated the phospholipids fraction and analyzed it for fatty acid composition, it was found that EPA and DHA content varied seasonally [62]. In particular, high amounts of EPA and DHA were found in mussel phospholipids in spring, which was attributed to increased availability of dietary PUFA-rich phytoplankton [62]. The authors thus concluded that the optimal time for the extraction of omega-3 from the mussels would be spring-summer [62]. These results were corroborated for the New Zealand green-lipped mussel (*Perna canaliculus*), with peak total n-3 content and mussel condition in spring for mussels collected at several sampling sites [51]. EPA levels, in particular, peaked in *P. canaliculus* samples during spring, which was attributed to increased diatom abundance coincident with the spring bloom [64]. Thus, it was suggested that spring would be the optimal harvesting period for omega-3 extracts from this species [51]. Conversely, in winter there was a marked decrease in n-3 PUFA content and a concomitant increase in saturated fatty acids (C14:0 and C16:0) with energetic functions [65], suggesting that the mussels had increased metabolic demands and limited PUFA-rich plankton as food [51].

Geographical variation in food availability and phytoplankton classes can also cause variation in the fat content and fatty acid composition of bivalves among different sites. *Perna canaliculus* is an important aquaculture species in New Zealand with production centered in the Marlborough Sounds. Comparison of fatty acid profiles of mussels from three sites within the Marlborough Sounds region showed no significant difference in the major fatty acid composition among sites [50]. However, when *P. canaliculus* samples were compared between regions (Marlborough Sounds: latitude 41 °S and Stewart Island: latitude 46.5 °S), there were significant latitudinal differences in individual fatty acids. In particular, the mussels collected from Stewart Island had significantly higher EPA levels than those from Marlborough Sounds, whereas the Marlborough mussels had elevated levels of DHA [51]. While the effect of temperature on fatty acid composition could not be ruled out, the dramatic decrease in EPA content in winter suggested that the latitudinal differences most likely reflected regional differences in phytoplankton biomass and community composition [51]. Despite these differences in individual fatty acids, there was no significant difference in total omega-3 PUFA content between regions, which implies that harvesting mussels for n-3 extracts would be driven more by logistical considerations [51].

4.2.2. Species-Specific Variation in Omega-3 Content in Bivalves

The same diet may give rise to varying fatty acid profiles in different species due to species-specific variation in how dietary fatty acids are incorporated and modified in lipids of the consumers. Two bivalve species, the oyster *Crassostrea gigas* and clam *Ruditapes philippinarium* were fed monospecific microalgal diets with varying PUFA composition for eight weeks and the resultant changes in their gill polar lipids analyzed [66]. The PUFA contents in oyster and clam gills were greatly influenced by the dietary conditioning, however there were intrinsic differences in the retention and modification of the dietary and synthesized fatty acids between the two species. Oysters maintained higher EPA levels (~18%) than clams, which reached a maximum (~10%) after four weeks and thereafter modulated the incorporation of EPA in gill lipids [66]. Contrary to the EPA results, oysters and clams presented similar levels of DHA according to the dietary supply [66]. Thus, although both species were fed the same monospecific diets and reflected a similar dietary imprint, the ultimate fatty acid composition recorded in the oysters and clams differed between the two species [66].

4.2.3. Variation in Omega-3 Content in Bivalves Due to Spawning Activity

Many invertebrates exhibit seasonal biochemical cycles [67]. Therefore, it is not surprising that there are often marked seasonal shifts in omega-3 content in bivalves that reflect changes in the gametogenic cycle. Many bivalves store energy in the form of glycogen, protein and lipid when food is abundant prior to gametogenesis, causing an increase in n-3 content when ripe and an associated decrease following spawning. For instance, seasonal variation in the biochemical composition of the oyster, *Crassostrea gigas*, showed changes in fatty acid levels that were closely related to reproductive activity [52]. High levels of total fatty acids in the gonad-visceral mass of the oyster were documented in the period of maximum ripeness, with peak PUFA levels especially DHA, recorded during oocyte maturation in spring [52]. The role of DHA in oogenesis and embryogenesis has been noted previously [68]. PUFA levels decreased dramatically during spawning, and thereafter variations in PUFA content appeared to be influenced by available food, with increases in PUFA reflecting increases in the chlorophyll *a* concentration in the seawater [52]. Similarly, seasonal fluctuations in lipid content and fatty acid levels have been found in relation to gametogenesis for a number of commercially important shellfish species, with an increase in total fatty acid content prior to spawning and a decrease associated with the post-spawning period [53, 67].

4.2.4. Variation in Omega-3 Content in Bivalves Due to Environmental Factors

Seasonal changes in environmental factors such as water temperature can influence the fatty acid composition of bivalves both (a) indirectly through altering the quantity and quality of its dietary sources [39, 43, 44], and (b) directly by affecting the lipid composition in their membranes.

(a) As discussed previously, the fatty acid composition of bivalves is influenced by food quality and quantity, which in turn, may be affected by environmental variability. The relationship between water temperature and fatty acid profiles of mussels (*Mytilus edulis*) and phytoplankton at two aquaculture sites showed that mussels and plankton were qualitatively similar, with EPA and DHA being among the most abundant fatty acids in both mussel and plankton samples [47]. Seasonal variation in the DHA:EPA ratio in mussels and plankton, which is often used as an indicator of the predominant phytoplankton class [36], showed strong seasonal variation corresponding to temperature fluctuations (Figure 3). The DHA:EPA ratio in plankton showed some temporal lag with the temperature profile but was more accentuated than the ratio in mussels; whereas the DHA:EPA ratio in mussels corresponded closely to fluctuations in temperature [47].

(b) Temperature also has a direct effect on bivalve fatty acid composition due to the role of phospholipids in maintaining membrane fluidity. A temperature decrease usually induces an ordering effect in membrane phospholipids, which can lead to membrane dysfunction [69]. Ectotherms (or 'cold-blooded' organisms) such as molluscs can counteract the ordering effects of reduced temperature by changing the structure of their membranes – a process known as homeoviscous adaptation [3]. In particular, the content of unsaturated fatty acids in the membranes of ectothermic animals is generally known to increase in response to cold. Correspondingly, studies have reported a negative correlation between water temperature and the levels of polyunsaturated fatty acids in, for example, scallops (*Pecten maximus)* [70], (*Placopecten magellanicus*) [3], clams (*Mercenaria mercenaria*) [69], mussels (*Mytilus edulis*) and oysters (*Crassostrea virginica*) [71]. A comparison of the lipid dynamics of blue mussels (*Mytilus edulis*), a cold-adapted species, and oysters (*Crassostrea virginica*), a warmer-water species, during simulated overwintering and passage to spring temperatures showed that both bivalves underwent a major remodelling of membrane phospholipids with decreases in EPA and DHA during the early stages of warming [71]. In addition, the mussels accumulated high levels of EPA in the digestive gland as triglyceride (TAG) stores prior to overwintering [71]. In contrast, oysters did not accumulate large TAG stores, thus emphasizing the contrasting use of lipid stores by oysters and mussels during colder temperature exposure [71].

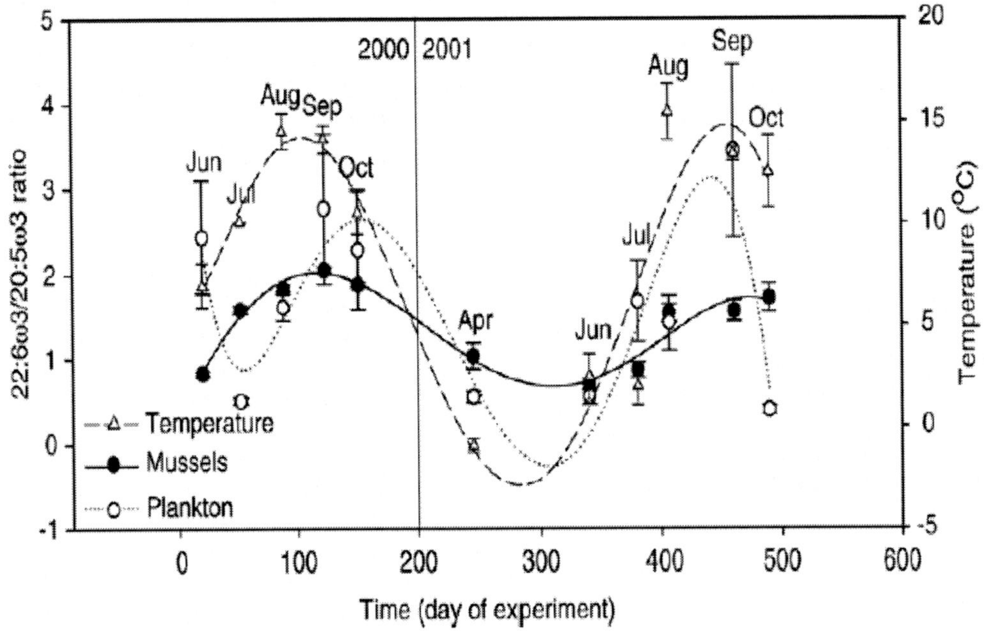

Figure 3. Relationship between water temperature and DHA:EPA ratio in cultured mussels (*Mytilus edulis)* and phytoplankton in Notre Dame Bay, Newfoundland, in 2000 and 2001 (From: [47]).

4.3. Finfish

Dietary intakes of fish and fish oil supplements are the most commonly promoted sources of very long-chain omega-3 polyunsaturated fatty acids for human health [72, 73]. Lipids and their constituent fatty acids play a major role as sources of metabolic energy in fish, thus total lipid content of fish can markedly exceed that of the protein content [74]. For example, capelin (*Mallotus vilosus*) can routinely contain 20% or more of their wet body weight as oil [74]. Fish are generally classified into three categories on the basis of their muscle fat content: lean – (<1% of muscle fat), intermediate – (5-10% of wet weight), and fatty fish (>10% muscle fat) [75]. In general, lean fish tend to accumulate lipids in their liver, have a higher water content and as a result have a whiter flesh, while fatty fish generally have elevated lipid levels in their flesh relative to lean fish, thus giving the flesh a yellow, grey or pink colour. For example, analyses of lipids in seafood showed that lean fish species, Northern cod (*Gadus morhua)* and rock cod (*Gadus ogac*) had <1% total lipid per wet weight, compared to herring (*Clupea harengus*) and capelin (*Mallotus vilosus*) – both fatty fish – with >8% lipid content [76]. The opposite trend was true for fish livers, with cod and herring containing 27% and 4% lipid in their livers, respectively [76]. Traditionally, it is these cold-water pelagic fish with fatty muscle tissue such as menhaden, herring, mackerel, and salmon, among others, which have been used for omega-3 extracts. As discussed above, these long-chain fatty acids such as EPA and DHA are produced mainly by phytoplankton and transferred through the food chain to secondary consumers such as marine fishes.

4.3.1. Variation in Omega-3 Content in Fish Due to Changes in Dietary Fatty Acids

The acquisition and accumulation of omega-3 fatty acids in fish is largely dependent on what the fish eats. Atlantic menhaden (*Brevoortia tyrannus*), which is extensively used for fish oil production, is a good example of direct transfer of fatty acids from primary producers to fish since adult fish are filter-feeders [17]. Similarly, some herbivorous fish that graze on macroalgae are directly linked to the plant material that synthesizes the fatty acids and therefore closely approximate their dietary fatty acid signatures. A comparison of the lipid profiles of two lean fish species from the same area that represented different trophic levels: the herbivorous rabbitfish (*Siganus rivulatus*) and the carnivorous white sea-bream (*Diplodus sargus*), showed seasonal variability in fatty acid composition for both species [77]. However, the herbivorous rabbitfish showed less temporal change in total fat content and the proportion of omega-3 fatty acids compared to the carnivorous white sea-bream, with the latter species varying in fat content by a factor of 10 seasonally [77]. This was attributed to less seasonal variation in the macroalgal diet of the rabbitfish as opposed to the diet of white sea-bream, which varies extensively with season [77].

Thus, in addition to dietary niche and variability of the component dietary sources, seasonal variation in food availability has also been noted to influence the total lipid content and PUFA content of fish (e.g., [78, 79]. The fatty acid composition of Pacific herring (*Clupea harengus*) caught between 2001 and 2003 in waters around Denmark showed strong seasonal variation in EPA content with significantly higher concentrations in summer than winter [79]. This seasonal variation was attributed to the lack of feeding during winter and consequently a depletion of storage lipids [80]. Conversely, no seasonal variation was noted for DHA [79]. In this study, omega-3 content was mainly related to catch season and there was no relationship with body length, which ranged from 195 to 343 mm (age 1 – 11 years) [79]. Conversely, another study on *Clupea harengus* showed a positive correlation between EPA and DHA fatty acid content and body length of herring [81]. The timing of these changes in fatty acids was consistent with known dietary shifts reported from stomach contents analyses when herring shifted from a zooplankton diet to piscivory [81].

Thus, dietary niche breadth can affect the seasonal variability in total fat content and omega-3 fatty acids among different species. It follows logically, therefore, that geographical differences in diet can also produce varying amounts of omega-3 fatty acids in the same fish species caught in different regions (for example, [82] *c.f.* [77]).

4.3.2. Species-Specific Variation in PUFA Composition in Fish

While it has long been recognized that lipid content and PUFA composition may vary among different fish species, with some being more suited for fish oil production, only in the past decade has attention been paid to the PUFA composition of commonly consumed species when selecting fish for diets [12, 24, 82-84]. These studies show that omega-3 PUFA generally dominate in marine fishes, yet individual fatty acid profiles vary considerably between species collected from the same area due to genetic differences in lipid metabolism. Nevertheless, some general trends are discernable (Table 2). The marine fish analyzed in the studies above were always richer in n-3 PUFA than n-6 PUFA, showing that they had higher proportions of the class of fatty acids associated with human health benefits. In addition, docosahexaenoic acid was the predominant PUFA in all fish and exceeded that of EPA for all marine species analyzed [12, 24, 82, 84].

4.3.3. Variation in Omega-3 Content in Fish Due to Spawning Periodicity

The lipid content and fatty acid composition of fish varies seasonally, which often corresponds to the spawning period [85-89]. The fatty acid composition from various organs of spawning and non-spawning Pacific herring (*Clupea harengus*) was compared to determine the relative effects of spawning on the nutritional value of herring [89]. The lipid content ranged from 1.9 to 3.4% (wet weight) in different tissues of spawning herring, to 10.5 to 16% in non-spawning individuals (Figure 4). The fatty acid profile exhibited many differences between the two fish groups, with spawning herring having significantly higher PUFA content in the gonads, with DHA representing the greatest proportion at more than twice that found in flesh [89]. Taken together, however, the total content of important n-3 fatty acids was low, due to the fact that herring eggs and milt contain a low amount of lipid. Moreover, due to the significant lipid loss during spawning, the absolute amounts of n-3 PUFA in the flesh of spawning fish will decrease even further. Thus non-spawning herring are a better nutritional food source than spawning herring [89]. It is also noteworthy that the n-3 fatty acid concentrations reported in herring during the spawning season were similar to those found in a lean fish species, such as walleye Pollock [89].

Table 2. Levels of total polyunsaturated fatty acids (ΣPUFA), omega-3 PUFA (ΣPUFA n-3), EPA, DHA, and the ratio of omega-3:omega-6 in selected finfish. Values are reported as % of total fatty acids and are based on muscle flesh tissue

Common name	Scientific name	ΣPUFA	ΣPUFA n-3	C20:5 n-3 (EPA)	C22:6 n-3 (DHA)	n-3/n-6	Notes	Reference
Garfish	*Belone belone*	50.3	46.3	6.1	34.9	12.2	n = 10	[84]
Atlantic menhaden	*Brevoortia tyrannus*	59.8	28.9	8.5	6.2	2.0	n = 4	[24]
Fringescale sardine	*Clupea fimbriata*	76.9	30.9	4.4	17.3	2.8	n = 4	[24]
Herring	*Clupea harenus*	20.3	17.2	7.0	8.8	18.8	n = 3	[76]
Yellow striped scad	*Delarides leptolejus*	85.5	41.3	4.0	27.3	3.5	n = 4	[24]
Sea bass	*Dicentrarchus labrax*	44.8	39.0	13.4	24.4	6.8	n = 10	[84]
Annular sea bream	*Diplodus annularis*	32.1	28.9	11.6	14.1	9.1	n = 10	[84]
Two-banded bream	*Diplodus vulgaris*	52.1	42.5	18.7	20.3	4.4	n = 10	[84]
Fourfinger threadfin	*Eleutheronema tradactylum*	89.5	29.7	5.9	9.9	1.5	n = 4	[24]
Anchovy	*Engraulis engrasicolus*	42.4	37.9	11.7	25.9	8.3	n = 10	[84]
Northern cod	*Gadus morhua*	58.3	52.5	19.1	32.6	15.6	n = 3	[76]
Rock cod	Gadus ogac	55.4	50.0	19.6	29.2	16.8	n = 3	[76]
Whiptail stingray	*Gymnura* spp.	91.5	38.9	5.2	17.5	2.2	n = 4	[24]
Hardtail scad	*Magalapsis cordyla*	86.9	48.4	5.8	28.6	4.1	n = 4	[24]
Capelin	Mallotus vilosus	20.0	15.2	9.3	4.1	17.6	n = 3	[76]
Grey mullet	*Mugil cephalus*	39.5	35.3	8.7	22.7	8.6	n = 10	[84]
Silver pomfret	*Pampus argenteus*	81.5	31.7	0.8	18.9	2.4	n = 4	[24]
Black pomfret	*Parastromateus niger*	74.5	30.5	5.2	9.4	2.3	n = 4	[24]
Striped sea catfish	*Plotosus* spp.	84.0	32.0	6.8	10.4	1.8	n = 4	[24]
Bluefish	*Pomatomus salatrix*	24.7	21.7	6.2	12.2	7.3	n = 10	[84]
Indian mackerel	*Rastrelliger kanagurta*	56.8	33.4	4.7	14.5	1.7	n = 4	[24]
Atlantic salmon	*Salmo salar*	35.8	29.7	6.0	20.7	5.0	n = 10	[84]

Table 2. (Continued)

Common name	Scientific name	ΣPUFA	ΣPUFA n-3	C20:5 n-3 (EPA)	C22:6 n-3 (DHA)	n-3/n-6	Notes	Reference
Atlantic bonito	*Sarda sarda*	33.5	30.0	8.2	19.6	9.9	n = 10	[84]
Mackerel	*Scomber scombrus*	33.6	28.9	10.2	12.7	5.6	n = 10	[84]
Spanish mackerel	*Scomberomorus commersonii*	86.7	47.9	5.9	23.3	4.3	n = 4	[24]
Gilthead sea bream	*Sparus aurata*	45.5	33.8	8.7	20.6	2.7	n = 10	[84]
Horse mackerel	*Trachurus trachurus*	22.9	21.2	7.5	10.6	12.6	n = 10	[84]

4.2.4. Variation in PUFA Content in Fish Due to Environmental Factors

As has been established for bivalves, studies on finfish have indicated that temperature is a major environmental factor influencing the fatty acid composition of fishes [90]. A review of seafood from the Indo-Pacific region found that with increased latitude there was a concomitant increase in omega-3 PUFA fatty acids, with DHA being particularly high in fish from cooler waters [90]. This was also supported in the survey of different species off the coast of South America, with Antarctic fish species, Icefish (*Chaenocephalus aceratus*) and Rock perch (*Notothenia neglecta*), having higher total n-3 PUFA content than the warmer water species caught off Brazil [12]. Thus, in general, cold-water fish present higher PUFA concentrations than those of tropical water fish. However, there are some exceptions, with some tuna species caught in warm water (e.g., frigate tuna, *Thunnus thynnus*) showing particularly high absolute values of DHA and therefore total omega-3 PUFA [12]. This latitudinal gradient is most likely because marine ectotherms can adjust their cell membrane fatty acid, sterol, and phospholipids compositions in relation to temperature [76], as indicated for bivalves.

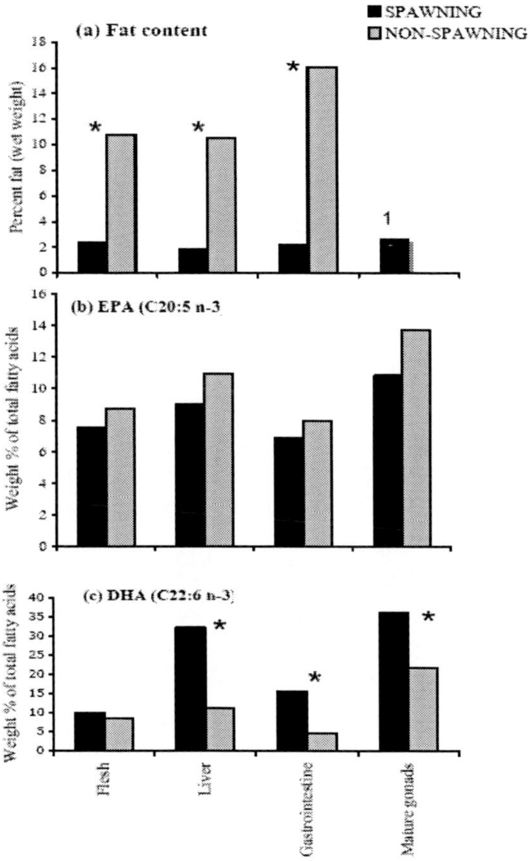

Figure 4. (a) Fat content and proportion of (b) EPA and (c) DHA in different tissues (muscle flesh; liver; gastrointestinal tract; mature gonads) from spawning and non-spawning herring (*Clupea harengus*). * = Significant difference ($p<0.05$) between the two groups of fish. [1] = average value for roe and milt. Modified from [89].

5. AQUACULTURE PROSPECTS

The lipid content and fatty acid composition of fish and bivalves are affected by various factors including genetic factors, seasonal and environmental factors, and the nutritional quality of each dietary component, as discussed previously. This can cause large variation in the nutritional quality of seafood caught from different areas and at different times of the year. Farming marine fish specifically for high PUFA content can, in principal, overcome the problem of variation in the quantity and quality of feed and therefore changes in the fatty acid composition due to dietary variance. Moreover, in aquaculture, one has greater control during the production and processing of the organisms to optimize omega-3 content. For example, one can select for genetic strains that have higher PUFA content, the location for the growout phase can be chosen to ensure optimal environmental conditions for omega-3 content of the culture species, and one has better control during harvesting to overcome any stability problems. Foods rich in PUFA are often highly susceptible to liquid peroxidation, which results in oxidative products that cause deterioration of food quality [91]. Since there is better control during the harvesting and processing in aquaculture facilities than capture fisheries, these food safety issues can be minimized. A further advantage of farming fish and shellfish for omega-3 content as opposed to harvesting wild stocks is the potential better control of environmental contaminants. Some fish and shellfish are a source of contaminants including methylmercury [21], dioxins and many other persistent organic pollutants [92]. Refining reduces the contaminant levels, especially heavy metals, and ensures that the fish oils conform to edible food standards [17]. Nevertheless, highly lipophilic contaminants may remain at low levels [17]. The approach advocated by some researchers is to choose fish and shellfish that maintain the benefits of n-3 fatty acids while minimizing the risks of exposure to contaminants [21], while other studies consider it prudent to source fish from a pristine environment [17].

However, perhaps one of the most important reasons for culturing marine organisms for omega-3 products as opposed to using capture fisheries is to ensure a sustainable resource. Fish oil remains the main source of omega-3 PUFA for human health supplements. However, diminishing wild fish stocks are forcing the industry to look at more sustainable alternatives to capture fisheries for food and related products. The supply of seafood from global capture fisheries is presently around 100 million tonnes per annum, with catches having reached a steady state since the mid-1980s, despite improvements in fishing techniques [93]. Furthermore, we are 'fishing down' our marine food webs, with the mean trophic levels of catches in decline in several important fisheries [94]. Thus, we are removing some of the most PUFA-rich fish species [17]. It is unlikely that this amount will increase in the future since most fisheries are heavily overexploited [95] and are currently at or above sustainable levels of production.

Aquaculture has been hailed as an alternative to capture fisheries and has grown at an average rate of 8.8 percent from 1950 to 2004 [93]. By 2004, aquaculture provided nearly 40 percent, or 59.4 million tonnes valued at US$ 70.3 billion, of the annual world fisheries production [93]. However, aquaculture is not always a net addition to global fish demands. As Naylor et al. [96] indicate, about 10 million tonnes of capture fisheries are used for fishmeal and fish oil to support aquaculture production. Farming of carnivorous fish such as salmon, cod, sea-bass, halibut and tuna, require large amounts of fishmeal in formulated feeds and this type of aquaculture is on the increase and expected to double by 2010 [97]. Yet if the supply of fishmeal and oil stays the same – as it has since the mid 1980s – then demand will outstrip supply of fish oil by 2010 and fishmeal by 2050 [97]. Consequently, the industry is starting to replace fish oils with vegetable oils, incorporating soybean, corn, and rapeseed (canola) into aquaculture feeds.

5.1. Replacement Oils in Aquaculture and Effects on Omega-3 Content

Feed is the largest production cost for commercial aquaculture. As the price of fish oil continues to escalate with increasing demand that exceeds supply, there are strong economic incentives to replace fishmeal and oil with vegetable oils. Apart from economic gains, there are environmental advantages of using vegetable oils instead of fish oil. Fish fed a diet containing a greater proportion of vegetable oils tend to excrete less nitrogen and phosphorus in their waste [97], the two nutrients often associated with coastal eutrophication [98]. However, there is also a downside to using large amounts of vegetable oils in aquaculture feeds. Digestion problems in fish have been noted, with fish suffering from irritated lower intestines, a depressed immune system and a diminished ability to absorb key minerals such as zinc and iron [97]. Moreover, for consumers there may be a trade-off in terms of the amount of omega-3 fatty acids in farmed fish flesh [99]. The nutritional quality of fish flesh in marine farms is largely determined by the feed and while vegetable oils impart a high concentration of oleic acid (18:1ω9), linoleic acid (18:2ω6) and α- linoleic acid (18:3ω3), they do not contain the long-chain C_{20} and C_{22} fatty acids most associated with human health benefits.

In an experiment in which turbot (*Psetta maxima*) were fed with various feeds composed of different lipid sources, the fatty acid composition of the farmed fish almost perfectly reflected that of their diet [100]. The control group were fed a feed containing fish oil with 8% EPA and 11% DHA, and the EPA and DHA concentration in the muscle of these fish was respectively 7% and 12% [100]. In contrast, turbot that received feeds rich in soybean oil and linseed oil, expressed significantly lower EPA and DHA concentrations. However, it was shown that the fatty acid composition of the fish fed vegetable oils could be altered by switching to a ‘finishing diet’ containing fish oil for a few months prior to being harvested. These latter fish subsequently exhibited fatty acid profiles very similar to those fish fed fish oils exclusively [100].

The use of vegetable oils in aquaculture feeds can therefore lower C_{20} and C_{22} omega-3 fatty acids in farmed fish compared to wild fish. However, farmed fish generally have a higher total lipid content due to their feeds that are formulated to maximize growth. It has thus been argued that the amounts of EPA and DHA provided by a given quantity of farmed fish may be higher than that produced by the same quantity of wild fish [75]. For example, some authors have stipulated that 100g fillets of farmed halibut provide 10-20 times more EPA and DHA than 100g of wild caught fish [101]. Therefore, provided they are grown under appropriate conditions, farmed fish can provide consumers with the same nutritional benefits as that provided by wild fish [75, 102].

In a study where Atlantic salmon (*Salmo salar*) parr were fed a control fish oil diet (FO) and one of three experimental diets containing 100% canola oil (CO), 100% stearidonic acid (SDA) oil (SO), and a 1:1 mix of CO and SDA (MX), the amounts of EPA, DHA and total omega-3 were comparable between fish fed the SO and FO diet [103]. SDA is a biosynthetic precursor of omega-3 long-chain PUFA and the authors argue that the EPA and DHA content were biosynthesized by the salmon fed SDA. However, the feeding trial was conducted for 42 days only and they admit that further research is needed to elucidate whether biosynthesis or greater retention occurred in the fish fed stearidonic acid [103]. It is also pertinent that this study was conducted on salmon parr, which were therefore in the freshwater stage of their lifecycle; (pre-smolt salmon exhibit increased fatty acid desaturation and elongation activities to build up reserves of essential fatty acids that are required to synthesize new cell membranes to facilitate adaptation to sea water [29, 104]). In two other studies where they used stearidonic acid oil as a replacement oil for feeding trials with marine fish, namely Arctic char (*Salvelinus alpinus*) and Atlantic cod (*Gadus morhua*), there was no evidence for increased levels of omega-3 long-chain PUFA, particularly EPA and DHA [104, 105].

Therefore, while the use of vegetable oils as a substitute for fish oils in aquaculture feeds has clear environmental benefits, there are still hurdles to overcome in terms of optimizing feed that does not compromise digestion in fish or the omega-3 content for consumers. In the future, microalgae may hold the key to producing high-quality PUFA as substitutes for fish oil in aquaculture operations [30, 44, 106-109]. Biotechnology approaches indisputably offer tremendous potential but are still undergoing development. The use of a fish oil finishing diet can increase n-3 PUFA levels in fish prior to harvesting and reduce the amount of fish oil used during a fish's life substantially – by as much as 85% for salmon [97]. However, it does not move away from reliance on ocean fish stocks for supporting aquaculture production. Consequently, it has been argued that neither wild nor farmed fish constitute a long-term sustainable source of n-3 fatty acids [72]. Other species that have been suggested as alternative sources of omega-3 for human consumption, such as krill (*Euphasia* spp.) [91], could cause widespread ecosystem perturbations if harvested in large amounts since they are the foundation of many food marine webs.

5.2. Use of Waste Products for Omega-3 Extracts

There is potential for utilization of fish-processing and aquaculture wastes as raw material for fish oil and omega-3 extracts [13, 110]. Various fish and shellfish materials such as skin, bone and internal organs, especially liver, are presently used to isolate health-promoting bioactive ingredients, including omega-3 extracts [110]. It should be cautioned however that some of the fatty deposits that are typically removed during processing (e.g., viscera) accumulate contaminants at higher concentrations than muscle tissue. Moreover, these tissues would exhibit high seasonal fluctuations, as discussed previously, thus being of inconsistent quality. Therefore, while the approach of turning waste into value-added products is alluring, special consideration needs to be given to the types of tissues used and where those tissues are sourced.

5.3. Bivalves as a Sustainable Alternative for Dietary Omega-3 Fatty Acids

Marine bivalves are grown in aquaculture farms for human consumption worldwide, with molluscs contributing 43% of global mariculture production [93]. The most important species economically is the Pacific cupped oyster (*Crassostrea gigas*), with a global production of 4.6 million tonnes (~10% of total aquaculture production by weight, excluding plants) valued at ~US$ 3 billion in 2005 [93]. Nearly 84 percent of global production of this species is in China. Other bivalves grown in aquaculture farms with global production exceeding 200 000 tonnes per annum include the Manilla clam (*Ruditapes philippinarum),* Japanese scallop (*Patinopecten yessoensis)*, blue mussel (*Mytilus edulis),* and green-lipped mussel *(Perna viridis)* [93]. Consumption of shellfish is common in Asia yet remains underutilized in the Western world. Due to overestimations in the measurement of cholesterol, shellfish have unnecessarily been excluded from low-cholesterol diets [111]. However, bivalves are actually low in cholesterol (~50 mg per 100 g wet weight) compared to other traditional protein sources, such as beef (~100 mg per 100 g wet weight). Moreover, they contain a complex mix of sterols in addition to cholesterol including the phytosterols that are believed to aid in lowering cholesterol in humans [48]. Thus, the amount of fat and the proportions of saturated and polyunsaturated fat in bivalves contribute to their nutritional value.

Sea-based culture of mussels, oysters, clams, etc. is a particularly attractive form of aquaculture as the animals utilize plankton and organic particles from the surrounding water and therefore do not require the addition of artificial food supplements [96]. Concerns have been expressed that intensive bivalve culture may compete with other filter-feeding organisms for plankton in the marine environment. However, provided the industry takes cognizance of the carrying capacity of an area [112], bivalves offer a more sustainable product to farm. In contrast, cage farming of carnivorous fish requires large amounts of fossil fuel inputs [113] and 2-5 times more fish protein than is supplied by the farmed product [96]. Further, there is tremendous potential for bivalves to be integrated with other culture organisms in a polyculture system, whereby the wastes from one type of culture are used as resources in the other [113]. In China, mussel culture ropes, scallop culture net cages and seaweed (*Laminaria* and *Undaria)* culture ropes are grown together. The seaweeds enhance production of the bivalves through the release of oxygen and cause hydrodynamic changes that stimulate plankton growth, while the bivalves produce metabolic by-products including nitrogen and CO_2 that act as fertilizers for seaweeds [114]. Although investment costs are higher, this mussel/kelp polyculture is estimated to earn 58% more market return than monoculture of each species [114]. There is potential to extend this idea to encompass finfish-bivalve-seaweed polyculture [113]. Such systems are aimed at increased efficiency, reduced resource use, less waste generation, and the recycling of nutrients and materials [113].

6. Conclusion

In conclusion, the long-chain omega-3 polyunsaturated fatty acids, specifically eicosapentaenoic acid (EPA; C20:5 n-3) and docosahexaenoic acid (DHA; C22:6 n-3), are essential fatty acids that promote human health and nutrition. These fatty acids have traditionally been supplied from dietary intakes of fish and fish oil supplements. Cold-water pelagic fish species, in particular, are rich in polyunsaturated fatty acids often with DHA predominating. However, the global decline in fisheries has pressurized the industry to identify sustainable alternatives for omega-3 sources. Farming marine organisms for high PUFA content may address this growing market for functional foods and nutraceuticals.

Marine bivalves, in particular, are an underutilized dietary omega-3 resource that generally has low fat content, elevated levels of long-chain highly unsaturated fatty acids, and phytosterols believed to lower cholesterol in humans. Moreover, in an aquaculture setting there is better control over the factors that cause variation in omega-3 content in organisms, both during the growout phase and harvesting. Aquaculture, by supplying a potential future source of omega-3 fatty acids, may in fact secure a healthier diet that more closely mirrors our original evolutionary seafood diet.

REFERENCES

[1] Burdge GC, Wootton SA. (2002). Conversion of α-linolenic acid to eicosapentaenoic, docosapentaenoic and docosahexaenoic acids in young women. *British Journal of Nutrition 88*: 411-420.

[2] Nettleton JA. (1995). *Omega-3 fatty acids and health.* New York, Chapman and Hall.

[3] Hall JM, Parrish CC, Thompson RJ. (2002). Eicosapentaenoic acid regulates scallop (*Placopecten magellanicus*) membrane fluidity in response to cold. *Biological Bulletin 202(3)*: 201-203.

[4] Valentine RC, Valentine DL. (2004). Omega-3 fatty acids in cellular membranes: a unified concept. *Progress in Lipid Research 43(5)*: 383-402.

[5] Dyerberg J, Bang HO, Stoffersen E, Moncada S, Vane JR. (1978). Eicosapentanoic Acid and Prevention of Thrombosis and Atherosclerosis. *Lancet 2(8081)*: 117-119.

[6] Astorg P, Guesnet P, Alessandri JM, Galan P, Lavialle M. (2006). Omega-3 polyunsaturated fatty acids and health: a mini-review of current knowledge. *Sciences Des Aliments 26(1)*: 8-28.

[7] Gibson SLM, Gibson RG. (1998). The treatment of arthritis with a lipid extract of *Perna canaliculus*: a randomized trial. *Complementary Therapies in Medicine 6(3)*: 122-126.

[8] Emelyanov A, Fedoseev G, Krasnoschekova O, Abulimity A, Trendeleva T, Barnes PJ. (2002). Treatment of asthma with lipid extract of New Zealand green-lipped mussel: a randomised clinical trial. *European Respiratory Journal 20*: 596-600.

[9] Luzon-Toro B, Schaefer S, Wirths O, Jimenez-Lopez J, Boza-Puerta J, Geerlings A, Falkai P, Bayer TA. (2004). Dietary fatty acids and brain mechanism of action throughout the lifespan. *Neurology Psychiatry and Brain Research 11(3)*: 149-160.

[10] Makrides M, Neumann MA, Byard RW, Simmer K, Gibson RA. (1994). Fatty-Acid Composition of Brain, Retina, and Erythrocytes in Breast-Fed and Formula-Fed Infants. *American Journal of Clinical Nutrition 60(2):* 189-194.

[11] Wen ZY, Chen F. (2003). Heterotrophic production of eicosapentaenoic acid by microalgae. *Biotechnology Advances 21(4):* 273-294.

[12] Visentainer JV, Noffs MD, Carvalho PDO, de Almeida VV, de Oliveira CC, de Souza NE. (2007). Lipid content and fatty acid composition of 15 marine fish species from the southeast coast of Brazil. *Journal of the American Oil Chemists Society 84(6):* 543-547.

[13] Shahidi F. (2007). Omega-3 Oils: Sources, Applications, and Health Effects In: Barrow CJ, Shahidi F, (Eds). *Marine Nutraceuticals and Functional Foods*. Boca Raton, FL, CRC Press, 23-61.

[14] Robson A. (2006). Shellfish view of omega-3 and sustainable fisheries. *Nature 444(7122)*: 1002-1002.

[15] Simopoulos AP. (2002). The importance of the ratio of omega-6/omega-3 essential fatty acids. *Biomedicine and Pharmacotherapy 56(8)*: 365-379.

[16] Burdge GC, Jones AE, Wootton SA. (2002). Eicosapentaenoic and docosapentaenoic acids are the principal products of α-linolenic acid metabolism in young men. *British Journal of Nutrition 88:* 355-363.

[17] Arts MT, Ackman RG, Holub BJ. (2001). "Essential fatty acids" in aquatic ecosystems: a crucial link between diet and human health and evolution. *Canadian Journal of Fisheries and Aquatic Sciences 58(1)*: 122-137.

[18] Kamal-Eldin A, Yanishlieva NV. (2002). N-3 fatty acids for human nutrition: stability considerations. *European Journal of Lipid Science and Technology 104(12):* 825-836.

[19] Newton I, Snyder D. (1997). Nutritional aspects of long-chain omega-3 fatty acids and their use in bread enrichment. *Cereal Foods World 42(3):* 126-131.

[20] Gerster H. (1998). Can adults adequately convert alpha-linolenic acid (18:3n-3) to eicosapentaenoic acid (20:5n-3) and docosahexaenoic acid (22:6n-3)? *International Journal for Vitamin and Nutrition Research 68(3)*: 159-173.

[21] Mahaffey KR. (2004). Fish and shellfish as dietary sources of methylmercury and the omega-3 fatty acids, eicosahexaenoic acid and docosahexaenoic acid: risks and benefits. *Environmental Research 95(3)*: 414-428.

[22] Kelley DS, Taylor PC, Nelson GJ, Mackey BE. (1998). Dietary docosahexaenoic acid and immunocompetence in young healthy men. *Lipids 33(6):* 559-566.

[23] Kelley DS, Taylor PC, Nelson GJ, Schmidt PC, Ferretti A, Erickson KL, Yu R, Chandra RK. (1999). Docosahexaenoic acid ingestion inhibits natural killer cell activity and production of inflammatory mediators in young healthy men. *Lipids 34(4)*: 317-324.

[24] Osman H, Suriah AR, Law EC. (2001). Fatty acid composition and cholesterol content of selected marine fish in Malaysian waters. *Food Chemistry 73(1)*: 55-60.

[25] Shahidi F, Wanasundara UN. (1998). Omega-3 fatty acid concentrates: Nutritional aspects and production technologies. *Trends in Food Science and Technology 9(6):* 230-240.

[26] Elvevoll EO, Barstad H, Breimo ES, Brox J, Eilertsen KE, Lund T, Olsen JO, Osterud B. (2006). Enhanced incorporation of n-3 fatty acids from fish compared with fish oils. *Lipids 41(12)*: 1109-1114.

[27] Kinsella JE. (1986). Food Components with Potential Therapeutic Benefits - the n-3-Polyunsaturated Fatty-Acids of Fish Oils. *Food Technology 40(2)*: 89-97.

[28] Piggot GM, Tucker BW. (1990). *Seafood: Effects of technology on nutrition*. Boca Raton, FL, CRC Press.

[29] Kanazawa A, Teshima SI, Ono K. (1979). Relationship between Essential Fatty-Acid Requirements of Aquatic Animals and the Capacity for Bioconversion of Linolenic Acid to Highly Unsaturated Fatty-Acids. *Comparative Biochemistry and Physiology B-Biochemistry and Molecular Biology 63(3)*: 295-298.

[30] Guschina IA, Harwood JL. (2006). Lipids and lipid metabolism in eukaryotic algae. *Progress in Lipid Research 45(2):* 160-186.

[31] Volkman JK, Jeffrey SW, Nichols PD, Rogers GI, Garland CD. (1989). Fatty-Acid And Lipid-Composition Of 10 Species Of Microalgae Used In Mariculture. *Journal Of Experimental Marine Biology And Ecology 128(3)*: 219-240.

[32] Dunstan GA, Volkman JK, Barrett SM, Leroi JM, Jeffrey SW. (1994). Essential Polyunsaturated Fatty-Acids from 14 Species of Diatom (Bacillariophyceae). *Phytochemistry 35(1)*: 155-161.

[33] Skerratt JH, Nichols PD, McMeekin TA, Burton H. (1995). Seasonal and Interannual Changes in Planktonic Biomass and Community Structure in Eastern Antarctica Using Signature Lipids. *Marine Chemistry 51(2)*: 93-113.

[34] Sushchik NN, Gladyshev MI, Makhutova ON, Kalachova GS, Kravchuk ES, Ivanova EA. (2004). Associating particulate essential fatty acids of the omega 3 family with phytoplankton species composition in a Siberian reservoir. *Freshwater Biology 49(9)*: 1206-1219.

[35] Shamsudin L. (1998). Seasonal variation of fatty acid content in natural microplankton from the Tumpat coastal waters of the South China Sea. *Archives of Physiology and Biochemistry 106(3)*: 253-260.

[36] Budge SM, Parrish CC. (1998). Lipid biogeochemistry of plankton, settling matter and sediments in Trinity Bay, Newfoundland. II. Fatty acids. *Organic Geochemistry 29(5-7)*: 1547-1559.

[37] Budge SM, Parrish CC, McKenzie CH. (2001). Fatty acid composition of phytoplankton, settling particulate matter and sediments at a sheltered bivalve aquaculture site. *Marine Chemistry 76(4):* 285-303.

[38] Thompson PA, Guo MX, Harrison PJ, Whyte JNC. (1992). Effects of Variation in Temperature II. On the Fatty-Acid Composition of 8 Species of Marine-Phytoplankton. *Journal of Phycology 28(4):* 488-497.

[39] Hu HH, Gao KS. (2006). Response of growth and fatty acid compositions of Nannochloropsis sp to environmental factors under elevated CO2 concentration. *Biotechnology Letters 28(13)*: 987-992.

[40] Thompson PA, Guo MX, Harrison PJ. (1994). Influence of Irradiance on the Nutritional-Value of 2 Phytoplankton Species Fed to Larval Japanese Scallops (Patinopecten-Yessoensis). *Marine Biology 119(1)*: 89-97.

[41] Wang KS, Chai TJ. (1994). Reduction in Omega-3-Fatty-Acids by UV-B Irradiation in Microalgae. *Journal of Applied Phycology 6(4)*: 415-421.

[42] Mortensen SH, Borsheim K, Y., Rainuzzo JR, Knutsen G. (1988). Fatty acid and elemental composition of the marine diatom *Chaetoceros gracilis* Schutt. Effects of silicate deprivation, temperature and light intensity. *Journal of Experimental Marine Biology and Ecology 122:* 173-185.

[43] Thompson PA, Guo M-X, Harrison PJ, Whyte JNC. (1992). Effects of variation in temperature. 2. On the fatty acid composition of eight species of marine phytoplankton. *Journal of Phycology 28(4):* 488-497.

[44] Jiang HM, Gao KS. (2004). Effects of lowering temperature during culture on the production of polyunsaturated fatty acids in the marine diatom *Phaeodactylum tricornutum* (Bacillariophyceae). *Journal of Phycology 40(4):* 651-654.

[45] Hochachka PW, Somero GN. (1984). *Biochemical Adaptations*. New Jersey, Princeton University Press.

[46] Chen F, Johns MR. (1991). Effect of C/N Ratio and Aeration on the Fatty-Acid Composition of Heterotrophic *Chlorella sorokiniana. Journal of Applied Phycology 3(3)*: 203-209.

[47] Alkanani T, Parrish CC, Thompson RJ, McKenzie CH. (2007). Role of fatty acids in cultured mussels, *Mytilus edulis*, grown in Notre Dame Bay, Newfoundland. *Journal of Experimental Marine Biology and Ecology 348(1-2)*: 33-45.

[48] Murphy KJ, Mooney BD, Mann NJ, Nichols PD, Sinclair AJ. (2002). Lipid, FA, and sterol composition of New Zealand green lipped mussel (*Perna canaliculus*) and Tasmanian blue mussel (*Mytilus edulis). Lipids 37(6):* 587-595.

[49] Chan KY, Gao QF, Yip KM, Wong WH, Shin PKS, Cheung SG. (2004). Lipid content and fatty acid composition in the green-lipped mussel *Perna viridis* (L.). *Journal of Food Lipids 11(2):* 123-130.

[50] Murphy KJ, Mann NJ, Sinclair AJ. (2003). Fatty acid and sterol composition of frozen and freeze-dried New Zealand Green Lipped Mussel (*Perna canaliculus*) from three sites in New Zealand. *Asia Pacific Journal Of Clinical Nutrition 12(1):* 50-60.

[51] Taylor AG, Savage C. (2006). Fatty acid composition of New Zealand green-lipped mussels, *Perna canaliculus:* Implications for harvesting for n-3 extracts. *Aquaculture 261(1):* 430-439.

[52] Dridi S, Romdhane MS, Elcafsi M. (2007). Seasonal variation in weight and biochemical composition of the Pacific oyster, *Crassostrea gigas* in relation to the gametogenic cycle and environmental conditions of the Bizert lagoon, Tunisia. *Aquaculture 263(1-4):* 238-248.

[53] Surh J, Ryu JS, Kwon H. (2003). Seasonal variations of fatty acid compositions in various Korean shellfish. *Journal of Agricultural and Food Chemistry 51(6):* 1617-1622.

[54] Freites L, Labarta U, Fernandez-Reiriz MJ. (2002). Evolution of fatty acid profiles of subtidal and rocky shore mussel seed (*Mytilus galloprovincialis*, Lmk.). Influence of environmental parameters. *Journal Of Experimental Marine Biology And Ecology 268(2)*: 185-204.

[55] Leonardos N, Lucas IAN. (2000). The use of larval fatty acids as an index of growth in *Mytilus edulis* L-larvae. *Aquaculture 184(1-2)*: 155-166.

[56] Abad M, Ruiz C, Martinez D, Mosquera G, Sanchez JL. (1995). Seasonal-Variations of Lipid Classes and Fatty-Acids in Flat Oyster, *Ostrea edulis,* from San-Cibran (Galicia, Spain). *Comparative Biochemistry and Physiology C - Pharmacology Toxicology and Endocrinology 110(2):* 109-118.

[57] Caers M, Coutteau P, Cure K, Morales V, Gajardo G, Sorgeloos P. (1999). The Chilean scallop *Argopecten purpuratus* (Lamarck, 1819): I. fatty acid composition and lipid content of six organs. *Comparative Biochemistry and Physiology B -Biochemistry and Molecular Biology 123(1)*: 89-96.

[58] De Moreno JEA, Moreno VJ, Brenner RR. (1976). Lipid-Metabolism of Yellow Clam, *Mesodesma mactroides* - 2- Polyunsaturated Fatty-Acid Metabolism. *Lipids 11(7)*: 561-566.

[59] Waldock MJ, Holland DL. (1984). Fatty-Acid Metabolism in Young Oysters, *Crassostrea gigas* - Polyunsaturated Fatty-Acids. *Lipids 19(5):* 332-336.

[60] Berntsson KM, Jonsson PR, Wangberg SA, Carlsson AS. (1997). Effects of broodstock diets on fatty acid composition, survival and growth rates in larvae of the European flat oyster, *Ostrea edulis*. *Aquaculture 154(2):* 139-153.

[61] Linehan LG, O'Connor TP, Burnell G. (1999). Seasonal variation in the chemical composition and fatty acid profile of Pacific oysters (*Crassostrea gigas*). *Food Chemistry 64(2)*: 211-214.

[62] Lin H, Jiang J, Xue CH, Zhang B, Xu JC. (2003). Seasonal changes in phospholipids of mussel (*Mytilus edulis* Linne). *Journal Of The Science Of Food And Agriculture 83(2)*: 133-135.

[63] Camacho AP, Albentosa M, Fernandez-Reiriz MJ, Labarta U. (1998). Effect of microalgal and inert (cornmeal and cornstarch) diets on growth performance and biochemical composition of *Ruditapes decussatus* seed. *Aquaculture 160(1-2):* 89-102.

[64] Mackenzie AL, Gillespie PA. (1986). Plankton Ecology And Productivity, Nutrient Chemistry, And Hydrography Of Tasman Bay, New-Zealand, 1982-1984. *New Zealand Journal Of Marine And Freshwater Research 20(3)*: 365-395.

[65] Freites L, Fernandez-Reiriz MJ, Labarta U. (2002). Fatty acid profiles of *Mytilus galloprovincialis* (Lmk) mussel of subtidal and rocky shore origin. *Comparative Biochemistry And Physiology B - Biochemistry and Molecular Biology 132(2):* 453-461.

[66] Delaporte M, Soudant P, Moal J, Kraffe E, Marty Y, Samain JF. (2005). Incorporation and modification of dietary fatty acids in gill polar lipids by two bivalve species *Crassostrea gigas* and *Ruditapes philippinarum.* Co*mparative Biochemistry and Physiology A - Molecular and Integrative Physiology 140(4):* 460-470.

[67] Kluytmans JH, Boot JH, Oudejans R, Zandee DI. (1985). Fatty-Acid Synthesis In Relation To Gametogenesis In The Mussel *Mytilus edulis* - L. *Comparative Biochemistry And Physiology B - Biochemistry and Molecular Biology 81(4):* 959-963.

[68] Soudant P, Marty Y, Moal J, Samain JF. (1996). Fatty acids and egg quality in great scallop. *Aquaculture International 4(3):* 191-200.

[69] Pernet F, Tremblay R, Gionet C, Landry T. (2006). Lipid remodeling in wild and selectively bred hard clams at low temperatures in relation to genetic and physiological parameters. *Journal of Experimental Biology 209(23):* 4663-4675.

[70] Pazos AJ, Sanchez JL, Roman G, Perez-Paralle ML, Abad M. (2003). Seasonal changes in lipid classes and fatty acid composition in the digestive gland of *Pecten maximus*. *Comparative Biochemistry and Physiology B - Biochemistry and Molecular Biology 134(2):* 367-380.

[71] Pernet F, Tremblay R, Comeau L, Guderley H. (2007). Temperature adaptation in two bivalve species from different thermal habitats: energetics and remodelling of membrane lipids. *Journal of Experimental Biology 210(17):* 2999-3014.

[72] Racine RA, Deckelbaum RJ. (2007). Sources of the very-long-chain unsaturated omega-3 fatty acids: eicosapentaenoic acid and docosahexaenoic acid. *Current Opinion in Clinical Nutrition and Metabolic Care 10(2):* 123-128.

[73] Kolanowski W, Laufenberg G. (2006). Enrichment of food products with polyunsaturated fatty acids by fish oil addition. *European Food Research and Technology 222(3-4)*: 472-477.

[74] Tocher DR. (2003). Metabolism and functions of lipids and fatty acids in teleost fish. *Reviews in Fisheries Science 11(2)*: 107-184.

[75] Cahu C, Salen P, de Lorgeril M. (2004). Farmed and wild fish in the prevention of cardiovascular diseases: Assessing possible differences in lipid nutritional values. *Nutrition Metabolism and Cardiovascular Diseases 14(1):* 34-41.

[76] Copeman LA, Parrish CC. (2004). Lipids classes, fatty acids, and sterols in seafood from gilbert bay, southern Labrador. *Journal of Agricultural and Food Chemistry 52(15)*: 4872-4881.

[77] Saoud IP, Batal M, Ghanawi J, Lebbos N. (2007). Seasonal variation in highly unsaturated fatty acid composition of muscle tissue of two fishes endemic to the eastern Mediterranean. *Ecology of Food and Nutrition 46(2)*: 77-89.

[78] Aidos I, van der Padt A, Luten JB, Boom RM. (2002). Seasonal, changes in crude and lipid composition of herring fillets, byproducts, and respective produced oils. *Journal of Agricultural and Food Chemistry 50(16):* 4589-4599.

[79] Jensen KN, Jacobsen C, Nielsen HH. (2007). Fatty acid composition of herring (*Clupea harengus* L.): influence of time and place of catch on n-3 PUFA content. *Journal of the Science of Food and Agriculture 87(4):* 710-718.

[80] Slotte A. (1999). Effects of fish length and condition on spawning migration in Norwegian spring spawning herring (*Clupea harengus* L.). *Sarsia 84(2):* 111-127.

[81] Iverson SJ, Frost KJ, Lang SLC. (2002). Fat content and fatty acid composition of forage fish and invertebrates in Prince William Sound, Alaska: factors contributing to among and within species variability. *Marine Ecology Progress Series 241*: 161-181.

[82] Soriguer F, Serna S, Valverde E, Hernando J, MartinReyes A, Soriguer M, Pareja A, Tinahones F, Esteva I. (1997). Lipid, protein, and calorie content of different Atlantic and Mediterranean fish, shellfish, and molluscs commonly eaten in the south of Spain. *European Journal of Epidemiology 13(4):* 451-463.

[83] Haliloglu HI, Aras NM, Yetim H. (2002). Comparison of muscle fatty acids of three trout species (*Salvelinus alpinus, Salmo trutta fario, Oncorhynchus mykiss*) raised under the same conditions. *Turkish Journal of Veterinary and Animal Sciences 26(5):* 1097-1102.

[84] Bayir A, Haliloglu HI, Sirkecioglu AN, Aras NM. (2006). Fatty acid composition in some selected marine fish species living in Turkish waters. *Journal of the Science of Food and Agriculture 86(1):* 163-168.

[85] Bandarra NM, Batista I, Nunes ML, Empis JM, Christie WW. (1997). Seasonal changes in lipid composition of sardine (*Sardina pilchardus). Journal of Food Science 62(1)*: 40-42.

[86] Uysal K, Aksoylar MY. (2005). Seasonal variations in fatty acid composition and the N-6/N-3 fatty acid ratio of pikeperch (*Sander lucioperca*) muscle lipids. *Ecology of Food and Nutrition 44(1):* 23-35.

[87] Cejas JR, Almansa E, Villamandos JE, Badia P, Bolanos A, Lorenzo A. (2003). Lipid and fatty acid composition of ovaries from wild fish and ovaries and eggs from captive fish of white sea bream (*Diplodus sargus*). *Aquaculture 216(1-4)*: 299-313.

[88] Bransden MP, Battaglene SC, Goldsmid RM, Dunstan GA, Nichols PD. (2007). Broodstock condition, egg morphology and lipid content and composition during the spawning season of captive striped trumpeter, *Latris lineata. Aquaculture 268(1-4)*: 2-12.

[89] Huynh MD, Kitts DD, Hu C, Trites AW. (2007). Comparison of fatty acid profiles of spawning and non-spawning Pacific herring, *Clupea harengus pallasi. Comparative Biochemistry and Physiology B - Biochemistry and Molecular Biology 146(4):* 504-511.

[90] Dunstan GA, Olley J, Ratkowsky DA. (1999). Major environmental and biological factors influencing the fatty acid composition of seafood from Indo-Pacific to Antarctic waters. *Recent Research Developments in Lipids Research 3:* 63-86.

[91] Tou JC, Jaczynski J, Chen YC. (2007). Krill for human consumption: Nutritional value and potential health benefits. *Nutrition Reviews 65(2)*: 63-77.

[92] Hites RA, Foran JA, Carpenter DO, Hamilton MC, Knuth BA, Schwager SJ. (2004). Global assessment of organic contaminants in farmed salmon. *Science 303(5655):* 226-229.

[93] FAO. (2006). *State of world aquaculture*. In FAO Fisheries Technical Paper vol 134. Rome, FAO Fisheries Department.

[94] Pauly D, Christensen V, Dalsgaard J, Froese R, Torres F. (1998). Fishing down marine food webs. *Science 279(5352)*: 860-863.

[95] Myers RA, Worm B. (2003). Rapid worldwide depletion of predatory fish communities. *Nature 423(6937):* 280-283.

[96] Naylor RL, Goldburg RJ, Primavera JH, Kautsky N, Beveridge MCM, Clay J, Folke C, Lubchenco J, Mooney H, Troell M. (2000). Effect of aquaculture on world fish supplies. *Nature 405(6790):* 1017-1024.

[97] Powell K. (2003). Eat your veg. *Nature 426(6965)*: 378-379.

[98] Nixon SW. (1995). Coastal marine eutrophication: a definition, social causes, and future concerns. *Ophelia 41:* 199-219.

[99] Karapanagiotidis IT, Bell MV, Little DC, Yakupitiyage A. (2007). Replacement of dietary fish oils by alpha-linolenic acid-rich oils lowers omega 3 content in tilapia flesh. *Lipids 42(6):* 547-559.

[100] Regost C, Arzel J, Robin J, Rosenlund G, Kaushik SJ. (2003). Total replacement of fish oil by soybean or linseed oil with a return to fish oil in turbot (*Psetta maxima*) - 1. Growth performance, flesh fatty acid profile, and lipid metabolism. *Aquaculture 217(1-4):* 465-482.

[101] Olsson GB, Olsen RL, Carlehog M, Ofstad R. (2003). Seasonal variations in chemical and sensory characteristics of farmed and wild Atlantic halibut (*Hippoglossus hippoglossus*). *Aquaculture 217(1-4)*: 191-205.

[102] Nichols PD, Brock M, Dunstan GA, Bransden MP, Goldsmid RM, Battaglene SC. (2005). Potential new Australian farmed seafood source of omega-3 oils: Oil and fatty acid composition of the finfish striped trumpeter (*Latris lineata*). *Food Australia 57(7):* 295-301.

[103] Miller MR, Nichols PD, Carter CG. (2007). Replacement of dietary fish oil for Atlantic salmon parr (*Salmo salar* L.) with a stearidonic acid containing oil has no effect on omega-3 long-chain polyunsaturated fatty acid concentrations. *Comparative Biochemistry and Physiology B - Biochemistry and Molecular Biology 146(2):* 197-206.

[104] Bell JG, Strachan F, Good JE, Tocher DR. (2006). Effect of dietary echium oil on growth, fatty acid composition and metabolism, gill prostaglandin production and macrophage activity in Atlantic cod (*Gadus morhua* L.). *Aquaculture Research 37(6):* 606-617.

[105] Tocher DR, Dick JR, MacGlaughlin P, Bell JG. (2006). Effect of diets enriched in Δ6 desaturated fatty acids (18:3n-6 and 18:4n-3), on growth, fatty acid composition and highly unsaturated fatty acid synthesis in two populations of Arctic charr (*Salvelinus alpinus* L.). *Comparative Biochemistry and Physiology B - Biochemistry and Molecular Biology 144(2):* 245-253.

[106] Benemann JR. (1992). Microalgae Aquaculture Feeds. *Journal of Applied Phycology 4(3):* 233-245.

[107] Carvalho AP, Malcata FX. (2005). Optimization of omega-3 fatty acid production by microalgae: Crossover effects of CO_2 and light intensity under batch and continuous cultivation modes. *Marine Biotechnology 7(4):* 381-388.

[108] Dunstan GA, Volkman JK, Barrett SM, Garland CD. (1993). Changes in the Lipid-Composition and Maximization of the Polyunsaturated Fatty-Acid Content of 3 Microalgae Grown in Mass-Culture. *Journal of Applied Phycology 5(1):* 71-83.

[109] de Swaaf ME, de Rijk TC, Eggink G, Sijtsma L. (1999). Optimisation of docosahexaenoic acid production in batch cultivations by *Crypthecodinium cohnii. Journal of Biotechnology 70(1-3)*: 185-192.

[110] Kim S-K, Mendis E, Shahidi F. (2007). Marine Fisheries By-Products and Potential Nutraceuticals: An Overview In: Barrow CJ, Shahidi F, (Eds.) *Marine Nutraceuticals and Functional Foods*. Boca Raton, FL, CRC Press, 1-22.

[111] Krzynowek J. (1985). Sterols and Fatty-Acids in Seafood. *Food Technology 39(2)*: 61-68.

[112] Gibbs MT. (2007). Sustainability performance indicators for suspended bivalve aquaculture activities. *Ecological Indicators 7(1)*: 94-107.

[113] Folke C, Kautsky N. (1992). Aquaculture with Its Environment - Prospects for Sustainability. *Ocean and Coastal Management 17(1):* 5-24.

[114] Brzeski V, Newkirk G. (1997). Integrated coastal food production systems - A review of current literature. *Ocean and Coastal Management 34(1):* 55-71.

In: Aquaculture Research Progress
Editor: Takumi K. Nakamura
ISBN 978-1-60456-247-7

Chapter 6

PROSPECTS AND DEVELOPMENT IN FISH SPERM AND EMBRYO CRYOPRESERVATION

Vanesa Robles[1], Elsa Cabrita[2], Vikram Kohli[3], and M. Paz Herráez[4]

[1] Instituto de Sanidad Animal y Desarrollo Ganadero, University of León, Spain
[2] Center for Marine Science-CCMAR, University of Algarve, Portugal
[3]Department of Electrical and Computer Engineering, University of Alberta, Edmonton, Alberta
[4]Department of Cell Biology, University of León, Spain

ABSTRACT

The Industry of Aquaculture has been profiting from fish sperm cryopreservation for many years. Sperm from more than two hundred fish species, most of them with commercial value, has been cryopreserved, allowing year round availability of sperm and the creation of sperm banks from those individuals with special interest. However, to date, fish embryo cryopreservation has not been successfully achieved in any teleost species. The benefits of fish embryo cryopreservation are numerous, not only for aquaculture but also for conservation purposes. This technique will allow for the maintenance of a constant supply of animals, reduce the facilities required on the fish farm, facilitate animal transportation between different farms and enable the preservation of valuable lines. These practical advantages will also report a cost reduction and minimize the impact of epidemics on productivity. This chapter presents an overview of the recent advances in this field and their impact on Aquaculture.

SPERM CRYOPRESERVATION

Cryopreservation is a technique with undoubted interest, not only for fish farming but also for the conservation and genetic improvement of broodstocks. This technique has been established in some freshwater fish species mainly, salmonids, sturgeons and carps, however, only in the last decade intensive work was done in marine species. The benefits of sperm cryopreservation include: 1- synchronization of gamete availability of both sexes: sperm can be stored and used when the eggs are available; 2- sperm economy: it allows the use of total volume of available sperm. This is useful when semen is difficult to obtain and also when low volume of semen can be stripped in captivity; 3- simplifies broodstock maintenance: off season spawning can be induced only in females and cryopreserved sperm can be used to fertilize the eggs, and 4- allows the transport of gametes from different fish farms, avoiding animal transportation and their requirements and costs. Moreover, due to the high domestication of cultured species, several inbreeding problems have been identified in broodstocks, being necessary more concerned about broodstocks genetic improvement and selection of specific strain lines.

Thus, the use of sperm cryobanks would be of undoubtedly interest not only from a productive point of view, but also can benefit the reposition of stocks from a genetic point a view. Sperm cryobanking has been used in conservation programs, genetic improving of wild and captive populations and in the selection of specific fish characteristics based on their reproductive performance. This last issue would certainly benefit aquaculture industry.

Successful attempt has been done to cryopreserve sperm from most commercialized fish species, such as turbot (*Scophthalmus maximus*) [1,2], gilthead seabream (*Sparus aurata*) [3,4], European seabass (*Dicentrarchus labrax*) [5,6], European and Japanese eel (*Anguilla anguilla, Anguilla japonica*) [7,8], striped bass (*Morone saxatilis*) [9], sharpsnout seabream (*Diplodus puntazzo*) [10], halibut (*Hippoglossus hippoglossus*) [11], Winter flounder (*Pleuronectes americanus*) [12], dusky and malaba grouper (*Epinephelus marginatus, Epinephelus malabaricus* [13, 14] cod and haddock (*Gadus morhua, Melanogrammus aeglefinus*) [15]. Indeed, regardless marine species, some good results have been achieved, but there is not a concern about the standardization of protocols including an exhaustive description of procedures and evaluation of sperm quality before and after freezing, giving special attention to initial sperm quality and to specific damage occurring during cryopreservation. From a production, conservation and specially, research point of view this is a crucial mater since sperm quality must be guarantee.

Standardized methods have been proposed in human spermatozoa, however, no such procedures have been developed to standardize sperm analysis of domestically species, especially fish. The development of standardized protocols for the analysis of fish sperm quality before and after cryopreservation would be crucial to guarantee reproducible results within species worldwide. Nonetheless, this is far from being achieved in fish. Although sperm from a wide range of fish species have been cryopreserved, few protocols were adapted and standardize for commercial scale [16]. This could be one of the principal bottlenecks in the cryopreservation of fish sperm and in the spreadness and commercialization of samples.

Semen characteristics vary among species, stocks and even within samples from the same animal, depending on the collection period during the reproductive season. Therefore, the assay of seminal quality and the development of tools to characterize this quality should be used extensively for quality control in research on seminal physiology and in commercial artificial breeding programs. Most of these quality parameters should allow to predict fertilizing potential as well as to establish stocks of good quality. It is well known that sperm samples have different liability for freezing, according to their quality. Thus, it is crucial an evaluation of sperm in order to select the best quality samples for freezing.

Cryopreservation produce several damage in cells, affecting plasma membrane, organelles (in particular the mitochondrion) and it can also produce damage at chromatin level. Cell damage can be produced by mechanical forces due to ice crystals formation, both within the cells and in the external medium or to osmotic stress [17]. The conservation of each cell structure as well as its functionality will depend on the viability of the cryopreservation protocol. Plasma membrane is the principal target of cell damage. In order to fertilize the eggs, sperm must conserve their integrity and cells must maintain the control of the osmotic regulation during the fertilization process. Thus, membrane integrity and functionality can not be compromised. Moreover, spermatozoa need to maintain its moving capacity when released into water in order to achieve the oocyte. Spermatozoa motility is crucial and depends on different aspects of the cell, such as mitochondria status, ATP production, plasma membrane channels and flagella structure. Any change in these aspects will compromise motility. Also, DNA integrity must be guaranteed to accomplish successful embryo development after fertilization.

The assessment of sperm quality using these tools will be decisive for the selection of good sperm samples and for the standardization of the designed cryopreservation protocols.

Embryo Cryopreservation

Fish embryo cryopreservation has yet to be achieved. Successful preservation of teleost embryos would undoubtedly impact the aquaculture industry. The ability to cryopreserve embryos would be important to genetic resource banking for the maintenance of transgenic fish lines and the re-population of exotic species for wildlife conservation. A resource bank of embryos would simplify the transfer of animals between fish farms and improve hatchery operations by reducing the spread of epidemics while maintaining storage of valuable genetic lineages. Despite these benefits, fish embryo preservation has been tedious, with active research being pursued in the cryopreservation of numerous teleost species including the winterflounder (*Pleuronectes americanus*), the Japanese flounder (*Paralichthys olivaceus*), the zebrafish (*Danio rerio*), the turbot (*Scophthalmus maximus*), the gilthead seabream (*Sparus aurata*) and the medaka (*Orizias latipes*). Zebrafish and medaka have received particular interest among vertebrate embryologists, zoologists and developmental biologists as important model systems in medical and biological research, while others like turbot and gilthead seabream have significant commercial value in the fisheries industry. The zebrafishmodel system has been used in the study of genetics, drug monitoring, human disease, cardiac function and blood disorders [18-25]. While the race to cryopreserve fish embryos continues, the field of cryobiology has expanded, eliciting the help of embryologists, developmental biologists, biophysicists, physicists and engineers, with the common intent of preserving teleost embryos. It is through this multidisciplinary pursuit that we hope to achieve successful fish embryo cryopreservation in the near future. Challenges hindering embryo cryopreservation include the embryo's low surface-to-volume ratio, low membrane permeability, differing osmotic properties of the cells and yolk, the large size of the yolk and cells, the high chilling sensitivity and the relatively large osmotically inactive water fraction volumes [26-34]. Preservation requires a proper balance between cooling rates, thawing rates and cryoprotectant concentration [35-37]. Intracellular cryoprotectants function to lower the intracellular ice formation temperature allowing cells to be cooled to lower temperatures while avoiding ice formation [38, 39]. Several membrane permeable and impermeable cryoprotectants have been used in the attempted cryopreservation of fish embryos including:

methanol (MeOH), dimethyl sulfoxide (DMSO), glycerol, propane-1,2-diol, propylene glycol (PG), ethylene glycol (EG), trehalose and sucrose [27, 32-34, 40-46]. Using such cryoprotectants several studies have examined the toxicity, permeability and chilling sensitivity of various species of teleost embryos.

The chilling sensitivity and survivability of embryos exposed to various cryoprotectants has been actively studied in order to determine the optimal parameters for embryo cryopreservation. For instance, in the zebrafish, Zhang et al. [27] examined the survival of blastula (3 hrs), epiboly (7 and 10 hrs), optic vesicle (15 hrs), auditory placode (20 hrs) and heartbeat stage (27 hrs) embryos in the absence of cryoprotectants. The embryos were held in a low temperature bath at 0°C for 24 hrs. Zhang and colleagues [27] found that early developmental stages were extremely sensitive to chilling. However, the chilling sensitivity was found to decrease as the embryo aged. To determine if cryoprotectants can improve the chilling sensitivity, Zhang and colleagues [27] examined the survival (% hatch) of heartbeat stage embryos stored at 0, -5 and -15°C in a cocktail of cryoprotectants. Among the different cryoprotectants tested 1.0 M MeOH provided the best protection with a maximum survival of 79 and 69% at 0°C for 18h and 24h of storage . A combination of MeOH and sucrose was also used, with embryo survival reaching a maximum of 84%. Using this combination of cryoprotectants had minor effects on survival when the storage temperature was decreased to –15°C. In another study by Zhang et al. [32], the effect of developmental arrest on chilling injury in zebrafish embryos was investigated. Embryos were exposed to methanol or anoxic conditions before they were cooled with different cooling rates: 1, 30 and 300°C/min. Zhang found that anoxia and developmental arrest had no effect on the chilling sensitivity of these embryos.. Hagedorn et al. [33] also performed chilling resistance studies on zebrafish embryos Similar to Zhang et al. [27] results, early stage embryos (1.25, 1.5 and 1.75 and 2 hrs), were found to be the most sensitive to chilling, while 50%, 75%, 100% epiboly and 3-somite staged embryos had higher survival rates. They also demonstrated that enzymatic removal of the chorion did not alter the pattern of sensitivity to chilling [33].

The above chilling resistance studies indicate that later embryonic developmental stages are more resistant to chilling injury. It is well known that chilling sensitivity observed at slow cooling rates might be avoided using vitrification. Therefore, cryobiologists have attempted to use this method for cryopreserving teleost embryos. However, the problem with vitrification is that rapid cooling can induce intracellular ice formation, not allowing sufficient water removal through cell dehydration. Formation of intraembryonic ice causes irreparable cellular damage leading to embryo death. The addition of high concentration cryoprotectants in vitifrication is meant to prevent ice formation. However, the high cryoprotectant concentrations required in vitrification make embryo toxicity an issue. Hagedorn et al. [33] showed that exposure of 3-somite embryos to 1.5 M glycerol or EG for up to 30min at RT was deleterious to zebrafish embryos. Glycerol caused 100% embryo mortality, and EG resulted in disassociation of the blastoderm from the yolk [33]. Suzuki et al. [47] measured the toxicity of DMSO in several fish embryos including the rainbow trout, medaka, pejerrey and carp. Suzuki and colleagues did not observe any deleterious effects on embryo survival using 2.0 M DMSO. However, prolonged exposure to 2.0 M and higher DMSO concentrations resulted in increased embryo mortality. Liu et al. [41] attempted to vitrify zebrafish embryos using MeOH in a two-step process. Early and late cleavage (1-cell and 64-cell), 50 % epiboly, 6-somite and prim-6 embryos were first exposed to 2.0 M MeOH at room temperature for 10 (1-cell), 20 (64-cell) and 30 min (50 %, 6-somite and prim-6). Afterward, embryos were re-exposed to MeOH at a concentration of 10.0 M for 1 (1-cell), 3 (64-cell), 10 (50% epiboly), 5 (6-somite) and 10 min (prim-6), respectively. Embryos were vitrified in liquid nitrogen or nitrogen slush for 30 sec. Following vitrification, Liu and colleagues [41] report that 80% of the later stage embryos (50% epiboly, 6-somite and prim-6) and 50% of the early stage embryos (1-cell and 64-cells) were morphologically intact immediately after thawing. However, within 20 min the plasma membrane collapsed and the yolk ruptured in all embryo stages [41]. Intraembryonic ice formation could not be avoided despite the high concentration of MeOH used in the study [41]. In similar work, Zhang et al. [48] attempted to vitrify 6-somite and heartbeat stage (27 hrs) embryos. Following thawing and dilution of the cryoprotectants, the vitelline membrane collapsed and the yolk ruptured with the observance of intraembryonic ice [48].

As previously mentioned, intraembryonic ice formation causes cellular damage and embryo death. Ice formation has been observed in both vitrification and slow cooling studies. Hagedorn et al. [49] recently reported that when zebrafish are cooled at a rate of 2 °C/min, intraembryonic ice forms within seconds after the appearance of external ice in the extracellular solution. This ice front propagates and causes the embryo to freeze. While intraembryonic ice should be avoided in the presence of cryoprotectants, the lack of conferred protection indicates a permeability barrier. Several studies have attempted to elucidate the permeability barrier hindering the intraembryonic accumulation of cryoprotectants and the movement of water. Harvey et al. [40] measured the permeability of chorionated and dechorionated 50% epiboly zebrafish embryos to labelled DMSO ([^{14}C]DMSO) and glycerol ([^{3}H]Glycerol). Using a scintillation counter, Harvey and colleagues [40] were able to theoretically estimate permeation or percentage equilibration of both labelled cryoprotectants. Harvey et al. [40] found that after 2 hrs of incubation, glycerol permeation reached a maximum of 8% while DMSO was slightly greater than 2%. Dechorionated 50% epiboly embryos showed increased permeability (~ 9%) indicating that the chorion could represent a barrier to the permeation of cryoprotectants [40]. Hagedorn et al. [28] determined the permeability and distribution of water in 40 % epiboly to 6-somite stage zebrafish embryos, using transport equations and volumetric measurements of the embryos suspended in nonpermeating hypertonic solutions. Hydraulic conductivitiy (L_p) measurements revealed that the permeability of water changed as a function of developmental stage, with L_p ranging from 0.022 to 0.040 $\mu m \bullet min^{-1} \bullet atm^{-1}$ for 75% epiboly to 3-somite stage L_p increased by two folds for embryos at 6-somite (0.049 to 0.100 $\mu m \bullet min^{-1} \bullet atm^{-1}$). However, Hagedorn and colleagues [28] are unaware if this twofold change in L_p was a result of an increase in water permeability or an artefact due to morphological complexity of the embryo stage. Water was also found to be unequally distributed within the embryo. For instance, in 6-somite embryos, the yolk contained 61% of the total volume and 42% water while the blastoderm contained a total volume of 39% and 82% water [28]. Further complicating permeability was the measurement of the osmotically inactive volume (V_b) fraction (volume not participating in osmosis), which was found to be 55% for the entire 6-somite stage embryos [28]. The small L_p values indicated that zebrafish have low membrane permeabilities (compare to Drosophila $L_p = 0.722$ $\mu m \bullet min^{-1} \bullet atm^{-1}$ [50]) at later developmental stages. The permeability of water and cryoprotectants to one-cell and 6-somite stage zebrafish embryos were also investigated by Zhang et al [34]. Using video microscopy and transport equations, the solute permeability (P_s)

of 2.0 M MeOH to both developmental stages was evaluated. P_s was found to be significantly higher for one-cell stage embryos, 0.45 + 0.19 µm/s, than 6-somite embryos, 0.04 ± 0.018 µm/s [34]. Zhang measured higher L_p and V_b values than Hagedorn for 6-somite embryos, 0.35 $\mu m \bullet min^{-1} \bullet atm^{-1}$ and 82.6% respectively. These discrepancies likely arise from differences in the volume and weight measurements. In a study by Susuki et al. [47], the authors used high-performance liquid chromatography (HPLC) to determine intraembryonic DMSO cryoprotectant concentrations in rainbow trout, pejerrey, medaka and carp. HPLC measurements in rainbow trout and medaka suspended in 4.0 M MeOH for 2 hrs reached a maximum permeating concentration of 1.25 M for the former and 3.0 M for the latter [47]. Only 1.0 and 1.5% permeation was measured in pejerrey and carp exposed to 3.0 M MeOH for 2 hrs [47]. These relatively low permeating concentrations, especially for rainbow trout, indicate that these embryos have similar cryoprotectant permeabilities to that of the zebrafish. Cabrita et al. [45] also used HPLC to estimate the DMSO concentration in the yolk, blastoderm and perivitelline space in turbot embryos. Cabrita and colleagues [45] found that DMSO had a low permeability to the yolk and blastoderm, but could readily permeate into the perivitelline space. Hagedorn and Zhang hypothesized that specific embryonic barriers in the zebrafish might hinder cryoprotectant permeability. Hagedorn et al. [51] used magnetic resonance (MR) microscopy and spectroscopy to determine the kinetics of cryoprotectant permeation in multiple compartments of 3 to 6-somite stage zebrafish embryos. Using chemical-shift (CS) imaging, these authors were able to generate a map of cryoprotectant distribution within the embryos [51]. The permeability of three cryoprotectants were studied, 2.0 M DMSO, 2.0 M PG and 2.25 M MeOH. From CS imaging of the embryos in each solution, DMSO and PG images appeared dark within the yolk, indicating an absence of cryoprotectants, for over 2 hrs, whereas MeOH maps appeared relatively bright within 15 min, indicating permeation of this cryoprotectant throughout the embryo [51]. To determine the compartmental barrier hindering cryoprotectant permeation, Hagedorn and colleagues [29, 51] microinjected DMSO and PG into the yolk of 6-somite stage zebrafish embryos. It was found that the cryoprotectants could not diffuse outside the yolk, but could diffuse freely within the yolk, indicating that this compartment was not a cryoprotectant barrier [29]. Through MR spectroscopy, the blastoderm was found not to be a cryoprotectant barrier as well. This lead Hagedorn et al. [29] to suggest that the yolk syncytial layer (YSL) (compartment between the yolk and blastoderm) was the primary barrier blocking cryoprotectant entry into the yolk cell. In addition to the YSL, Zhang and colleagues [34] suggested that the yolk mass may also represent a barrier that could prevent cryoprotectant penetration. Both hypotheses are currently being actively pursued.

The combination of chilling sensitivity and permeability barrier has been the main difficulty hindering successful fish embryo cryopreservation. While the chilling sensitivity can be reduced using later stage embryos, the development of the YSL during development (in the case of the zebrafish beginning at mid blastula) has been implicated as a permeability barrier to cryoprotectants. Liu et al. [52] reported that yolk-reduced embryos could improve embryo chilling sensitivity in zebrafish. Prim-6 and high-pec after partial removal of the yolk were less sensitive to chilling injury at 0 °C than control embryos [35]. Since the yolk provides nutrients to the developing blastomeres, partial yolk removal may be detrimental to embryo development. Lower embryo survival was observed in yolk-reduced embryos for all developmental stages studied [52]. As described by Harvey et al. [40], dechorionating embryos does improve the permeability of cryoprotectants, however, protease digestion of the chorion may not be suitable in vitrification and slow-cooling studies, as dechorionated embryos are very sensitive to handling. Several reports have addressed methods for overcoming the membrane permeability problems in fish embryos. In an attempt to increase both the permeability of water and cryoprotectants, Hagedorn et al. [30] microinjected early cleavage (1 to 4-cell) stage zebrafish embryos with mRNA encoding the aquaporin-3 (AQP3) water channel protein. These embryos were subsequently cultured to 50% epiboly. L_p and P_s measurements revealed that the AQP3 treated embryos had a hydraulic conductivity and solute permeability that was 6 and 2.5 times greater than those values for control embryos (cryoprotectant used was 2.0 M ^{14}C PG) [30]. However, while this approach to solving the permeability barrier was interesting, the addition of AQP3 in the zebrafish membrane did not improve the embryo freezing response, with control embryos behaving similar to modified embryos [49]. As an alternative approach, Janik et al. [53] microinjected PG and DMSO/amide into the yolk of 100% epiboly zebrafish embryos. To determine if direct injection into the yolk confers protection during vitrification and dehydration, Janik and colleagues [53] microinjected 30 nl of PG (final concentration 5.9 M) into the yolk of 100% epiboly embryos. Post-vitrification and fixing, electron micrographs revealed damaged mitochondria, the disorganization of ribosomes and the disruption of the YSL [53]. Similar results were observed in control embryos. These findings indicated that bypassing the permeability barrier by direct injection into the yolk was not sufficient to obtain normal embryo morphology and high survival post-vitrification. Kohli et al. [54] recently demonstrated the application of femtosecond (fs) laser pulses as method for introducing cryprotective carbohydrates (i.e. sucrose) into mammalian cells. When fs laser pulses were focused onto the plasma membrane, transient pores (reversible permeabilization) were formed exposing the extracellular solution to the intracellular environment. The pores, as measured through volumetric responses, were found to seal within less than 270 ms [54]. Theoretical estimates of the delivered intracellular concentration were estimated to be 72.3 and 66% for 0.2, 0.3 and 0.5 M sucrose to a diffusion length of 10 μm [54]. For the highest delivered concentration (0.2 M; 72.3% loading efficiency) cell survival was found to be 91.5 ± 8% (as determined by membrane integrity assays), indicating that the application of fs laser pulses was non-invasive [54]. Kohli and colleagues [55] further demonstrated the application of fs laser pulses by introducing a fluorescent reporter molecule, streptavidin-conjugated quantum dots or DNA into the blastomere cells of developing zebrafish embryos. Transient pores were formed in the embryonic cells of chorionated and dechorionated embryos; in the chorionated embryos, pore formation in the blastomeres occurred without damage to the chorion [55]. Using this method of delivery, perivitelline captured reporter molecules were introduced into the blastomere cells of chorionated embryos without disrupting this layer. As reported by the authors, fs laser pulses may benefit the field of cryobiology, as a technique for introducing cryoprotectants. Perivitelline captured cryoprotectants could be introduced directly into the blastomeres, or the solute permeability of blastomere-permeable cryoprotectants could be increased by transient pore formation in the cells [55].

In addition to the above techniques, cryobiologists have explored the use of antifreeze proteins (AFP) as a cryprotectant to reduce intraembryonic ice formation and improve chilling sensitivity. AFPs function by binding to ice crystals preventing ice growth and recrystallization. The proteins are known to lower the freezing temperature and potentially protect cellular membranes from damage due to chilling [56-58]. Several types of AFP exist [59] (AFP I, II, III; there is also a sub-class of antifreeze proteins called antifreeze glycoproteins (AFGP)), however, AFP I from the winter flounder has received particular attention in cryopreservation studies [60-64]. Robles et al. [42] found that when 11-day-old winter flounder embryos were vitrified and thawed, a small percentage survived with intact morphology and continued pigmentation for 6 days [42]. Robles and colleagues hypothesized that survival likely arose from the presence of AFPs in the winter flounder [42]. Encouraged by these results, Robles et al. [65-66] microinjected AFP I and III into different embryonic compartments of turbot and seabream embryos. The authors found that the presence of AFP I improved the chilling sensitivity of seabream embryos microinjected at 2-cell stage with a hatching rate approaching 100% after chilling exposure.

Despite over 50 years of research, the cryopreservation of teleost embryos has still been unsuccessful. As an alternative approach to embryo cryopreservation, cryobiologists have considered other options for fish genome cryobanking, such as the storage of somatic fish cells suggested by Mauger et al. [67]. Cryopreservation of primordial germ cells (PGCs) and blastomeres have already been achieved [68, 69]. Recently, Kobayashi et al. [68] reported that cryopreserved/tawed PGCs could be transplanted into the peritoneal cavities of allogenic trout hatchlings and differentiate into mature spermatozoa and eggs in the recipient gonads. These authors have also reported that the fertilization of eggs derived from cryopreserved PGCs by cryopreserved spermatozoa resulted in the development of fertile F1 fish. [68]. This PGC cryopreservation technique represents an important tool for aquaculture industry but also for re-population of fish species and wildlife conservation.

REFERENCES

[1] Suquet, M., Dreanno, C., Petton, B., Normant, Y., Omnes, M.H., and Billard, R. (1998). Long-term effects of the cryopreservation of turbot (*Psetta maxima*) spermatozoa, *Aquat. Living Resour. 11*, 45- 48.

[2] Chereguini, O., Garcia de la Banda, I., Herrera, M., Martinez, C., and De la Hera, M. (2003). Cryopreservation of turbot *Scophthalmus maximus* (L.) sperm: fertilization and hatching rates, *Aquac. Res. 34*, 739-747.

[3] Cabrita, E., Robles, V., Cuñado, S., Wallace, J.C., Sarasquete, C., Herráez, M.P. (2006). Evaluation of gilthead seabream, S*parus aurata*, sperm quality after cryopreservation in 5 ml macrotubes. *Cryobiology 50*, 273-284.

[4] Fabbrocini, A., Lavadera, L., Rispoli, S., and Sansone, G., (2000). Cryopreservation of sea bream (*Sparus aurata*) spermatozoa. *Cryobiology 40*, 46-53.

[5] Fauvel, C., Suquet, M., Dreanno, C., Zonno, V., and Menu, B. (1998). Cryopreservation of sea bass (*Dicentrarchus labrax*) spermatozoa in experimental and production simulating conditions. *Aquatic Living Resources 11,* 387-394.

[6] Sansone, G., Fabbrocini, A., Ieropoli, S., Langellotti, A., Occidente, M., and Matassino, D. (2002). Effects of extender composition, cooling rate, and freezing on the motility of sea bass (*Dicentrarchus labrax*, L.) spermatozoa after thawing, *Cryobiology 44*, 229-239.

[7] Asturiano, J.F., Pérez, L., Marco-Jiménez, F., Olivares, L., Vicente, J.S., and Jover, M. (2003). Media and methods for the cryopreservation of European eel (*Anguilla anguilla*) sperm. *Fish Physiol. Biochem. 28*, 501-502.

[8] Tanaka, S., Zhang, H., Horie, N., Yamada, Y., Okamura, A., Utoh, T., Mikawa, N., Oka, H.P., and Kurokura, H. (2002). Long-term cryopreservation of sperm of Japanese eel. *J. Fish Biol. 60*, 139-146.

[9] He, S. and Woods III, L.C. (2003). Effects of glycine and alanine on short-term storage and cryopreservation of striped bass (*Morone saxatilis*) spermatozoa, *Cryobiology 46*, 17-25.

[10] Taddei, A.R., Barbato, F., Abelli, L., Canese, S., Moretti, F., Rana, K.J., Fausto, A.M., and Mazzini, M. (2001). Is cryopreservation a homogeneous process? Ultrastructure and motility of untreated, prefreezing and postthawed spermatozoa of *Diplodus puntazzo* (Cetti), *Cryobiology 42*, 244-255.

[11] Babiak, I., Ottesen, O., Rudolfsen, G.,and Johnsen, S.(2006). Chilled storage of semen from Halibut, Hippoglossus hippoglossus: Optimizing the protocol. *Theriogenology 66*, 2025-2035.

[12] Rideout, R.M., Litvak, M.K., and Trippel, E.A. (2003). The development of a sperm cryopreservation protocol for winter flounder *Pseudopleuronectes americanus* (Walbaum): evaluation of cryoprotectants and diluents, *Aquac. Res. 34*, 653-659.

[13] Cabrita, E., Engrola, S., Conceição, L.E.C., Pousão-Ferreira, P., Dinis, M.T. Successeful cryopreservation of sex-reversed sperm from *Epinephelus marginatus. Aquaculture (in press)*

[14] Gwo, J-C. (1993). Cryopreservation of black grouper (*Epinephelus malabaricus*). *Theriogenology 39*, 1331-1342.

[15] Rideout, R.M., Trippel, E.A., and Litvak, M.K. (2004). The development of haddock and Atlantic cod sperm cryopreservation techniques and the effect of sperm age on cryopreservation success, *J. Fish Biol. 65*, 299-311.

[16] Tiersch, T.R. (2000). Introduction. In: Cryopreservation in Aquatic Species. Tiersch, T.R. and Mazik, P.M.(Eds). World aquaculture Society, Baton Rouge, Louisiana, XIX-XXVI.

[17] Watson, P.F. and Morris, G.J. (1987). Cold shock injury in animal cells. *Symp. Soc. Exp. Biol. 41*, 311-40.

[18] Sar, A.M.v.d., Appelmelk, B.J., Vandenbroucke-Grauls, C.M.J.E., and Bitter, W. (2004). A star with stripes: zebrafish as an infection model. *Trends Microbiol 12*, 451-457.

[19] Barut, B.A. and Zon, L.I. (2000). Realizing the potential of zebrafish as a model for human disease. Physiol. *Genomics 2*, 49-51.

[20] Warren, K.S., Wu, J.C., Pinet, F., and Fishman, M.C. (2000). The genetic basis of cardiac function: dissection by zebrafish (*Danio rerio*) screens. *Phil. Trans. R. Soc. Lond.* B. *355*, 939-944.

[21] Dooley, K. and Zon, L.I. (2000). Zebrafish: a model system for the study of human disease. *COGD 10*, 252-256.

[22] Hill, A.J., Teraoka, H., Heideman, W., and Peterson, R.E. (2005). REVIEW: Zebrafish as a Model Vetebrate for Investigating Chemical Toxicity. *Toxicol. Sci. 86*, 6-19.

[23] Jagadeeswaran, P., and Sheenhan, J.P. (1999). Analysis of Blood Coagulation in the Zebrafish. *Blood Cells Mol. Dis. 25*, 239-249.

[24] Nasevicius, A., and Ekker, S.C. (2000). Effective targeted gene 'knockdown' in zebrafish. *Nat. Genet. 26*, 216-220.

[25] Thisse, C., and Zon, L.I. (2002). Organogenesis-Heart and Blood Formation from the Zebrafish Point of View. *Science 295*, 457-462.

[26] Kasai, M. (1996). Simple and efficient methods for vitrification of mammalian embryos. *Anim.Reprod. Sci. 42*, 67-75.

[27] Zhang, T. and Rawson, D.M. (1995). Studies on Chilling Sensitivity of Zebrafish (*Brachydanio rerio*) Embryos. *Cryobiology 32*, 239-246.

[28] Hagedorn, M., Kleinhans, F.W., Freitas, R., Liu, J., Hsu, E.W., Wildt, D.E., and Rall, W.F. (1997). Water Distribution and Permeability of Zebrafish Embryos, *Brachydanio rerio*. *J. Exp. Zool. 278*, 356-371.

[29] Hagedorn, M., Hsu, E., Kleinhans, F.W., and Wildt, D.E. (1997). New Approaches for Studying the Permeability of Fish Embryos: Toward Successful Cryopreservation. *Cryobiology 34*, 335-347.

[30] Hagedorn, M., Lance, S.L., Fonseca, D.M., Kleinhans, F.W., Artimov, D., Fleischer, R., Hoque, A.T.M.S., Hamilton, M.B., and Pukazhenthi, B.S. (2002). Altering Fish Embryos with Aquaporin-3: An Essential Step Toward Successful Cryopreservation. *Biol. Reprod. 67*, 961-966.

[31] Harvey, B., and Chamberlain, J.B. (1982). Water Permeability In The Developing Embryo Of the Zebrafish, *Brachydanion Reiro*. *Can. J. Zoolog. 60*, 268-270.

[32] Zhang, T., Liu, X.-H., and Rawson, D.M. (2003). Effects of methanol and developmental arrest on chilling injury in zebrafish (*Danio rerio*) embryos. *Theriogenology 59*, 1545-1556.

[33] Hagedorn, M., Kleinhans, F.W., Wildt, D.E., and Rall, W.F. (1997). Chill Sensitivity and Cryoprotectant Permeability of Dechorionated Zebrafish Embryos, *Brachydanio rerio*. *Cryobiology 34*, 251-263.

[34] Zhang, T., and Rawson, D.M. (1998). Permeability of Dechorionated One-Cell and Six-Somite Zebrafish (*Brachydanio rerio*) Embryos to Water and Methanol. *Cryobiology 37*, 13-21.

[35] Diller, K.R. (1975). Intracellular Freezing: Effect of Extracellular Supercooling. *Cryobiology 12*, 480-485.

[36] Fahy, G.M., MacFarlane, D.R., Angell, C.A., and Meryman, H.T. (1984). Vitrification as an Approach to Cryopreservation. *Cryobiology 21*, 407-426.

[37] Mazur, P. (1984). Freezing of living cells: mechanisms and implications. *Am J Physiol Cell Physiol 247*, C125-C141.

[38] Hubalek, Z. (2003). Protectants used in the cryopreservation of microorganisms. *Cryobiology 46*, 205-229.

[39] Rall, W.F., Mazur, P., and McGrath, J.J. (1983). Depression of the ice-nucleation temperature of rapidly cooled mouse embryos by glycerol and dimethyl sulfoxide. *Biophys*. J. *41*, 1-12.

[40] Harvey, B., Kelley, R.N., and Ashwood-Smith, M.J. (1983). Permeability of Intact and Dechorionated Zebra Fish Embryos to Glycerol and Dimethyl Sulfoxide. *Cryobiology 20*, 432-439.

[41] Liu, X.-H., Zhang, T., and Rawson, D.M. (1998). Feasibility of vitrification of zebrafish (*Danio rerio*) embryos using methanol. *Cryo Letters 19*, 309-318.

[42] Robles, V., Cabrita, E., Fletcher, G.L., Shears, M.A., King, M.J., and Herráez, M.P. (2005). Vitrification assays with embryos from a cold tolerant sub-arctic fish species The first report of fish embryo survival following vitrification. *Theriogenology 64*, 1633-1646.

[43] Robles, V., Cabrita, E., Paz, P.d., Cunado, S., Anel, L., and Herráez, M.P. (2004). Effect of a vitrification protocol on the lactate dehydrogenase and glucose-6-phosphate dehydrogenase activities and the hatching rates of Zebrafish (*Danio rerio*) and Turbot (*Scophthalmus maximus*) embryos. *Theriogenology 61*, 1367-1379.

[44] Robles, V., Cabrita, E., Real, M., Alvarez, R., and Herráez, M.P. (2003). Vitrification of turbot embryos: preliminary assays. *Cryobiology 47*, 30-39.

[45] Cabrita, E., Robles, V., Chereguini, O., Paz, P.d., Anel, L., and Herráez, M.P. (2003). *Dimethly sulfoxide influx in turbot embryos exposed to a vitrification protocol. Theriogenology 60*, 463-473.

[46] Cabrita, E., Robles, V., Chereguini, O., Wallace, J.C., and Herráez, M.P. (2003). Effect of different cryoprotectants and vitrificant solution on the hatching rate of turbot embryos (*Scophthalmus maximus*). *Cryobiology 47*, 204-213.

[47] Suzuki, T., Komada, H., Takai, R., Arii, K., and Kozima, T.T. (1995). Relation between Toxicity of Cryoprotectant DMSO and Its Concentration in Several Fish Embryos. *Fish. Sci. 61*, 193-197.

[48] Zhang, T., and Rawson, D.M. (1996). Feasibility Studies on Vitrification of Intact Zebrafish (*Brachydanio rerio*) Embryos. *Cryobiology 33*, 1-13.

[49] Hagedorn, M., Perterson, A., Mazur, P., and Kleinhans, F.W. (2004). High ice nucleation temperature of zebrafish embryos: slow-freezing is not an option. *Cryobiology 49*, 181-189.

[50] Lin, T.-T., Pitt, R.E., and Steponkus, P.L. (1989). Osmometric Behavior of *Drosophila melanogaster* Embryos. *Cryobiology 26*, 453-471.

[51] Hagedorn, M., Hsu, E.W., Pilatus, U., Wildt, D.E., Rali, W.F., and Blackband, S.J. (1996). Magnetic resonance microscopy and spectroscopy reveal kinetics of cryoprotectant permeation in a multicompartmental biological system. *Proc. Natl. Acad. Sci USA 93*, 7454-7459.

[52] Liu, X.-H., Zhang, T., and Rawson, D.M. (1999). The Effect of Partial Removal of Yolk on the Chilling Sensitivity of Zebrafish (*Danio Rerio)* Embryos. *Cryobiology 39*, 236-242.

[53] Janik, M., Kleinhans, F.W., and Hagedorn, M. (2000). Overcoming a Permeability Barrier by Microinjecting Cryoprotectants into Zebrafish Embryos (*Brachydanio rerio*). *Cryobiology 41*, 25-34.

[54] Kohli, V., Acker, J.P., and Elezzabi, A.Y. (2005). Reversible permeabilization using high-intensity femtosecond laser pulses: Applications to biopreservation. *Biotechnol. Bioeng. 92*, 889-899.

[55] Kohli, V., Robles, V., Cancela, M.L., Acker, J.P., Waskiewicz, A.J., and Elezzabi, A.Y. (2007). An alternative method for delivering exogenous material into developing zebrafish embryos. *Biotechnology and Bioengineering 98, 1230-1241.*

[56] Fletcher, G.L., Hew, C.L., and Davies, P.L. (2001). Antifreeze proteins of teleost fishes. *Annu. Rev. Physiol. 63*, 359-390.

[57] Tomczak, M.M., Vigh, L., Meyer, J.D., Manning, M.C., Hincha, D.K., and Crowe, J.H. (2002). Lipid unsaturation determines the interaction of AFP type I with model membranes during thermotropic phase transitions. *Cryobiology 45*, 135-142.
[58] Wang, J.H. (2000). A comprehensive evaluation of the effects and mechanisms of antifreeze proteins during low-temperature preservation. *Cryobiology 41*, 1-9.
[59] Yeh, Y., and Feeney, R.E. (1996). Antifreeze Proteins: Structures and Mechanism of Function. *Chemical Reviews 96*.
[60] Shaw, J.M., C.Ward, and Trounson, A.O. (1995). Evaluation of propanediol, ethylene glycol, sucrose and antifreeze proteins on the survival of slow-cooled mouse pronuclear and 4-cell embryos. *Human Reproduction 10*, 396-402.
[61] Carpenter, J.F. and Hansen, T.N. (1992). Antifreeze protein modulates cell survival during cryopreservation: Mediation through influence on ice crystal growth. *Proc. Natl. Acad. Sci USA 89*, 8953-8957.
[62] Chao, H., Davies, P.L., and Carpenter, J.F. (1996). Effects of antifreeze proteins on red blood cell survival during cryopreservation. *The Journal of Experimental Biology 199*, 2071-2076.
[63] Tomczak, M.M., Hincha, D.K., Estrada, S.D., Wolkers, W.F., Crowe, L.M., Feeney, R.E., Tablin, F., and Crowe, J.H. (2002). A mechanism for stabilization of membranes at low temperatures by an antifreeze protein. *Biophysical J. 82*, 874-881.
[64] Arav, A., Rubinsky, B., Fletcher, G., and Seren, E. (1993). Cryogenic protection of oocytes with antifreeze proteins. *Molecular reproduction and development 36*, 488-493.
[65] Robles, V., Barbosa, V., Herraez, M.P., Martinez-Paramo, S., and Cancela, M.L. (2007). The antifreeze protein type I (AFPI) increases seabream (*Sparus aurata*) embryos tolerance to low temperatures. *Theriogenology 68*, 284-289.
[66] Robles, V., Cabrita, E., Anel, L., and Herraez, M.P. (2006). Microinjection of the antifreeze protein type III (AFPIII) in turbot embryos: toxicity and protein distriubtion. *Aquaculture 261*, 1299-1306.
[67] Mauger, P.-E., Le, P.-Y.B., and Labbe, C. (2006). Cryobanking of fish somatic cells: Optimizations of fin explant culture and fin cell cryopreservation. *Comp. Biochem. Physiol B Biochem Mol. Biol. 144*, 29-37.
[68] Kobayashi, T., Takeuchi, Y., and Takeuchi, T. Yoshizaki, G., (2007). Generation of viable fish from cryopreserved primordial germ cells. *Mol. Reprod. Dev. 74*, 207-213.
[69] Strüssmann, C.A., Nakatsugawa, H., Takashima, F., Hasobe, M., Suzuki, T. and Takai, R. (1999). Cryopreservation of isolated fish blastomeres: Effects of Cell Stage, Cryoprotectant Concentration, and cooling rate on postthawing survival. Cryobiology 39, 252-261.

In: Aquaculture Research Progress
Editor: Takumi K. Nakamura

ISBN 978-1-60456-247-7

Chapter 7

REARING BRAZILIAN PINK SHRIMP *FARFANTEPENAEUS PAULENSIS* IN CAGES USING ALTERNATIVE FEEDING MANAGEMENT: A TENTATIVE APPROACH FOR PROMOTING SUSTAINABLE AQUACULTURE IN OFFSHORE AREAS

***Julio Vicente Lombardi*[1], *Hélcio Luis de Almeida Marques*[1], *and Elisabete Maria Macedo Viegas*[2]**

[1]Fishery Institute - Instituto de Pesca –Address : Av. Francisco Matarazzo, 455 - São Paulo -Brazil 05001-900

[2] FZEA/USP – University of São Paulo – Av. Duque de Caxias Norte, 225 - Pirassununga – São Paulo, Brasil 13635-900.

ABSTRACT

The aim of the study presented in this chapter was to develop a new concept in rearing shrimp in cages, by testing different stocking densities and an alternative feed management. Juveniles of the Brazilian pink shrimp *Farfantepenaeus paulensis* were reared in experimental floating cages (1.00m long x 1.00m wide x 1.00m deep), constructed with a polyester/PVC 5mm-mesh and fixed in PVC floating frames, installed in a offshore area of São Paulo's marine coast, in Brazil. The study was carried out according to two different phases. Phase I involved testing three nursery densities of 500, 1,000 and 2,000 postlarvae/m^2, and four grow-out stocking densities of 100, 200, 400 and 800 juveniles/m^2. Commercial pellets of 35 to 40% crude protein was supplied twice a day, and shrimp were harvested after 37 and 54 days of nursery and grow-out stages, respectively. The best results in terms of growth and survival rate were obtained for the densities of 500 postlarvae/m^2 (nursery), and 100 and 200 shrimp/m^2 (grow-out). Phase II was carried out in order to determine growth performance, as well as composition and sensory aspects of shrimp meat from studies of an alternative feed management. In this phase, shrimp were reared at a density of 150 juveniles/m^2, and fed three different diet

[1] E-mail:lombardi@pesca.sp.gov.br.

compositions: diet I = exclusively commercial pellets consisting of 34% crude protein, diet II = exclusively mussel flesh (*Perna perna*), and diet III = mixed diet (50% diet I + 50% diet II). After 122 days of growth, diet II showed the best results in terms of shrimp weight gain, as well as the best chemical and sensory evaluation of shrimp meat. In conclusion, nursery stocking density of 500 postlarvae/m^2 and grow-out stocking densities of 100 to 200 juveniles/ m^2, showed the best results for rearing shrimp in cages, and feed based exclusively on mussel flesh seemed to be the best feed management for producing shrimp within the concepts of organic production and ecological sustainability.

1. INTRODUCTION

The culture of marine shrimp has grown substantially around the world. As with any other commercial activity, it should be accompanied by technological advances in production, especially with regard to economic and environmental sustainability. In the traditional system of shrimp production, the construction of ponds that are excavated directly in the ground represents a limiting factor for the implementation and expansion of shrimp farms, because this can conflict with legal measures aimed at the preservation of natural areas, for example, wetlands and mangrove forests. At the same time, consumers, in general, are demanding that more attention be paid to current questions concerning the future of our planet. The most optimistic projects, based on public opinion research, point to the emergence of a strong tendency in the market toward the consumption of healthier foods with less impact on the environment. This situation leads to strong pressure on producers to make alterations in traditional systems to meet the new expectations of the consumer market.

Floating cage systems offer excellent alternatives for the rearing of aquatic organisms, because they dispense with the use of soil and can represent an important strategy in the sustained development of aquaculture. This technique is being practiced in various countries for marine shrimp production on a small scale (Li and Chen, 1987; Walford and Lam, 1987; Avault Jr, 1988; Martinez-Cordova, 1988; Sinha and Sengupta,1992; Angell, 1993; Rodriguez et al., 1993; Srikrishnadhas and Sundararaj, 1993; Shanmugam et al., 1995; Stuck et al., 1996; Paquotte et al., 1998). In Brazil, excellent results, in terms of productive performance (yields over 2.5 kg/m^2/yr), have been stirring interest among producers who already use this technique in other sectors of marine farming. However, some challenges still merit the attention of specialists in that area. The high expense of purchasing commercial feed is the main item which contributes to high production costs in the shrimp industry. At the same time, feeding shrimp with commercial pellets can also lead to a potential risk for environmental pollution, especially when this feed management is carried out in cages installed directly in the environment (offshore areas). A major challenge facing shrimp farmers is the reduction of production costs as well as the implementation of alternative techniques to make the activity more environment friendly.

Organic agriculture has recently revealed a new trend for producing natural foods, and the main concepts of this technique can also be employed by aquaculturists. The suppressed use of manufactured chemicals and artificial compounds, regularly used for rearing aquatic organisms, as well as the substitution of commercial feed (ration) by natural ingredients, are essential factors in bringing aquaculture within the standards of organic production.

Mussel culture is one of the most developed sectors of mariculture along the southeast Brazilian coastline. This activity could be integrated with shrimp culture through the use of mussel flesh as shrimp meal, since the cost usually sustained for producing mussels is lower than that of purchasing commercial food for shrimp. Therefore, this kind of integration could establish a form of management within the standards of ecological and economic sustainability, and could also make it possible to produce shrimp within the concepts of organic farming.

Some trials for rearing the Brazilian pink shrimp *Farfantepenaeus paulensis* in cages were first reported by Wasielesky Jr. (1999), who demonstrated that such species has a great potential for aquaculture management. However, there are many aspects of this species that should be investigated further to determine its suitability for the rearing in cages. Studies focusing on the use of alternative feed management and the integration of shrimp culture with other activities are very important for the establishment of caged shrimp technology.

The study presented in this chapter was carried out in order to establish the technology for rearing the Brazilian pink shrimp *F. paulensis* in cages, especially concerning the determination of the best stocking densities (nursery and grow-out), as well as to test the use of mussel flesh *Perna perna* as an alternative feed for shrimp, with the intention of promoting ecological and economic sustainability, and to bring the activity within the standards of organic aquaculture.

2. Material and Methods

2.1. Location and Main Structure

The experiments were carried out at the Center for Marine Research of the Fishery Institute in Ubatuba-SP (24°26'S; 45°04'W). Experimental cages were constructed according to the specifications listed in Table 1 and fixed to a floating-deck (6.0 x 8.0m) set up in an offshore area whose depth varied from 3.0 to 4.0 m.

2.2. Phase I – Experiments with Stocking Densities

2.2.1. Acclimation

Postlarvae of the Brazilian pink shrimp *Farfantepenaeus paulensis* were purchased from a commercial hatchery and transported to the experimental site. The organisms aged 25 days from hatching (PL_{25}) had a mean weight of 0.017 ± 0.002 g. They were kept in two acclimation cages (Table 1a) for 12 days at a stocking density of 12,500 postlarvae/cage. This acclimation phase was carried out to adapt postarvae to the local environmental conditions. During this period, the shrimp were fed small commercial pellets (40% crude protein), which was supplied twice a day based on a body weight proportion of 5% of total biomass/day.

2.2.2. Testing Nursery Densities

After the 12 days of acclimation, postlarvae (0.023 ± 0.004g) were counted and transferred to experimental nursery cages (Table 1b). They were randomly distributed according to the following treatments: stocking densities of 500, 1,000, and 2,000 postlarvae/m^2), with four simultaneous replicates, totaling the occupation of 12 nursery cages for this phase.

Table 1. Framework description of experimental cages

Cage by stage	Net description	dimension (WxLxH) (m)
(a) (acclimation)	Nylon (1.0 mm mesh)	(1.00 x 1.00 x 1.40)
(b) Nursery	Polyester coated with PVC (2.0 mm mesh)	(1.00 x 1.00 x 1.40)
(c) Grow-out	Polyester coated with PVC (5.0 mm mesh)	(1.00 x 1.00 x 1.40)

During 37 days of nursing, shrimp were fed commercial pellets (40% crude protein), supplied twice a day in a proportion varying from 50% of total biomass/day (at the beginning) to 10% (at the end of the experiment). During this period, three measurements were performed for body weight records, and survival was calculated only at the end of the experiment. Juveniles produced in this phase were kept in a large cage (2.0 x 2.0 x 1.40m) for one day before being transferred to the next experimental phase (grow-out).

2.2.3. Testing Grow-out Densities

Juveniles (total weight ≥ 0.5 g) coming from the previous phase were transferred to grow-out experimental cages (Table 1c). They were randomly distributed according to the following stocking densities: 100, 200, 400 and 800 juveniles/m^2, with four simultaneous replicates, totaling the occupation of 16 cages.

During 54 days of grow-out, shrimp were fed commercial pellets (varying from 40% crude protein - at the beginning to 35% at the end). It was supplied twice a day in a proportion varying from 10% to 5% of total biomass/day. During this period, measurements of body weight were taken each fortnight, and survival was calculated only at the end of the experiment.

2.2.4. Statistical Treatments

Results were evaluated in terms of final body weight (g), and survival rate (%) performances. Curves for describing weight gain were plotted as a graph and fitted according to the equation: **Y = A + BX**, where **Y** = variable related to the body weight, **X** = variable related to the time of rearing, and **A** and **B** = fitted curve constants. Data for survival rates were transformed (arcsin X) before running ANOVA.

2.3. Phase II – Experiments with Alternative Feed Management

The location and main structures used in these experiments were the same as described in item 2.1. This experimental phase was based only on the grow-out stage, while juveniles of

0.752 ± 0.178 g were obtained from preceding phases performed for 22 days (postlarvae acclimation), and 47 days (nursing).

2.3.1. Testing Grow-out with Alternative Diets for Shrimp

The use of fresh flesh of the brown mussel *Perna perna* was tested as alternative feed management versus the traditional one, which is practiced using commercial pellets. Therefore, three treatments were tested as follows:

Diet I – feed based exclusively on commercial pellets
Diet II – feed based exclusively on brown mussel flesh
Diet III – mixed feed (50% diet I + 50% diet II)

The experiment was carried out for 122 days, according to four simultaneous replicates for each diet treatment. Every cage was kept with a stocking density of 150 juveniles/m^2.

Commercial pellets with 35% crude protein (label information) were used in diets I and III. The brown mussels used to make diets II and III were obtained from the ropes of an experimental culture located at the same site used for this study. The mussels were submitted to a pre-cooking process to remove the mussel flesh from the shell. This procedure was previously studied and selected for use in this experiment because it assures the integrity of mussel flesh, avoiding losses during the feeding of the shrimp.

Suitable quantities of wet weight of mussel flesh were adjusted according to corresponding dry weight, in order to follow the same protocol used for feeding shrimp with commercial pellet, where quantities are based on feed dry weight. Therefore, an experiment were performed by drying 50 mussels, where the wet weight was recorded before drying (wet weight), and after drying (dry weight). These data were used to determine the following expression ($\mathbf{y} = 0.292x - 0.1749$; $R^2 = 0.97$), were $\mathbf{y}$ = mussel dry weight, and $\mathbf{x}$ = mussel wet weight (Figure 1). This expression was used in the present study to estimate the actual quantity of dry weight supplied to shrimp, when they were fed moist mussel flesh.

The feed (pellet and/or mussel flesh) was supplied twice a day, according to the body proportion, which varied from 10% of total biomass/day (at the beginning of the experiment) down to 5% and 3% (at the end of the experiment). For diet III (mixed diet), pellet was supplied in the morning, and mussel flesh in the afternoon.

Feed was offered in polyethylene trays, which avoided losses and allowed observations of possible food remains, permitting constant adjustment of the feed quantities for the subsequent supplies.

Growth was monitored during the entire experiment by weight measurements performed each fortnight. Data of mean individual weight obtained at the end of the experiment were statistically evaluated by ANOVA.

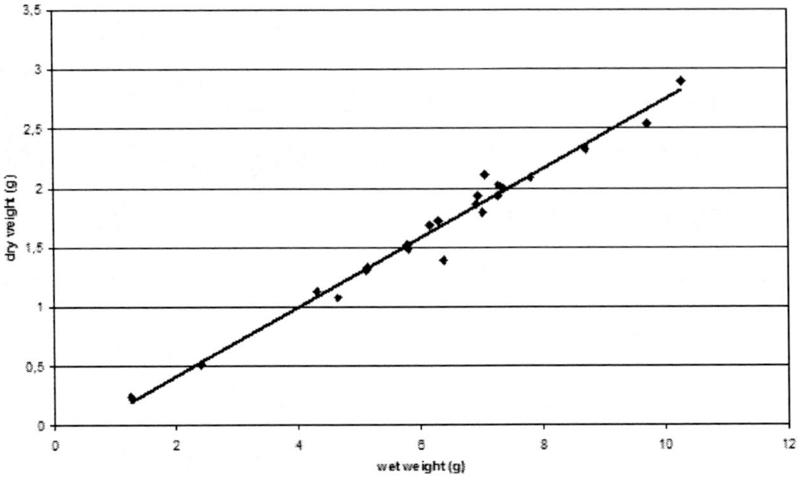

Figure 1. Relashionship between dry weight x wet weight for the pre-cooked flesh of the mussel *Perna perna*. ($y = 0.292x + 0.1749$; $R^2=0.97$).

2.3.2. Determination of Feed Attractiveness

An assessment of feed attractiveness was carried out to determine the preference of shrimp for the two kinds of feed (pellets and mussel flesh) tested in the present study. It was performed in an aquarium internally divided into two chambers, according to a "Y" format (Figure 2).

The test was developed with shrimp juveniles (3.05 ± 0.9g), previously acclimated for 72h under the same environmental conditions used for conducting the experiment. During this period, shrimp were not given any feed.

The two different kinds of feed (pellets and mussel flesh) were added separately to the respective chambers in the aquarium (Figure 2 E). Shrimp were then taken from the acclimation aquarium, in samples of 10 individuals, and immediately placed inside the neutral chamber in the test aquarium (Figure 2 F), where they were kept by a polyethylene mesh-wall, which was taken out after one minute of shrimp introduction. Five minutes later, the number of shrimp located in each chamber was recorded. In order to test only the chemical attractiveness, the whole experiment was carried out in the dark to avoid interferences by other aspects such as light attractiveness, shrimp stress, etc. Lights were turned on only after the five minutes of testing, for permitting instantaneous observations.

The experiment was replicated 20 times, each time with 10 different samples of shrimp. After each observation the rest of feed was siphoned out. Feed samples (pellets and mussel flesh) were replaced after every five observations (replicates). Data recorded for the number of shrimp observed in each chamber (neutral, pellets and mussel flesh) were evaluated statistically by ANOVA.

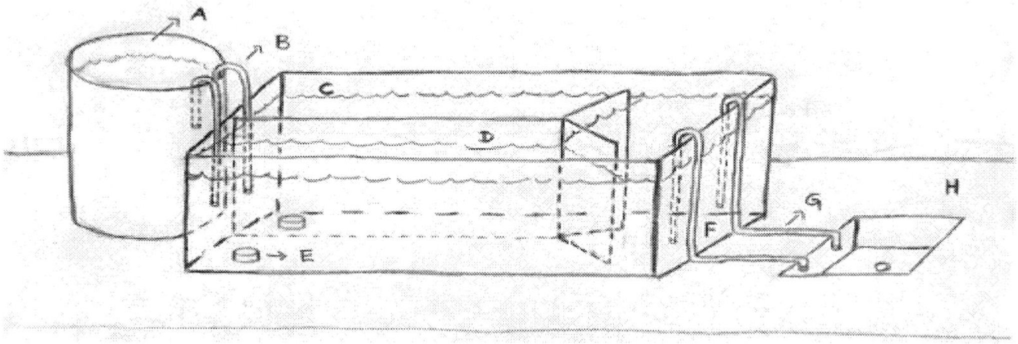

Figure 2. Lay-out of the aquarium used for testing attractiveness .A = containner filled with sea water; B = inlet pumps; C = glass aquarium; D = glass division in "Y" format; E = glass dishes for feed disposal; F = neutral chamber (site for initial shrimp releasing); G = outlet pipes; H = sink for waste water.

2.3.3. Chemical and Biochemical Analysis

Samples of food (ration and mussel) were sent to the FZEA-USP Aquaculture Laboratory (Faculdade de Zootecnia e Engenharia de Alimentos – Campus de Pirassununga – SP) to determine the following percentage values of nutritional parameters: moisture, crude protein, crude fat, ash, and non protein nitrogen (NPN) according AOAC (1990) methods. Chemical composition of shrimp meat was also carried out (AOAC,1990) only after ending the feeding experiment. Total volatile base nitrogen (TVB-N) was determined using the modified method of Malle and Tao (1987)

2.3.4. Sensory Analysis

Samples of shrimp coming from the three different treatments (diets I, II and III) were submitted to the sensory analysis to determine the influence of feed on the aspects of appearance (color), odor and flavor of the final product. This study was also performed at the FZEA-USP Sensory Analysis Laboratory (Faculdades de Zootecnia e Engenharia de Alimentos – Campus de Pirassununga – SP), which has facilities adequate for performing this kind of test.

Prawns were prepared for assessment as follows. Each batch of prawns was washed, peeled and cooked by immersion in boiling fresh water for 4 min. One percent salt was added to water for cooking. The sensory assessments were carried out within 10 min of cooking.

To determine the sensory attributes, a non-trained panel (n=50) was used aimed at representing the common consumer. The panelist received the samples randomly and evaluated the flavor, odor and appearance (mainly color) in a hedonic scale of 9 points (from 9= I liked very much to 1= I disliked very much). Results were submitted to ANOVA analysis to determine significant variations among treatments.

2.5. Oceanic Variables Monitored

Oceanic variables were monitored daily in all experiments. The data recorded were maximum and minimum temperatures (^{o}C), dissolved oxygen (mg L^{-1} and % saturation –

inside and outside the cages), salinity, pH (inside and outside the cages), and transparency (cm).

3. Results and Discussion

3.1. Phase I – Experiment with Stocking Densities

3.1.1. Testing Nursery Densities

Survival rates observed at the end of the nursery phase were satisfactorily high, and were not statistically different among treatments (Table 2). Curves adjusted for describing growth in this phase (Figure 3) permitted the estimation of the suitable time for nursing shrimp, in order to reach an average weight between 0.4 and 0.5g, considered suitable for transferring juveniles to grow-out cages (data based on local observations). According to these observations, the results indicated the following time-density recommendations for maintaining shrimp in nursery cages: 45 days at a density of 500 postlarvae/m^2, 70 days at a density of 1,000 postlarvae/m^2, and 120 days at a density of 2,000 postlarvae/m^2. Therefore, the higher densities tested in this study (1,000 and 2,000 postlarvae/m^2) did not seem to be technically suitable for nursing shrimp, as it demands a long time of maintenance.

The results of final mean weight obtained in this experimental phase can be compared to those observed by Ballester et al. (2003) and Preto et al. (2005), who carried out similar studies focusing on nursing postlarvae of *F. paulensis* in cages, and who reported the following results: 0.72g at a density of 300 postlarvae/m^2 (after 30 days of nursing), and 0.61g at a density of 500 postlarvae/m^2 (after 35 days of nursing). Vaz et al. (2004) reported a mean weight of 0.88g after 26 days of nursing *F.paulensis* at a density of 200 postlarvae/m^2, also for studies carried out in the southern part of Brazil.

On the other hand, final mean weights observed in the present study were lower than those reported in other studies which were conducted with other shrimp species and under similar conditions of stocking density.

Table 2 – Individual mean weights (g) and survival rates (%) observed in the experiment developed with nursery cages, according to different stocking densities for *F. paulensis*

trial	mean weight (g) time (days)			survival
(postlarvae/ m^2)	0	24	37	(%)
500	0.023 ± 0.004	0.23 ± 0.06	0.41 ± 0.16	78.90 ± 18.93^{ns}
1,000	0.023 ± 0.004	0.17 ± 0.05	0.27 ± 0.08	81.95 ± 11.44^{ns}
2,000	0.023 ± 0.004	0.12 ± 0.02	0.16 ± 0.03	82.63 ± 8.76^{ns}

ns = not significantly different (*P>0.05*).

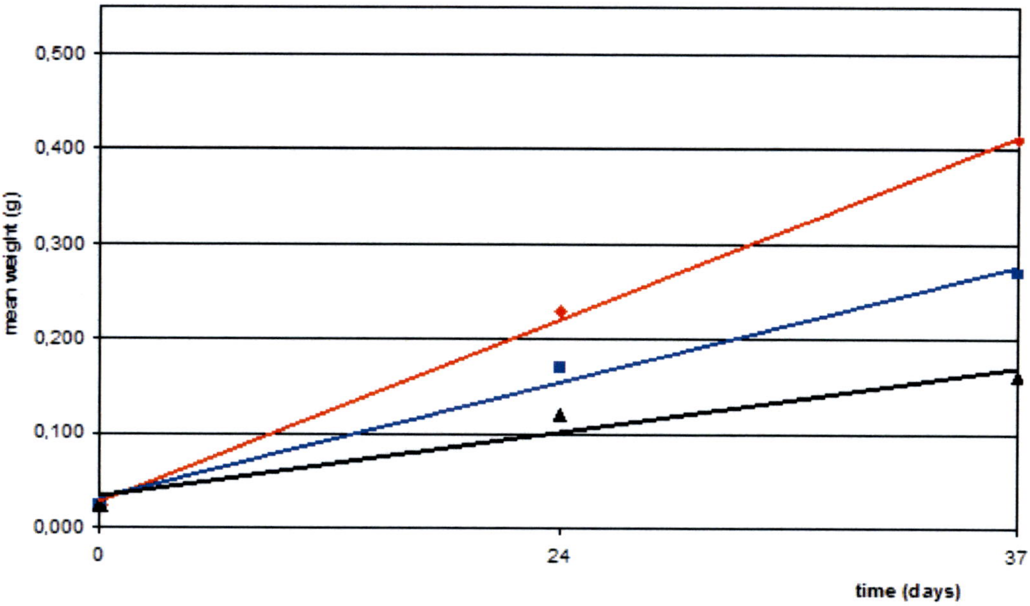

Figure 3. Lines adjusted for describing the growth of *F.paulensis* in nursery cages, according to different stocking densities .

500 / m^2	Y = 0.0130 + 0.0102 X	R^2 = 0.9916
1,000 / m^2	Y = 0.020 + 0.0066 X	R^2 = 0.9982
2,000 / m^2	Y = 0.0249 + 0.0037 X	R^2 = 0.9978

Siddharaju et al. (1980) reported a final mean weight varying from 0.8 to 1.0g for two species of shrimp (*Penaeus monodon* and *Fenneropenaeus indicus*) with stocking densities of 200 and 500 postlarvae/m^2. Natarajan et al. (1987) reported a final mean weight of 2.5g for rearing *Fenneropenaeus indicus* at a density of 100 postlarvae/m^2. However, time of nursing was not reported by these authors.

Walford and Lan (1987) observed a final mean weight of 0.48g for nursing *F. indicus* at a density of 2,100 postlarvae/m^2, for 14 days. Bose (1987) studied the growth of postlarvae of *Penaeus indicus* in cages at three stocking densities (66, 75 and 100 postlarvae/m^2), and the final mean weights reported were 1.87 and 2.34g after 30 and 40 days, respectively. Zarain-Hersberg et al. (2006) reported 0.5g as the final mean weight obtained after nursing *Litopenaeus vannamei* in cages at a density of 700 postlarvae/m^2, for 30 days. The latter species was also studied by Lombardi et al. (2006), where the results revealed a final mean weight of 2.30g after 73 days of nursing shrimp at a density of 500 postlarvae/m^2, under general management very similar to that used in the present study.

The lack of statistical differences in survival rate among the stocking densities tested in this study, may be explained by the absence of cannibalism, which is very common in this early phase of rearing, even at high densities. Vaz et al. (2004) emphasize that good survival rates in the nursery phase is essential for promoting the success of growing shrimp in the following phases. Other authors also observed good results in terms of survival of *F. paulensis* nursed in cages. Ballester et al. (2003) and Vaz et al. (2004) reported 96 and 92%

survival, respectively. While Preto et al. (2005) reported a lower survival rate (61%) for the stocking density of 500 postlarvae/m^2. The results reported in a study carried out with *P.monodon* mentioned a survival of 41% for nursing shrimp at a density of 4 to 17 postlarvae/m^3, for 2 to 3 weeks (Pena et al. 1984). Natarajan et al. (1987), and Walford and Lam (1987) reported a survival of 83% for *Fenneropenaeus indicus* nursed at densities varying from 100 to 1,700 postlarvae/m^2. In the study conducted by Zarain-Herberg et al. (2006), survival rate was about 90% for a 30-day nursery phase carried out with *L. vannamei* at a density of 700 postlarvae/m^2.

The good results in terms of survival observed in the present study encourages the development of new studies focusing on the use of artificial substrates inside nursery cages, aimed at improving viability with the use of higher stocking densities with no negative interference in the growth of postlarvae. This technique can also be beneficial by increasing the colonization by decomposition organisms, such as bacterial biofilm, algae, microcrustaceans, and other small organisms, which is a natural event observed in net cages, and can be considered as an important source of natural food for shrimp. In accordance with this, Zarain-Herzberg et al. (2006) reported that the addition of artificial substrates inside cages resulted in better results for rearing *L. vannamei* at high densities. While Ballester et al. (2007) reported that postlarvae of *F. paulensis* reared in the presence of artificial substrates achieved significantly higher survival. Nevertheless, Preto et al. (2005) did not observe significant differences in data for growth and survival recorded for *F. paulensis* reared in cages when comparing the presence and absence of artificial substrates, at stocking densities varying from 100 to 500 postlarvae/m^2.

3.1.2. Testing Grow-out Densities

Survival rates observed at the end of the grow-out phase were relatively low, suggesting the existence of an inversely proportional relationship between density and survival (Table 3). The results indicated that densities of 100 and 200 shrimp/m^2 were more suitable for growing *F. paulensis* in cages. In addition, the fitted plotted lines in Figure 4, as well as the respective mathematical expressions, permitted the estimate of 120 days as the necessary time for shrimp to reach minimum marketable sizes (approximately from 9.00 to 11.00g), when reared at these densities. On the other hand, higher densities tested in this study could be suitable for producing small shrimp, demanding less time for rearing (60 days). It could meet another segment of the market, especially to meet the needs for small organisms to be used as live bait shrimp. However, the development of new studies is also necessary in order to test techniques which could improve survival rates for higher stocking densities.

The final mean weight recorded in the present study can be compared to those obtained in experiments carried out in cages and pens with the same species in the southern part of Brazil. Wasielesky et al. (1995), reported initial and final mean weight as 2.6 and 9.5g, respectively, for eight weeks of rearing *F. paulensis* at a density of 80 shrimp/m^2. Wasielesky et al. (1999 and 2001) reported a final mean weight of 6.13g for 90 days at a density of 60 shrimp/m^2, and 6.39g for 90 days at densities varying from 30 to 120 shrimp/m^2. Jorgensen and Bemvenuti (2001) reported a final mean weight of 7g for 60 days at a density of 60 shrimp/m^2. Krummenauer et al. (2006) observed a final mean weight of 6.9 to 8.2g for 60 days, at densities of 20 to 120 shrimp/m^2, while for pen culturing, Wasielesky et al. (2004) reported final mean weights from 7.35 to 9.35g, after 100 days of grow-out phase.

Table 3. Individual mean weights (g) and survival rates (%) observed in the experiment developed with grow-out cages, according to different stocking densities for *F. paulensis*

Trial Shrimp / m^2	mean weight (g) Time (days) 0	15	29	40	54	Survival (%)
100	0.37 ± 0.04	1.33 ± 0.27	2.64 ± 0.24	3.61 ± 0.27	4.77 ± 0.36	$67.50^a \pm 5.32$
200	0.37 ± 0.04	1.15 ± 0.31	2.14 ± 0.26	3.08 ± 0.44	4.05 ± 0.62	$66.38^a \pm 14.27$
400	0.37 ± 0.04	0.81 ± 0.17	1.62 ± 0.12	2.27 ± 0.37	3.02 ± 0.08	$55.50^{a,b} \pm 4.38$
800	0.35 ± 0.05	0.75 ± 0.15	1.34 ± 0.24	1.89 ± 0.12	2.21 ± 0.15	$44.14^b \pm 4.30$

Values with at least one equal superscript letter are not significantly different ($P>0.05$).

Although the results reported by other authors were better than those obtained in the present study, in terms of final mean weight, it must be taken into account that lower stocking densities were used for carrying out these studies.

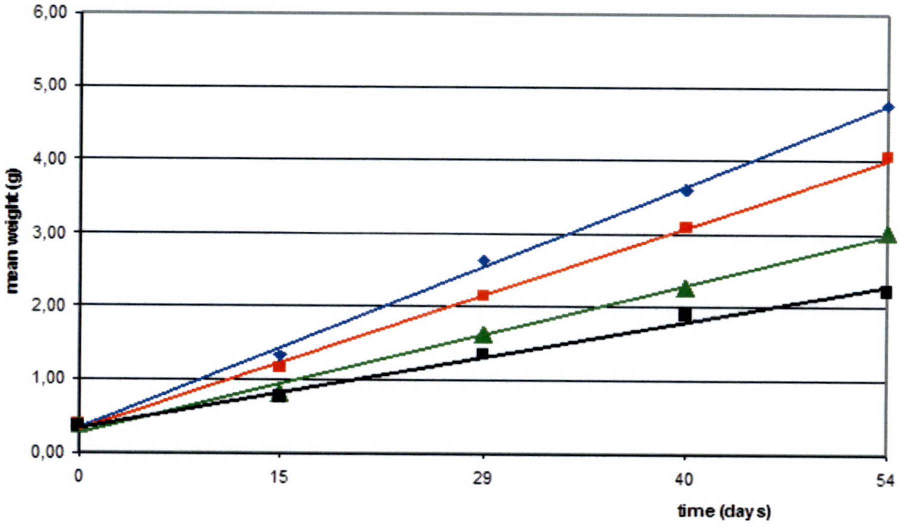

Figure 4. Lines adjusted for describing the growth of *F.paulensis* in grow-out cages, according to different stocking densities

100 / m^2	$Y = 0.2494 + 0.0831 X$	$R^2 = 0.9982$
200 / m^2	$Y = 0.2372 + 0.0696 X$	$R^2 = 0.9968$
400 / m^2	$Y = 0.2214 + 0.0506 X$	$R^2 = 0.9934$
800 / m^2	$Y = 0.3033 + 0.0364 X$	$R^2 = 0.9931$

Survival rates recorded in the present study were also satisfactory when compared to the results reported by other authors for *F. paulensis*. Jorgensen and Bemvenuti (2001) reported a survival of 25.6% after 60 days of grow-out at a density of 60 shrimp/m^2. Wasielesky et al.

(2001, 2004), reported survival rates of 85.8 and 95% after 90 days, and 82.6 and 90.3% after 100 days of grow-out phase, with no statistical differences between densities of 30 and 120 shrimp/m^2. According to Preto et al. (2006), stocking densities up to 200 shrimp/m^2 produce satisfactory results in terms of survival. While Krummenauer et al. (2006) reported survival varying from 43.3 to 65% after 60 days of grow-out at densities of 20 to 120 shrimp/m^2.

Natural live food items are known to play an important role in the growth of shrimp, even when artificial feed is provided (D'Abramo, 1997). Culturing shrimp in cages, without contact with the bottom, as done in the present study, can reduce the availability of an important item for feeding shrimp, which are the benthic organisms. Wasielesky et al. (1999) observed that shrimp grow better in pens when compared to cages, because pen enclosures allow shrimp access to the benthic source. Nevertheless, according to Abreu et al. (1998) and Ballester et al. (2003), the biofilm that naturally promotes incrustations on net cages, can supply part of the natural food for shrimp during the grow-out phase. Again, the use of artificial substrates inside cages could stimulate incrustation, increasing the availability of natural food, as stated by Zarain-Herzberg et al. (2006).

Farfanetepenaeus paulensis is a species whose behavior in the natural environment includes the habit of burrowing into sandy and muddy bottoms to hide their entire body in order to protect itself from predation (Iwai, 1978). Thus, culturing such species inside cages can cause stress to the organisms, reducing growth and survival, especially at high densities. On the other hand, the colonization of the cage bottom by natural organisms may play the role of providing a natural substrate for shrimp, thereby promoting a kind of shelter and possibly providing better results in terms of production.

From the data reported in many studies conducted with *L. vannamei,* it can be inferred that this species produces the best results in terms of survival rates compared to those obtained for *F. paulensis*. Zarain-Hersberg et al. (2006) observed that the final mean weight for *L. vannamei* reared at a density of 100 shrimp/m^2 ranged from 6.9 to 9.3g after 2 months of culturing, with survival of 90%. Casillas and Villarreal (1995) also reported survival near 90% for the same species reared for 102 days at densities of 50 and 66 shrimp/m^2, where final mean weights were 10.4 g and 10.8 g, respectively. Lombardi et al. (2006) carried out another study, in the same location where the present study was performed, but using the species *L. vannamei* at a cage density of 100 shrimp/m^2, and observed survival rates ranging from 81 to 100%, with maximum final mean weight of 6.73g. Peixoto et al. (2003) observed with *L. vannamei* a significantly higher increase in mean weight, reaching at harvest a size of 12.52g compared to 11.17g for the pink shrimp *F. paulensis*, after 102 days of culturing at a density of 15 shrimp/m^2 in earthen ponds.

Other species of marine shrimp reared in cages have also demonstrated a good growth potential with this technique. Walford and Lam (1987) reported that *Fenneropenaeus indicus,* reared in large net cages (25 m^2) at a density of 80 shrimp/m^2, reached a mean size of 9.8g after 113 days, with an average survival rate of 32%. Ruenreungdee and Mahasawasde (2003) observed that *Penaeus monodon* reared in cages for 120 days reached final mean weights ranging from 18 to 21g, with survival rates from 67 to 79%. Martinez-Cordova (1988) reared the shrimp *Penaeus stylirostris* for 60 days in cages at a density of 100 shrimp/ m^2, obtaining mean weights ranging from 5.65 to 8.34 g with survival rates from 31.8 to 45.3 %.

3.1.3. Oceanic Variables Monitored

All oceanic variables, monitored during all experimental periods (acclimation, nursery, and grow-out), were observed to be in accordance with the standards recommended by Lawrence et al. (1992) for rearing marine shrimp, and by Wasielesky (2000) especially for the species *F. paulensis*. In addition, no significant variations were observed among the different experimental treatments, nor between levels found inside and outside cages (Table 4).

Table 4. Mean of oceanic variables registered during the experiments of phase I

oceanic variables	experimental stages				
	acclimation	Nursery		grow-out	
		inside cage	outside cage	inside cage	outside cage
Minimum temperature (°C)	23.56 ± 2.93	26.88 ± 0.46	*	25.56 ± 0.98	*
Maximum temperature (°C)	25.44 ± 2.04	28.00 ± 0.61	*	26.64 ± 1.01	*
Salinity (ppt)	33.56 ± 0.76	35.25 ± 1.89	*	35.00 ± 0.88	*
pH	*	8.15 ± 0.02	8.09 ± 0.07	8.15 ± 0.24	8.13 ± 0.30
Dissolved Oxigen (mg/L)	*	5.82 ± 0.20	5.80 ± 0.40	6.01 ± 0.32	6.19 ± 0.35
Dissolved Oxigen (% sat)	*	91.90 ± 3.15	92.77 ± 8.27	91.33 ± 5.88	94.56 ± 6.52
Water transparency (m)	3.06 ± 0.93	*	3.73 ± 0.78	*	2.86 ± 1.09

(*) = variables not monitored.

3.2. Phase II – Experiment with Alternative Feed Management

3.2.1. Testing Grow-out with Alternative Diets for Shrimp

Shrimp fed mussel flesh (diets II and III) demonstrated better growth compared to those fed exclusively commercial pellets (diet I) (Table 5).

3.2.2. Oceanic Variables Monitored

Similar to the previous phase, all oceanic variables monitored during the experimental phase II, were observed to be in accordance with the standards recommended by Lawrence et al. (1992) for rearing marine shrimp, and by Wasielesky (2000) especially for the species *F. paulensis*. In addition, no significant variations were observed among the different experimental treatments, nor between levels determined inside and outside cages (Table 6).

Table 5. Individual mean weights (g) after 122 days of rearing *F.paulensis*, according to the following feed managements: Diet I = shrimp fed exclusively commercial pellet; diet II = shrimp fed exclusively mussel flesh; diet III = mixed diet (50% diet I + 50% diet II)

Treatments	Mean weight (g)
Diet I	7.21[a]
Diet II	12.05[b]
Diet III	13.17[b]

Values with equal superscript letter are not significantly different (*P>0.05*).

Table 6. Mean of oceanic variables registered during the experiment of phase II. Diet I = shrimp fed exclusively commercial pellet; diet II = shrimp fed exclusively mussel flesh; diet III = mixed diet (50% diet I + 50% diet II)

	treatments			
Variable	Diet I	Diet II	Diet III	outside cages
Temperature (°C)	25,80 ± 0.70	25.85 ± 0.86	25.84 ± 0.80	25.90 ± 0.72
Salinity (ppt)	*	*	*	35.08 ± 0.34
pH	8.14 ± 0.09	8.21 ± 0.06	8.20 ± 0.05	*
Dissolved Oxigen (mg/L)	5.62 ± 0.24	5.56 ± 0.32	5.55 ± 0.27	5.86 ± 0.21
Dissolved Oxigen (% sat)	85.28 ± 1.58	84.15 ± 2.51	84.47 ± 1.67	89.74 ± 1.02
Transparency (m)	*	*	*	1.99 ± 0.85

(*) = variables not monitored.

3.2.3. Feed Attractiveness Assessment

Observations made during the attractiveness assessment demonstrated that 42.5% of the organisms preferred mussel flesh compared to 20% that entered the chamber containing commercial pellet ($p<0.01$ – Student's t-test). The rest of the organisms remained in the neutral chamber (37.5%).

3.2.4. Chemical and Biochemical Analysis

The analysis of two types of feed offered to shrimp showed that the commercial food has a better nutritional value in comparison to the mussel flesh, mainly regarding crude protein (Table 7). However, some of these biochemical parameters when analyzed in the shrimp muscle, showed a slight advantage in the nutritional value of the organisms given mussel flesh (with exclusive or mixed feeding), mainly in regard to crude protein and ash content (Table 8). Crude protein levels observed in this study are slightly higher than those found by Contreras-Guzmán (2002) with the shrimp *Penaeus schmitii* (16.8-17.1%).

Table 7. Proximate composition of commercial feed and mussel flesh

Composition (%)	commercial pellet	mussel flesh
Moisture	10.4	72.5
Crude protein	33.95	19.44
Crude lipid	7.49	2.49
Ash	7.77	3.47
NFE[1]	40.39	2.10

[1] NFE= Nitrogen Free Extract calculated by difference (100 – others components).

Table 8. Chemical composition of shrimp flesh after feeding period. Diet I = shrimp fed exclusively commercial pellet; diet II = shrimp fed exclusively mussel flesh; diet III = mixed diet (50% diet I + 50% diet II)

Composition	shrimp fed diet I	shrimp fed diet II	shrimp fed diet III
Moisture (%)	79.4	79.3	79.2
Crude protein (%)	17.59[a]	18.21[b]	18.42[b]
Crude lipid (%)	0.28	0.31	0.33
Ash (%)	1.09[a]	1.23[b]	1.19[b]
NPN (mg/100g)[1]	613.88[a]	697.24[b]	681.39[b]
TVB-N (mg/100g)[2]	8.06	7.68	7.43

[1] NPN=Nonprotein nitrogen; [2]TVB-N= Total volatile base nitrogen. Values with equal superscript letter, in the same line, are not significantly different ($P>0.05$).

Emokpae (1983), observed changes in chemical composition according to the season of the year for pink shrimp (*Penaeus duorarum*). The water content varied 78-80%, crude protein 12-20%, ash 1-5%, and fat 0.5-1.0%.

The NPN content in aquatic animals comprises a variety of compounds of low molecular weight, which take part in various biochemical and physiological functions in the living muscle and in the post-mortem changes, besides contributing importantly to the odor and flavor of fishes, mussels and crustaceans. In this work, shrimp fed diets containing mussel flesh showed higher levels ($P<0.05$) of NPN. This fact may signify better odor and flavor, increasing the sensory acceptance of these animals (Table 9). According to Contreras-Guzmán (2002), 840 mg N/100g of NPN was found in *Penaeus schmitii* caught in the wild.

TVB-N is the most useful index for spoilage in fresh and slightly preserved seafood. In the current study, the TVB-N values (mg N/100 g muscle) in the shrimp flesh analyzed ranged from 7.43 to 8.06, showing that animals were in a state of excellent freshness. Shamshad et al. (1990) reported that the upper limit for TVB-N levels was 28.5 mgN/100g for *Penaeus merguiensis*. Brazilian law allows values up to 30mg/100g for the sale of fishery flesh.

3.2.5. Sensory Analysis

According to the results of sensory evaluations (Table 9), no significant difference was determined among the three groups (P>0.05). However, a preference tendency was observed among the panelists for shrimp fed mussels (be it exclusive or mixed feeding). It was also observed that feeding based on mussel flesh gives the shrimp a stronger pigmentation (aspect observed in product processing). It may have contributed to the large difference observed among the grades related to the appearance attribute (color).

Table 9. Sensory evaluations of shrimp flesh. Diet I = shrimp fed exclusively commercial pellet; diet II = shrimp fed exclusively mussel flesh; diet III = mixed diet (50% diet I + 50% diet II)

	Scores (1 - 9)		
Attributes	Diet I	Diet II	Diet III
Appearance (color)	7.3	7.8	7.7
Odor	7.1	7.4	7.2
Flavor	7.4	7.7	7.5

4. CONCLUSION

A density of 500 postlarvae/m^2 seemed to be suitable for the maintenance of the marine shrimp *F. paulensis* in nursery cages for about 45 days.

Densities ranging from 100 to 200 shrimp/m^2, showed the best results for the maintenance of the marine shrimp *F. paulensis* in grow-out cages for about 120 days.

The best results for rearing the shrimp *F. paulensis* in cages were observed when mussel flesh was introduced as an alternative feed (supplied separately and/or combined with commercial pellet). Such feed management also enhanced a little the nutritional values and organoleptical properties of shrimp meat.

5. ACKNOWLEDGMENTS

We thank FAPESP for funding part of the study related to this issue (Proc.n 00/11073-9). We are grateful to all our colleagues who contributed to this work, especially Vivian C. S. Rodigues, Denise C. Verreschi, and Peter G. Kirschnik. We are also grateful to Dr. A. Leyva for English editing of the manuscript.

6. REFERENCES

Abreu, PC; Thompson, FL; Wasielesky-Jr, W; Cavalli, RO (1998). New perspectives in the use of microorganisms in shrimp culture: food source, water quality and diseases control. In: Proceedings of Aqüicultura Brazil'98, Recife, PE, Nov, 2-6, v. 2, p. 703-709.

AOAC (1990). Official Methods of Analysis of the Association of Analytical Chemistry . AOAC , Washington, DC, USA, 1018 pp .

Angell, C (1993). Can "trickle down" economics work in the shrimp culture industry? Bay of Bengal News, 51: 21-24.

Avault Jr, JW (1988). "Confinement" aquaculture in cages. *Aquacult. Mag.*, 14(3): 60-62

Ballester, ELC; Wasielesky-Jr, W; Cavalli, RO; Santos, MHS; Abreu, PC (2003). Influência do biofilme no crescimento do camarão rosa *Farfantepenaeus paulensis* em sistem de berçário. Atlantica (Rio Grande), 25(2): 117-122.

Ballester, ELC; Wasielesky-Jr, W; Cavalli, RO; Abreu, PC (2007). Nursery of the pink shrimp *Farfantepenaeus paulensis* in cages with artificial substrates: biofilm composition and shrimp perfomance. *Aquaculture,* 269: 355-362.

Bartlett, P; Hodgson, A; Bonilla, P (1993). Growth of *Penaeus vannamei* without feed in cages of plastic netting placed in ponds. In: World Aquaculture '93 Int. Conf., Torremolinos (Spain), 26-28 May 1993. Book of Abstracts.

Beveridge, MCM (1996). Cage Aquaculture. 2nd ed. Fishing New Books, Oxford

Bose, SVC (1987). Nursery rearing of seed of *Penaeus indicus* in velon net cages. In: National Symposium on Research and Development in Marine Fisheries, Mandapam Camp, India, 16 sept 1987. Book of Abstracts .

Casillas, R; Villarreal, H (1995). Effect of density on the cage culture of the white shrimp *Penaeus vannamei* in Bahia de Guasimas, Sonora, Mexico. In: *Aquaculture '95*, California (USA), 1-4 Feb 1995. Book of Abstracts.

Contreras-Guzmán, E (2002). Bioquímica de Pescados e Invertebrados. CECTA-USACH Ed., Santiago, Chile, 309p.

Emokpae, AO (1983). Preliminary studies on the chemical and weight composition of some commercially important species of fish and shrimp caught in the Nigerian inshore waters. *J. Food Technol.*,18: 271-283.

D'Abramo, LR (1997). Triacylglycerols and fatty acids. In: D'Abramo, LR; Conklin, DE; Akyama, DM (eds). Crustacean Nutrition. World Aquaculture Society, *Baton. Rouge,* USA, p. 71-84.

Iwai, M (1978). Desenvolvimento larval e pós-larval de *Penaeus (Melicertus) paulensis* Pérez-Farfante, 1967 (Crustacea, Decapoda) e o ciclo de vida dos camarões do gênero *Penaeus* da região centro-sul do Brasil. PhD Thesis, Zoology Department, São Paulo State University, 137 p.

Jorgensen, P; Bemvenuti, CE (2001). Cultivo intensivo de juvenis do camarão rosa *Farfantepenaeus paulensis* (Pérez-Farfante, 1967) em cercados: avaliação experimental do sistema de engorda numa enseada estuarina da Lagoa dos Patos. Atlantica, 23: 47-58

Krummenauer, D; Wasielesky-Jr, W; Cavalli, RO; Peixoto, S; Zogbi, PR (2006). Viabilidade do cultivo do camarão rosa *Farfantepenaeus paulensis* (Crustacea, Decapoda) em gaiolas sob diferentes densidades durante o outono no sul do Brasil. Ciência Rural, 36(1): 252-257.

Lawrence, AL; Mcvey, JP; Hunter, JV; (1992). Penaeid shrimp culture. In: Crustacean and mollusk aquaculture in the United States. Huner, JV; Brown, EE (eds). Van Nostrand Reinhold, New York, p.127-157.

Li, H; Chen, B (1987). Experiment on culturing Chinese prawn in cage. Mar.Sci. Haiyang Kexue 6: 48-49.

Lombardi, JV; Marques, HLA; Pereira, RTL; Barreto, OJS; Paula, EJ (2006). Cage polyculture of the Pacific white shrimp *Litopenaeus vannamei* and the Philippines seaweed *Kappaphycus alvarezii*. *Aquaculture,* 258: 412-415.

Malle, P; Tao, SH (1987). Rapid quantitative determination of trimethylamine using steam distillation, *J. Food Protection*, 50: 756–760..

Martinez-Cordova, LR (1988). Culture of blue shrimp (*Penaeus stylirostris*) in floating cages. Progressive Fish-Culturist, 50(1): 36-38.

Natarajan, MV; Nalluchinnappan, I; Rajappan, G; Mohamed, SM (1987). Cage rearing of early juveniles of white prawn *Penaeus indicus* (H.Milne Edwards) for stocking in grow out net pens. Journal of the Marine Biological Association of India, 29(1-2): 387-390

Paquotte, P; Chim, L; Martin, JLM; Lemos, E; Stern, M; Tosta, G (1998). Intensive culture of shrimp *Penaeus vannamei* in floating cages: zootechnical, economic and environmental aspects. *Aquaculture,* 164: 151-166.

Peixoto, S; Wasielesky-Jr, W; Louzada Jr, LR (2003). Comparative analysis of pink shrimp, *Farfantepenaeus paulensis* and pacific white shrimp, *Litopennaeus vannamei,* cultured in extreme southern Brazil. *Journal of Applied Aquaculture,* 14(1): 101-111.

Pena, D; Young, AT; Prospero, OQ; Taki, Y; Primavera, JH; Llobrera, JA (1984). Floating cage nursery culture system for *Penaeus monodon*. In: Int Conf on the Culture of Penaeid Prawns/Shrimps (FICCPPS), Iloilo City (Philippines), 4-7 Dec 1984, Book of Abstracts

Preto, AL (2004). Efeito da densidade de estocagem e biofilme sobre o desempenho do camarão-rosa *Farfantepenaeus paulensis* cultivado em gaiolas nas fases de berçário e de produção de iscas vivas. Msc Dissertation, Oceanology Department, Rio Grande University, Brazil, 42 p.

Preto, AL; Cavalli, RO; Pissetti, T; Abreu, PC; Wasielesk Jr, W (2005). Efeito da densidade de estocagem sobre o biofilme e o desempenho de pós-larvas do camarão-rosa *Farfantepenaeus paulensis* cultivadas em gaiolas. Ciência Rural, 35(6): 1417-1423 .

Preto, AL; Cavalli, RO; Pissetti, TL; Poersch, LH; Wasielesky-Jr, W (2006). Cage culture of *Farfantepenaeus paulensis* as live bait-shrimp: the effect of stocking density. In: AQUA 2006 – World and European Aquaculture Society Meeting, Firenze (Italy), 9-13 May 2006. Book of Abstract, vol. 01, p.147 .

Rodriguez, EM; Bombeo-Tuburan, I; Fukumoto, S; Ticar, RB (1993). Nursery rearing of *Penaeus monodon* (Fabricius) using suspended (hapa) net enclosures installed in a pond. *Aquaculture,* 112 (1): 107-111

Ruenreungdee, K; Mahasawasde, C (2003). Monoculture of black tiger prawn (*Penaeus monodon*, Fabricius) by changing of net cages. In: Proceedings of 41st. Kasetsart University Annual Conference, Subject: Fisheries, Kasetsart, Thailand, 3-7 Feb 2003, p. 404-410.

Shanmugam, A; Rajamanickam, S; Kannupandi, T (1995). Cage culture of the *Penaeus indicus* in Vellar Estuary. *J. Mar. Biol .Assoc. India*, 37 (1-2): 166-170.

Shamshad, SI; Nisa, KU; Riaz, M; Zuberi, R; Qadri, RB (1990). Shelf life of shrimp (*Penaeus merguiensis)* stored at different temperatures. *Journal of Food Science*, 55 (5): 1201-1205.

Siddharaju, S; Sultan, KMM; Menon, VR (1980). Nursery rearing of postlarvae of two commercially important penaeid prawns in nylon cages. In: Symp on Coastal *Aquaculture,* Cochin (India), 12-18 Jan 1980, Book of Abstracts.

Sinha, I; Sengupta, S (1992). Nurturing shrimp in Ramnagar. An NGO's experience. Bay of Bengal News, 46: 24-27.

Srikrishnadhas, B; Sundararaj, V (1993). Studies on the growth of marine shrimps in floating cages and pen. In: Natarajan, P; Jayaprakas, V (eds.). Proceedings of the National Seminar on Aquaculture Development in India – Problems and Prospects, 27-29 November 1990, Trivandrum (India), p. 53-58.

Stuck, KC; Watts, SA; Wang, SY (1996). Biochemical responses during starvation and subsequent recovery in postlarval Pacific white shrimp, *Penaeus vannamei. Mar. Biol.,* 125 (1): 33-45.

Vaz, LJ; Wasielesky-Jr, W; Cavalli, RO; Peixoto, S; Santos, MHS; Ballester, E (2004). Growth and survival of pink shrimp (*Farfantepenaeus paulensis*) postlarvae in cages and pens enclosures. *Sci. Agricola*, 61(3): 332-335.

Walford, J; Lam, TJ (1980). Floating hatchery and net cage culture of *Penaeus indicus* in the Straits of Johore, Singapore. *Aquaculture,* 62 (1): 11-32.

Wasielesky-Jr, W (1999). Produção do camarão marinho *Penaeus paulensis* no sul do Brasil: cultivo em estruturas alternativas. In: Prêmio Jovem Cientista 1997. CNPq – Conselho Nacional de Desenvolvimento Científico e Tecnológico, Fundação Roberto Marinho, Grupo Gerdau, Rio de Janeiro, p. 53-106.

Wasielesky-Jr, W (2000). Cultivo de juvenis do camarão-rosa *Farfantepenaeus paulensis* (Decapoda, Penaeidae) no estuário da Lagoa dos Patos: efeito dos parâmetros ambientais. PhD Thesis, Oceanology Department, Rio Grande University, Brazil, 199 p.

Wasielesky-Jr, W; Cavalli, RO; Dolci, D; Silva, TMA (1995). Crescimento do camarão rosa *Penaeus paulensis* (Crustacea: Decapoda) cultivado em gaiolas e cercados, no estuário da Lagoa dos Patos. In: 6 Encontro Rio-Grandense de Técnicos em Aquacultura, Ibiruba (RS, Brazil), 13-15 Oct 1995. Anais, vol 1: p. 14-25

Wasielesky-Jr, W; Poersch, LH; Bianchini, A (1999). Comparação entre a sobrevivência e o crescimento do camarão rosa *Farfantepenaeus paulensis* cultivado em gaiolas e cercados. *Nauplius* (Rio Grande), 7: 173-177.

Wasielesky-Jr, W; Poersch, LH; Jensen, LV; Bianchini, A (2001). Effect of stocking density on pen reared pink shrimp *Farfantepenaeus paulensis* (Pérez-Farfante, 1967) (Decapoda, Penaeidae). Nauplius, 9 (2): 163-167.

Wasielesky-Jr, W; Bianchini, A; Sanchez, CC; Poersch, LH (2003). The effect of temperature, salinity and nitrogen products of food consumption of pink shrimp *Farfantepenaeus paulensis*. Brazilian Archives of Biology and Technology, 46(1): 135-141.

Wasielesky-Jr, W; Peixoto, S; Jensen, L; Poersch, LH; Bianchini, A (2004). Estudo preliminar do cultivo do camarão-rosa *Farfantepenaeus paulensis* em cercados no estuário da Lagoa dos Patos. Boletim do Instituto de Pesca, 30 (1): 63-70.

Zarain-Herzberg, M; Campa-Córdova, AI; Cavalli, RO (2006). Biological viability of producing white shrimp *Litopenaeus vannamei* in seawater floating cages. *Aquaculture,* 259: 283-289.

In: Aquaculture Research Progress
Editor: Takumi K. Nakamura
ISBN 978-1-60456-247-7

Chapter 8

STUDY OF AQUATIC BIODIVERSITY OF SOMESWARI AND NETHAI RIVER; GHARIA AND NIDAYA BEEL IN NORTHERN BANGLADESH

B. K. Chakraborty [1]
Fisheries Officer, Directorate of Fisheries, Bangladesh

ABSTRACT

A total number of 79 species of fishes, four species of prawn, one species of crabs, one species of snail and five species of reptiles were identified from the Someswari and Nethai river; Gharia and Nidaya beel in Bangladesh. About 11 types of fishing methods were found in operation. Increasing rate of current jal (gill net), Kapuri jal (seine net) and FAD (Fish aggregating device) were identified as detrimental gear killing different species during whole the year. The increasing rate in exploitation of the water bodies a thread to aquatic biodiversity of the open and closed water bodies. The fisheries production of these water bodies declined significantly over the last five (2001-2005) years. The total production of the Someswari and Nethai river; Gharia and Nidaya beel has decreased from 95.79±6.89 to 38.61±2.83; 175.20±11.83 to 76.63±5.59; 104.65±5.50 to 48.03±2.97 and 112.87±5.98 to 50.58±2.98 mt within five years and the percentage of total production had sharply decreased from 19.61% to 59.68%; 22.07% to 56.66%; 21.88% to 59.25% and 19.84% to 56.20% over the same period. As a result, a number of commercially important fish species such as, mohasher (*Tot tor* and *Tor putitora*) nandina (*Labeo nandina*), sharpunti (*Puntius sarana*), Gajar (*Channa marulius*) and reptiles (*Kachuga kachuga* and *Morenia petersi*) were extinct, 52-58 species were at extremely high risk and 7-18 species were in the verge of extinction.

1. INTRODUCTION

The study of biodiversity has become a major concern to the fishing biologists against the back drop of rapid decline in the natural population of fish and aquatic biota across all the continents of the world.

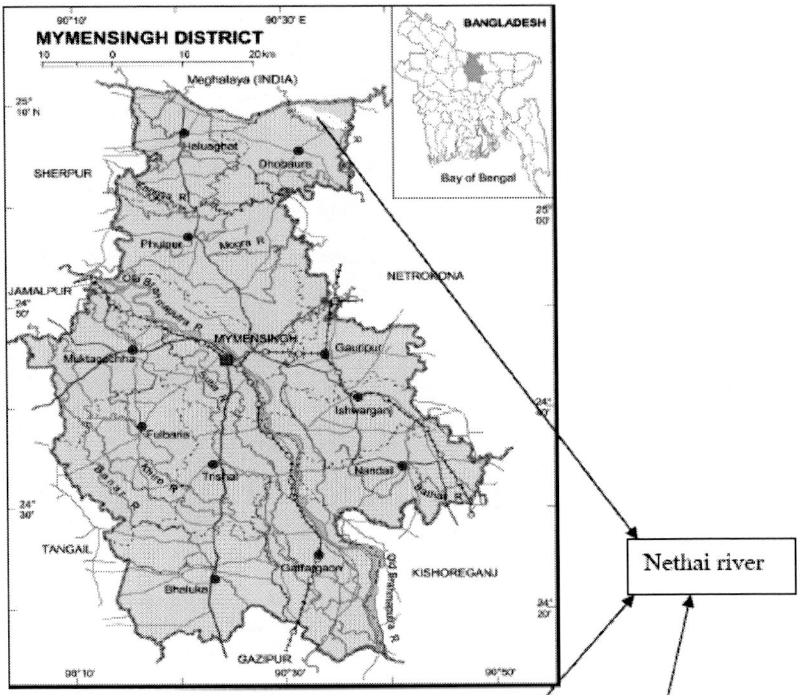

Figure 1. Map of Bangladesh and Mymensingh district showing the location of Nethai River.

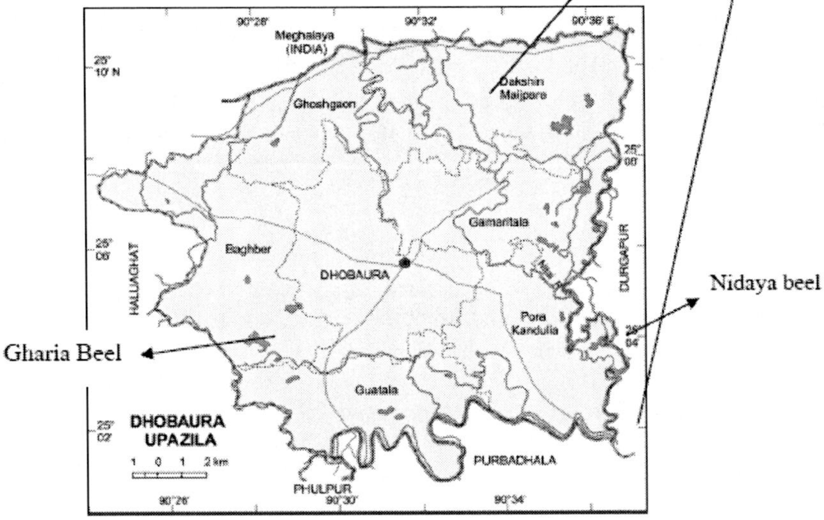

Figure 2. Map of Dhubaura Upazilla showing the study area of Nethai River. area and indiscriminate use of pesticides. Pollution from domestic, industrial and agrochemicals wastes and run off have resulted in extinction of a considerable amount of aquatic biota in same stretches of the open water system (Diaster, 1990; Mazid and Hossain 1995; Alam 1996; Hussain1977).

Biodiversity encompasses genetic species, assemblage, ecosystem and landcape levels of biological organization with structural, compositional and functional components (Noss, 1990; Cairns and Lackey, 1992). About 20% of the of commercially important fresh water fish species are now extinct or endangered; the main treats being habit degradation, over exploitation, pollution and competition for water. The worldwide loss of biodiversity is widely accepted as a major problem (Moyle and Williams, 1990). Though loss of aquatic species has been occurring rapidly, the aquatic organisms have received comparatively little attention from conservation biologists (Allendrof, 1988). A rich diversity of fish species is critical to the ecology and sustainable productivity of the flood plains. Fisheries resources in Bangladesh are under severe threat due to environmental degradation which include human interventions through construction of flood control embankments, drainage system and sluice gates, conversion of inundated land to crop land thereby reducing water

Someswari and Nethai River

Someswari and Nethai river are the two important river in the northern part of Bangladesh. Which originate from the hilly area of Nokhrak Habri of Baghmara and Thura district of Maghaloya, India. The former one with 24-26 km long course flows across the Bijoypur and Ranikong from northern to southern Jaria-Jhanjyle through the Sibgonj bazaar and Durgapur Sadar before joining the Kangshow River (Figure 3 and 4). During November 1991, a small abandoned channel north of Durgapur with a flow direction towards the east has been rejuvenated after 1990 (Alam, 2004). This small meandering channel locally named Atrakhali channel. But no evidence of water flow was observed in the original Someswari River and Atrakhali channel except rainy season. The Nethi river with its 32-35 km long course runs across the Rangsingpur from northern to southern Roykandulia through the Dhubaura Upazilla, Mymensingh before merging with Kangshow River (Figure 3 and 4). The water flow in both the river is continuous. During monsoon, the water flow comes down from the upper region of hilly Meghaloya, India and causes serious flood in the northern areas of Mymensingh district.

Gharia and Nidaya Beel

Gharia and Nidaya beel receives surface runoff water by rivers and canals, and consequently this beel becomes very extensive water body in the monsoon and retain small amount of water in the post-monsoon period. These beels are of tectonic origin and connected with the Nethai and Kongsha River (Figure 4). The beel basin comprised the flood plains tributaries with abundant aquatic vegetation. But through gradual sedimentation, the basin becomes shallower leading to the formation of reeds and sedges. This resulted in providing enough food and shelter for fish and other aquatic fauna and added to the fertility of the water bodies by their excreta promoting rich growth of phytoplankton and macrophytes thus partly contributing to the process of eutrophication. The basin of the beel supports a large variety of wetland bio-diversity.

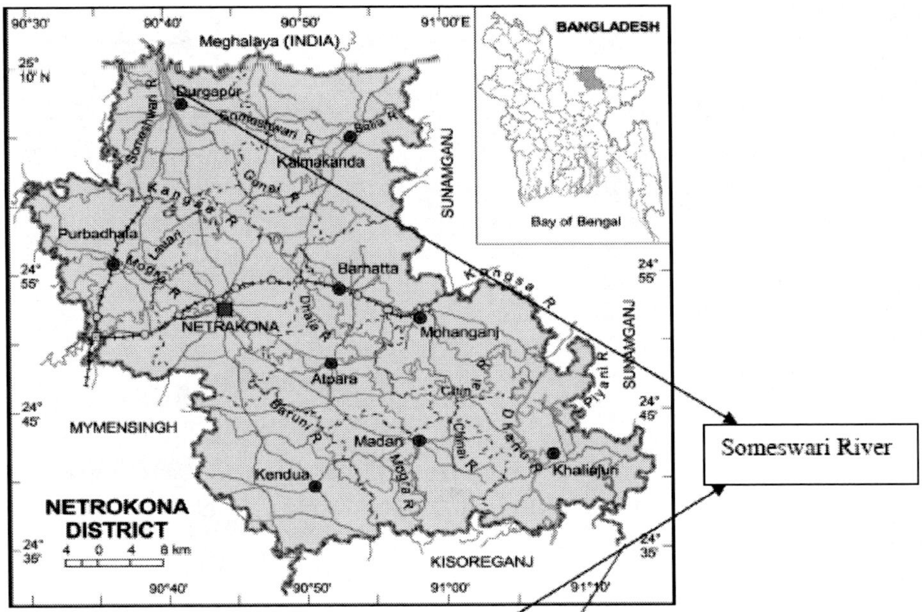

Figure 3. Map of Bangladesh and Netrokona district showing the location of Someswari River.

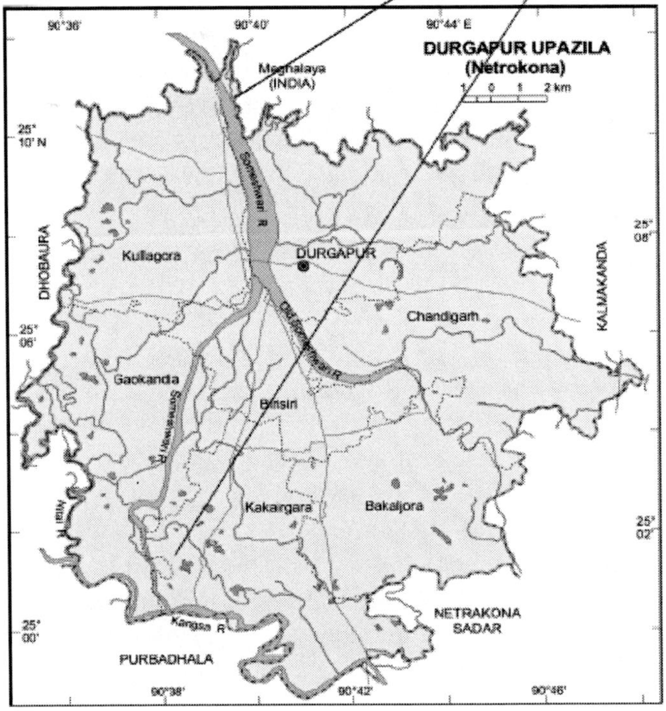

Figure 4. Map of Durgapur Upazilla showing the study area of Someswari River.

There are two flood phases: the early flood phase and the deep flood phase. The early flood phase (April to early June) occurs in the early monsoon season when the water level in rivers and basin is relatively low. The water level in the floodplain rises and falls in accordance with the water level in adjacent rivers. The deep flood phase (June to September) begins when the water level in the Someswari, Nethai and Kangshow River, causing deep flooding in the eight unions of Dhubaura and Durgapur Upazilla, Netrokona. Floodwater in flood plains starts receding in the post-monsoon season (October to December). The water recession starts at shallow areas and water surface area shrinks, fishes and other aquatic organism move with water flow into deep water area of the river.

The beels are also important fishing grounds of the country. Once these rivers and beels had abundance of native wild fish species, crabs and reptiles. Due to over-exploitation and various ecological changes of the rivers and beels, some important fish species and reptiles have disappeared. Like other floodplains the feeding and breeding grounds of fishes in and around the rivers have been reducing drastically from various human created problems. Indiscriminate destructive fishing practices, soil erosion, siltation, construction of flood control and drainage structures and agro-chemicals have caused havoc to the aquatic biodiversity in Bangladesh.

A well planned and systematic study is required to assess the present status of biodiversity in the

rivers and beel with a view to undertake appropriate measures to conserve and manage the aquatic resource. The present study focuses on the abundance, species combination, catch statistics and related aspects of rivers and beel.

2. Materials and Methods

Experimental Procedure

Survey of the Someswari and Nethai River were conducted during 2001-2005 with particular emphasis on soil and water quality, biological productivity and biotic communities and status of fishery exploitation. For the purpose of the study of river course was divided into upper and lower regions. The river courses of Bijoypur and Ranikong (boarder of India) to Shibgonj and Birishiri constitute the Upper region and Shibgonj and Birishiri to Jariya-Ganjail had constituted the lower region, in which the Someswari River joins with the Kangshow River and the river course of Nethai was also divided into upper and lower stretches. The river courses of Bhuynpara and Dhigolbadh (boarder of India) to Kalshindur constitute the Upper stretch and Kalshindur to Roykandulia village via Rangsingpore had constituted the Lower stretch, in which the Nathai joins with the Kangshow River.

The Gharia beel comprises an average area of 62.0 ha with an average depth 1.82±0.82 m. The beel is surrounded by Tarikandi, Haripur and Shanandakhila villages (Figure 4). The beel Nidaya comprising an average are of 82.4 ha with an average depth 1.68±0.62 m. The beel is surrounded by Purakandulia, Raykandulia, Udaypur, Baharbhita and Manikpur Atam villages (Figure 4). The research was based on both primary and secondary data, comprehensive literature review and extracts of local knowledge and information. Collection of primary data was made by field observation and different experimentations Viz. experimental fishing in the beel, survey of different fishing methods, survey of fish markets adjacent to beel, monitoring of hydrological, meteorological, physico-chemical and biological characteristics of beel and fishers' perception as well. Secondary data were collected from the fishers, local administration and department of Fisheries (DoF).

Study of Meteorological and Physico-Chemical Parameters and Plankton

Water depth was measured by means of a graduated bamboo stick. Water temperature was recorded using a Celsius thermometer and transparency was measured by using a Secchi disc of 20 cm diameter. Dissolved oxygen and pH were measured directly using a digital electronic oxygen meter (YSI Model 58) and an electronic pH meter (Jenway Model 3020). Alkalinity was determined by titrimetric method (Clesceri *et al.*, 1989).

The plankton sample was collected at every fortnight using 40 μ blotting silk plankton net and later analyzed numerically with the help of Sedgewick-Rafter counting cell (SR-cell) under a compound microscope (Clesceri *et al.*, 1989). Calculation of the abundance of plankton was done by using the formula of Stirling (1985).

Data Collection

An organized sampling program spread over a reasonably long time was needed to get a true picture of the catch and composition. The experimental rivers and beels were sampled simultaneously during winter (mid November to mid February), premonsoon (mid February to April), monsoon (May to August) and post monsoon (September to mid November) for assessment of fish abundance and availability. The present study, being a rapid survey, gives only a broad picture of a stock of fishes, crabs, snail and reptiles that could be obtained through market survey, collection of different species directly from fishers' catch, khata fishing, kua fishing, fish landing center and interaction with fisherman in the beel and riverside; and even in the beel and river. Aquatic weeds were collected and identified.

Analysis of Experimental Data

The data were analyzed through one way ANOVA using MSTAT. Duncan's Multiple Range Test was done to find out whether any significant difference existed among treatment means (Zar 1984). Standard deviation in each parameter was calculated and expressed as mean ±S.D.

3. Results

3.1. Physical Characteristics of Someswari and Nethai River

Soil texture of Someswari and Nethai river bed varied from sandy to loam sand. In the upper region, structure of the bed appeared to have predominantly sandy and in the lower region the soil was found to be sandy to sandy loam (Table 1).

3.2. Physical Characteristics of Gharia and Nidaya Beel

Soil texture of Gharia beel was 2.2±0.12% sandy, 58.2±5.66% loam sand and 39.6±2.48% clay and Nidaya beel was 1.8±0.11% sandy, 57.5±4.77% loam sand and

40.7±2.48% clay in which highest percentage of loam sand was recorded. The soil structure of the deeper bed appeared to have predominantly clay and in the surrounding area of the wet land was loamy to clay.

Table 1. Physical features of sediment of the Someswari and Nethai river

Name of the river	Location	Soil texture of the river bed (%)		
		Sandy	Loam sand	Clay
Someswari river	Upper region	91.4±5.62[A]	7.2±4.22[B]	1.2±2.32[C]
	Lower region	60.3±6.11[A]	24.5±3.88[B]	15.2±4.80[C]
Nathi river	Upper stretch	89.2±5.62[A]	8.6±4.22[B]	2.2±2.32[C]
	Lower stretch	62.0±6.11[A]	18.5±3.88[B]	19.5±4.80[C]

Figures with different superscripts in the same row varied significantly ($P>0.05$). Figures in the parenthesis indicate the range

3.3 Morphometry and Hydrodynamics of Experimental Rivers and Beels

The Kongsha and Nethai River passes through the east and southern side of the beel. This beels are connected with the Nethai and Kongsha River by two cannels locally called khal. In the dry season, almost 80% areas of the beel become dried up except the canals, and khata and kua fishing area where water remains during January to mid-April. Flooding originates from the Nethai and Kongsha River, the east and south of the beel. Surface run-off and increase the area of the beel due to rain and inflow of water between beel and river. The more water gain or exchange of water takes place during southwest monsoon when floodplains are flooded. The early flood phase (April to early June) occurs in the early monsoon season when the water level in basin is relatively low. The water level in the floodplain rises and falls even deep flood phase (June to September) begins when the water level in the Nethai and Kongsha river, causing deep flooding. Floodwater in flood plains starts receding in the post-monsoon season (October to December). After recession of flood, water level in the beel decreases snapping the beel connection with the river. The beel gets almost dried up through evapo-transpiration and seepage. Except deeper portion of the beel, the people use most of the area for crop production by extracting water from the beel. The water loss by various means causes shrinkage of the effective water area and lowering of depth in the beel.

Table 2. Physico-chemical parameters of Someswari River

Parameters	Years				
	2001	2002	2003	2004	2005
Temperature (°C)	26.29±7.88 (16.22-32.75)	26.58±6.85 (14.88-32.80)	26.22±7.47 (15.40-33.20)	26.28±7.72 (15.18-32.85)	26.90±7.82 (14.95-33.18)
Transparency (cm)	39.68±8.44[d] (28.52-50.22)	45.82±6.02[a] (33.02-52.62)	43.55±6.48[e] (34.38-51.41)	42.87±7.74[c] (32.30-48.44)	48.22±6.74[b] (31.18-50.72)
pH	7.50± 3.07 (6.60-8.85)	7.86±2.6 6 (6.51-8.75)	7.55± 2.58 (6.50-8.88)	7.77±1.86 (6.65-8.67)	7.89±2.26 (6.66-8.86)
Dissolved oxygen (mg/L)	5.44±1.74 (4.05-8.12)	5.77±1.12 (4.54-7.84)	5.64±1.22 (4.38-7.75)	5.15±1.28 (4.44-7.85)	5.25±1.66 (4.04-7.78)
Alkalinity (mg/L)	132.22±10.04[c] (110.24-145.45)	117.41±10.22[b] (100.88-138.02)	126.25±9.44[e] (107.25-137.35)	121.02±9.57[d] (111.80-135.12)	114.76±12.42[a] (102.26-137.85)

Figures with different superscripts in the same row varied significantly ($P>0.05$). Figures in the parenthesis indicate the range.

The results of the physico-chemical parameters of the Someswari River, Nethai River, Gharia beel and Rivers and beels were are furnished in Table 3, 4, 5 and 6 which included temperature, transparency, pH, dissolve oxygen and alkalinity of water were found to be more and less in a normal range.

It is evident from the Table (3-6) that the mean water temperatures of the four aquatic environments were not statistically significant (P>0.05). Mean Secchi disk transparency differed significantly (P<0.05), during the year 2001-2005. Higher values occurred during post monsoon and summer months due to reduced flow and relatively stable conditions. pH of the experimental rivers and beels did not differ significantly (P>0.05). A significant rise in pH during pre-monsoon followed by a drop in winter was noted. The mean dissolved Oxygen (DO) of the experimental rivers and beels also did not differ significantly (P>0.05). Total alkalinity of the experimental rivers and beels were differed significantly (P<0.05). Lowest value of alkalinity was recorded in the in the winter during 2005.

Table 3. Physico-Chemical Parameters Of Experimental Nethi River

Parameters	Years				
	2001	2002	2003	2004	2005
Temperature (°C)	26.42±7.88 (15.15-33.05)	26.28±6.85 (14.95-32.85)	26.32±7.47 (15.45-33.10)	26.38±7.72 (15.20-32.95)	26.50±7.82 (14.25-33.10)
Transparency (cm)	38.88±7.14^{d} (30.52-50.55)	45.12±6.22^{a} (33.02-52.66)	40.54±6.18^{e} (34.38-51.28)	40.87±7.64^{e} (32.30-50.14)	47.22±6.14^{b} (31.18-52.55)
pH	7.55± 3.07 (6.50-8.88)	7.81±2.6 6 (6.54-8.55)	7.50± 2.58 (6.55-8.68)	7.80±1.86 (6.45-8.65)	7.89±2.26 (6.60-8.80)
Dissolved oxygen (mg/L)	5.04±1.54 (4.05-8.22)	5.27±1.22 (4.54-7.55)	4.98±1.25 (4.38-7.65)	5.05±1.35 (4.44-7.75)	5.42±1.65 (4.04-7.75)
Alkalinity (mg/L)	122.22±10.04^{c} (108.24-144.35)	115.41±10.22^{b} (101.88-135.22)	126.25±9.44^{e} (117.25-157.55)	129.02±9.57^{d} (112.80-135.14)	115.76±12.42^{a} (112.26-138.25)

Figures with different superscripts in the same row varied significantly (P>0.05). Figures in the parenthesis indicate the range.'

Table 4. Physico-chemical parameters of Gharia beel

Parameters	Years				
	2001	2002	2003	2004	2005
Temperature (°C)	26.22±6.88 (16.28-33.44)	27.58±7.85 (17.11-32.80)	26.02±8.47 (16.2-33.20)	25.88±7.72 (16.08-32.85)	25.09±8.82 (15.95-33.08)
Transparency (cm)	34.68±8.44^{d} (28.52-50.22)	45.82±6.02^{a} (32.02-52.62)	30.55±6.48^{e} (24.38-47.41)	39.87±7.74^{c} (28.30-48.44)	42.22±6.74^{b} (27.18-50.72)
pH	7.48± 3.07 (6.70-8.80)	7.56±2.6 6 (6.55-8.70)	7.55± 1.58 (6.55-8.68)	7.48±1.86 (6.65-8.60)	7.60±2.06 (6.50-8.40)
Dissolved oxygen (mg/L)	5.14±1.54 (4.18-8.02)	5.44±1.08 (4.04-7.74)	5.84±1.42 (4.48-7.45)	4.85±1.28 (4.24-7.14)	5.05±1.66 (4.24-7.88)
Alkalinity (mg/L)	122.28±10.24^{c} (100.24-135.45)	127.02±10.02^{b} (99.88-138.02)	110.28±9.21^{e} (101.25-137.35)	116.02±9.32^{d} (102.80-135.12)	133.76±12.22^{a} (98.26-137.85)

Figures with different superscripts in the same row varied significantly (P>0.05). Figures in the parenthesis ndicate the range.

Table 5. Physico-chemical parameters of Nidaya beel

Parameters	Years				
	2001	2002	2003	2004	2005
Temperature (°C)	25.61±6.88 (15.88-32.84)	26.08±6.65 (16.01-32.80)	25.82±7.07 (16.11-32.50)	25.88±5.02 (16.08-32.55)	25.54±6.12 (15.92-32.18)
Transparency (cm)	35.14±6.24[d] (28.12-52.32)	42.34±5.42[a] (29.02-52.40)	30.25±5.41[e] (27.15-50.32)	39.06±6.44[c] (28.32-47.75)	33.28±5.74[b] (28.88-50.72)
pH	7.88± 2.17 (6.70-8.80)	7.67±3.06 (6.66-8.70)	7.85±2.08 (6.55-9.01)	7.78±1.99 (6.65-8.67)	8.08±3.06 (6.55-8.88)
Dissolved oxygen (mg/L)	5.33±1.54 (4.08-8.22)	5.14±1.08 (4.14-7.77)	5.44±1.42 (4.18-7.82)	4.92±1.28 (4.22-8.04)	5.65±1.66 (4.54-8.18)
Alkalinity (mg/L)	128.28±10.24[c] (110.24-140.45)	117.02±10.02[b] (109.88-135.12)	130.28±9.21[e] (111.25-139.05)	136.02±9.32[d] (112.80-145.32)	140.76±12.22[a] (120.16-147.85)

Figures with different superscripts in the same row varied significantly ($P>0.05$). Figures in the parenthesis indicate the range.

3.4. Plankton Population

Plankton abundance in different river and beel is presented in Table 6, 7, 8 and 9 which are discussed below:

Someswari and Nethai river

The quantity of phytoplankton and zooplankton was higher in the lower region of the Someswari and Nethai River where Someswari and Nethai met with the Kangsha River. Plankton population was much less in the upper region of the river where running water flows whole the year. The phytoplankton consisted of 27 genera in the rivers in four broad groups viz., Chlorophyceae, Bacillariophyceae, Cyanophyceae and Euglenophyceae. Chlorophyceae was represented by the genera were *Protococcus, Mougeotia, Microspora, Mesotenium, Clasterium, Eremesphaera, Chlorococcum, Ophiocytium, Penium, Spyrogyra, Zygnema, Kirchneriella, Gonotozygon, Pediastrum, Oocystis, Tetraedron* and *Volvox.* Bacillariophyceae included various species belonging to genera *Melosira*, *Diatoma, Fragilaria* and *Navicula.* Cyanophyceae included the genera of *Anabaena, Chroococcus, Merismopedia, Mycrocystis* and *Oscillatoria.*

Euglenophyceae represented by the genera *Euglena.* Euglenophyceae occurred in the lower region which may be attributed to the modification of the river course from fast flowing continuous system into discontinuous, shallow water level. Chlorophyceae was the dominant group followed by Bacillariophyceae which differed significantly ($P<0.05$) between 2001 and 2005.

Among zooplankton the represented genera were *Bosmina, Brachionus, Cyclops, Filinia, Keratella, Trichocerca, Lecane, Diaptomus, Daphnia, Moina, Nauplius* and *Oicomonas* belonging to two groups. The zooplankton population consisted of 12 genera including nauplii in two groups viz., Rotifera, Crustacea and other groups. Rotifera and Crustacea were significantly higher ($P<0.05$) in lower region followed by upper region. The mean abundance of total phytoplankton and zooplankton

Table 6. Mean variation of phytoplankton (individual/ml) and zooplankton (organism/ml) population in the Someswari river

Plankton group ($\times10^3$ cells/L)	Years				
	2001	2002	2003	2004	2005
Chlorophyceae	17.05 ± 4.88^{b} (14.82-25.41)	14.77 ± 4.22^{e} (12.04-21.12)	16.05 ± 4.28^{c} (13.25-23.48)	15.17 ± 4.22^{d} (12.22-22.45)	18.25 ± 4.25^{a} (14.80-26.33)
Bacillariophyceae	11.42 ± 3.24^{d} (8.08-18.47)	13.08 ± 3.46^{b} (9.28-18.30)	14.14 ± 3.44^{a} (10.38-18.22)	12.81 ± 2.46^{c} (9.18-18.14)	10.24 ± 3.44^{e} (8.18-13.32)
Cyanophyceae	6.27 ± 2.28^{b} (4.13-9.28)	5.64 ± 2.15^{c} (4.22-9.32)	7.12 ± 2.08^{a} (4.8 5-9.65)	5.11 ± 2.05^{d} (3.88-9.25)	6.22 ± 2.26^{b} (4.15-9.87)
Euglenophyceae	0.77 ± 0.00^{d} (0.60-1.02)	1.10 ± 0.02^{a} (0.88-1.32)	0.78 ± 0.01^{c} (0.34-1.10)	1.04 ± 0.22^{b} (0.40-1.02)	0.40 ± 0.01^{e} (0.20-0.82)
Total Phytoplankton ($\times10^3$ cells/L)	35.51 ± 6.97^{c}	37.59 ± 7.29^{b}	38.09 ± 6.98^{a}	34.13 ± 6.59^{e}	35.11 ± 7.50^{d}
Rotifera	4.07 ± 1.22^{e} (2.10-5.76)	5.26 ± 1.82^{d} (3.21-7.07)	5.38 ± 1.25^{c} (4.10-7.26)	6.08 ± 1.22^{a} (4.22-7.98)	5.88 ± 1.15^{b} (3.78-7.16)
Crustaceae	3.58 ± 1.01^{d} (2.12-4.74)	4.24 ± 1.14^{b} (3.06-6.52)	3.78 ± 1.28^{c} (2.62-5.08)	4.24 ± 1.20^{b} (3.11-6.12)	4.58 ± 1.26^{a} (3.22-6.10)
Others	1.07 ± 0.20^{d} (0.80-2.13)	1.22 ± 0.82^{b} (1.00-2.88)	1.06 ± 0.48^{d} (0.88-2.03)	1.11 ± 0.32^{c} (0.68-2.28)	1.55 ± 0.28^{a} (0.59-2.35)
Total Zooplankton ($\times10^3$ Organisms/L)	8.72 ± 1.50^{e}	10.72 ± 2.10^{c}	10.22 ± 2.19^{d}	11.43 ± 2.51^{b}	12.0 ± 2.22^{a}

Figure in the same row having the same superscript are not significantly different ($P>0.05$). Figures in the parenthesis indicate the range.

Table 7. Mean variation of phytoplankton (individual/ml) and zooplankton (organism/ml) population in the Nethai river

Plankton group ($\times10^3$ cells/L)	Years				
	2001	2002	2003	2004	2005
Chlorophyceae	16.11 ± 3.25^{a} (12.28-20.22)	14.77 ± 3.02^{c} (09.04-15.72)	12.52 ± 3.28^{e} (09.44-17.08)	15.17 ± 3.22^{b} (10.32-17.33)	13.28 ± 4.05^{d} (09.14-16.03)
Bacillariophyceae	10.22 ± 2.04^{d} (8.18-14.24)	12.14 ± 2.16^{a} (9.20-16.14)	11.64 ± 2.42^{b} (9.38-16.32)	9.88 ± 1.86^{e} (9.78-14.24)	11.23 ± 2.14^{c} (9.28-13.88)
Cyanophyceae	5.24 ± 1.28^{c} (3.13-7.28)	5.88 ± 1.15^{b} (4.02-7.32)	4.92 ± 1.08^{d} (3.06-6.25)	4.11 ± 1.32^{e} (3.48-7.22)	6.02 ± 1.06^{a} (4.35-9.67)
Euglenophyceae	0.37 ± 0.01^{d} (0.20-0.74)	0.45 ± 0.02^{c} (0.28-0.66)	0.58 ± 0.01^{b} (0.44-0.96)	0.78 ± 0.22^{a} (0.50-0.85)	0.25 ± 0.01^{e} (0.15-0.53)
Total Phytoplankton ($\times10^3$ cells/L)	31.94 ± 6.74^{b}	33.24 ± 6.43^{a}	29.66 ± 6.38^{e}	29.94 ± 6.35^{d}	30.78 ± 5.83^{c}
Rotifera	6.38 ± 1.02^{a} (3.11-6.55)	4.06 ± 1.22^{e} (3.55-6.24)	4.88 ± 1.15^{d} (4.10-7.66)	5.02 ± 1.32^{c} (4.12-7.88)	6.12 ± 1.10^{b} (5.02-7.16)
Crustaceae	4.61 ± 1.04^{a} (2.02-4.44)	4.16 ± 1.17^{c} (3.06-6.02)	3.67 ± 1.02^{e} (3.12-5.18)	4.58 ± 1.10^{b} (3.22-6.62)	4.05 ± 1.06^{d} (2.88-5.12)
Others	1.04 ± 0.20^{c} (0.80-2.01)	1.08 ± 0.42^{b} (0.95-2.18)	1.00 ± 0.28^{d} (0.85-1.58)	1.05 ± 0.33^{c} (0.72-1.95)	1.15 ± 0.22^{a} (0.99-2.01)
Total Zooplankton ($\times10^3$ Organisms/L)	12.03 ± 2.18^{a}	10.00 ± 2.01^{d}	9.55 ± 2.45^{e}	10.65 ± 2.18^{c}	11.32 ± 2.49^{b}

Figure in the same row having the same superscript are not significantly different ($P>0.05$). Figures in the parenthesis indicate the range.differed significantly ($P<0.05$) in the river in different year.

Table 8. Mean variation of phytoplankton (individual/ml) and zooplankton (organism/ml) population in the Gharia beel

Plankton group ($\times10^3$ cells/L)	Years				
	2001	2002	2003	2004	2005
Chlorophyceae	26.05±5.18[a] (17.88-29.43)	22.77±7.22[d] (16.04-30.62)	20.05±6.28[e] (16.69-29.48)	25.17±5.22[b] (17.54-30.72)	24.05±6.85[c] (17.80-31.45)
Bacillariophyceae	12.42±3.04[d] (8.48-18.47)	14.28±4.46[a] (10.28-20.30)	13.48±3.44[b] (8.38-20.92)	13.88±2.46[b] (9.18-20.14)	12.44±3.44[d] (9.88-19.40)
Cyanophyceae	7.22±2.08 (6.15-11.08)	8.07±2.55 (7.11-13.12)	7.82±1.88 (6.05-12.48)	8.11±2.05 (7.01-12.12)	7.77±3.02 (6.15-12.21)
Euglenophyceae	0.00±0.00	1.0±0.02 (0.80-1.22)	0.88±0.00 (0.24-1.40)	1.01±0.82 (0.40-1.32)	0.00±0.00
Total Phytoplankton ($\times10^3$ cells/L)	45.69±11.00[c]	46.12±9.25[b]	42.23±8.16[e]	48.17±10.21[a]	44.26±10.06[d]
Rotifera	5.28±1.22[a] (3.10-5.76)	4.66±1.12[c] (3.21-6.58)	5.08±1.05[b] (4.00-7.16)	4.22±1.22[c] (3.22-6.98)	6.38±1.15[a] (3.88-7.16)
Crustaceae	3.78±1.06[d] (2.62-4.28)	4.04±1.12[c] (3.36-6.72)	3.68±1.26[e] (2.62-5.18)	4.14±1.22[b] (3.06-6.62)	4.28±1.46[a] (2.90-5.18)
Others	1.10±0.20 (1.80-2.03)	1.17±0.82 (1.01-2.88)	1.15±0.48 (10.88-2.13)	1.07±1.32 (0.88-2.18)	1.05±0.28 (0.99-2.30)
Total Zooplankton ($\times10^3$ Organisms/L)	10.16±2.12[b]	9.87±1.86[c]	9.91±1.99[c]	9.43±1.80[d]	11.71±2.25[a]

Figure in the same row having the same superscript are not significantly different (P>0.05). Figures in the parenthesis indicate the range.

Gharia and Nidaya beel

The quantity of phytoplankton and zooplankton of Gharia and Nidaya beel were particularly dominant in the month of June and July and lowest count was obtained in December and January. The phytoplankton consisted of 27 genera in the Gharia beel in four broad groups' viz., Chlorophyceae, Bacillariophyceae, Cyanophyceae and Euglenophyceae. Chlorophyceae contributed the genera were Protococcus, Mougeotia, Microspora, Mesotenium, Clasterium, Eremesphaera, Chlorococcum, Ophiocytium, Penium, Spyrogyra, Zygnema, Kirchneriella, Gonotozygon, Pediastrum, Oocystis, Tetraedron, Volvox. Bacillariophyceae included various species belonging to genera Melosira, Diatoma, Fragilaria and Navicula. Cyanophyceae included the genera of Anabaena, Chroococcus, Merismopedia, Mycrocystis and Oscillatoria.

The phytoplankton consisted of 26 genera in the Nidaya beel in four broad groups' viz., Chlorophyceae, Bacillariophyceae, Cyanophyceae and Euglenophyceae. Only *Pediastrum* (Chlorophyceae) was not found in Nidaya beel. Bacillariophyceae included various species belonging to genera *Diatoma, Fragilaria, Melosira* and *Navicula.* Cyanophyceae included the genera of *Anabaena, Chroococcus, Merismopedia, Mycrocystis* and *Oscillatoria.* Euglenophyceae included only the genera of *Euglena.* Chlorophyceae was the dominant group which was significantly higher ($P<0.05$) during five years study. The abundance of Bacillariophyceae differ significantly ($P<0.05$) in different years. The mean abundance of total phytoplankton was differed significantly ($P<0.05$) during investigation period. Among zooplankton the represented 12 genera were *Bosmina, Brachionus, Cyclops, Filinia, Keratella, Trichocerca, Lecane, Diaptomus, Daphnia, Moina, Nauplius* and *Oicomonas* belonging to Rotifera and Crustacea groups including nauplii. Rotifera and Crustacea were significantly differed ($P<0.05$) in the two beels during investigation periods. The mean abundance of total zooplankton differed significantly ($P<0.05$) during different years.

Table 9. Mean variation of phytoplankton (individual/ml) and zooplankton (organism/ml) population in the Nidaya beel

Plankton group ($\times10^3$ cells/L)	Years				
	2001	2002	2003	2004	2005
Chlorophyceae	24.05 ± 4.24^a (19.22-28.74)	23.55 ± 6.21^d (17.04-27.42)	20.85 ± 4.22^e (18.57-26.52)	24.11 ± 5.02^b (18.04-29.02)	22.47 ± 5.02^c (17.11-29.40)
Bacillariophyceae	14.82 ± 3.34^d (10.48-17.41)	12.25 ± 4.26^a (10.88-16.44)	16.18 ± 3.54^b (11.31-18.22)	13.08 ± 2.16^b (11.18-19.12)	13.01 ± 3.04^d (10.48-18.90)
Cyanophyceae	7.88±1.08 (6.15-10.48)	8.27±2.05 (7.11-10.82)	8.42±2.22 (6.05-11.48)	8.12±2.15 (7.01-10.12)	7.77±3.21 (6.15-11.01)
Euglenophyceae	0.00±0.00	0.05±0.02 (0.01-0.08)	0.08±0.00 (0.04-1.10)	0.04±0.02 (0.01-0.5)	0.02±0.01 (0.01-0.05)
Total Phytoplankton ($\times10^3$ cells/L)	46.75 ± 10.63^c	44.12 ± 9.77^b	45.53 ± 9.11^e	45.35 ± 10.07^a	43.27 ± 9.42^d
Rotifera	5.08 ± 1.82^a (4.10-6.76)	5.61 ± 1.12^c (4.21-6.58)	5.11 ± 1.05^b (4.10-6.16)	4.77 ± 1.02^c (4.02-6.98)	5.33 ± 1.05^a (4.08-7.36)
Crustaceae	4.28 ± 1.26^d (3.62-5.27)	4.34 ± 1.12^c (3.51-6.72)	3.89 ± 1.22^e (3.01-5.58)	3.44 ± 1.22^b (3.05-5.62)	4.25 ± 1.46^a (3.20-5.18)
Others	1.88±0.50 (1.10-2.11)	1.72±0.62 (1.01-2.48)	1.65±0.42 (1.08-2.10)	1.08±0.86 (0.98-2.12)	1.57±0.78 (1.02-2.01)
Total Zooplankton ($\times10^3$ Organisms/L)	11.24 ± 1.67^b	11.67 ± 1.98^c	10.65 ± 1.89^c	9.29 ± 1.87^d	11.15 ± 1.94^a

Figure in the same row having the same superscript are not significantly different ($P>0.05$). Figures in the parenthesis indicate the range.

3.5 Macrophytes

A total number of 13 species belonging 12 genera and 11 families of aquatic weeds were identified from Gharia and Nidaya beel (Table 10). The weeds usually grow along the beels margins and only Hizal, *Barringotonia acutangula* area of the beel, which gives shelter and additional nutrients of aquatic wild animal. *Barringotonia acutangula* grows in the deeper regions. The weeds are usually used as human consumption and cattle food. *Najas najas* species was the most dominant group among the identified weeds. The eggs of prawn and miscellaneous fish species were found to be present onto the *N. najas* and water hyacinth (*Eichhornia crassipes*) during summer to winter. Water hyacinth usually covered a layer on the surface of Khua in the deep.

3.6. Uses of Fishing Craft and Gears

Fishermen used locally made country boat for transport of fishes, nets and related materials during fishing. For catching fish they used seine net or ber jal, komor jal, thela jal, dharma jal, bua jal, lift net, cast net, current jal and various type fish trap, hook and line and FAD (Fish aggregating device) according to season and availability of different species of aquatic lives (Table 11). The fishermen used a variety of fish trap, hook and line to capture the different groups of aquatic animals.

Craft and Gears Used in the Someswari River

It is noted that every year increased fishing effort is being increased by using illegal fishing gear like Gill net (current jal) and Ber jal (kaperi jal).

Table10. Aquatic weeds of Gharia and Nidaya beel

Family	Local Name	Scientific Name	Type
Lemnaceae	Edurkanipana	Wolffia Arrhiza	Floating
Pontederiaceae	Kachuripana	Eichhornia Crasssipes	Floating
Gramineae	Dal	Hudroryza Aristota	Emergent
Marsiliaceae	Shusnishak	Marsileaquadrifolia	Emergent
Najadaceae	Najas	Najas Najas	Submerged
Compositaceae	Helencha	Enhydra Flucktuans	Spreading
Gramineae	Arail	Leersia Hexandra	Spreading
Commelinaceae	Kanaibashi	Commelina Bengalensis	Spreading
Convolvulaceae	Kalmilata	Ipomoea Aquatica	Spreading
Nymphaceae	Shapla	Numphaea Nouchali	Rooted Plants With Floating Leaves
Nymphaceae	Padma	Nelumbo Nucifera	Rooted Plants With Floating Leaves
Myrtaesae	Hizal	Barringotonia Acutangula	Rooted Plants With Floating

The percentage of catch statistics by using current jal and ber jal (kaperi jal) were 29.80%, 30.70%, 31.70%, 32.60% and 34.40%; and 17.60%, 18.80%, 21.30%, 22.20% and 23.30% in the year 2001, 2002, 2003, 2004 and 2005, respectively (Figure 5) and using of current jal and ber jal (kaperi jal) differed significantly ($P<0.05$). Catch statistics by using of Komor jal were17.80%, 15.40%, 16.50%, 17.30% and 17.50% in the year 2001, 2002, 2003, 2004 and 2005, respectively but use of Komor jal did not differ significantly ($P>0.05$). The catch statistics by use of thela jal, dharma jal, bua jal, lift net, cast net, fish trap and hook and line was decreasing and differed significantly ($P<0.05$) in the year 2001, 2002, 2003, 2004 and 2005. As a result, a significant reduction in fish abundance was noted in the Someswari River every year.

Craft and Gears Used in the Nethai River

There were great similarity in the types and intensity of use of different crafts and gears in case of Someswari and Nethai river. Catch composition profile by different gears in the Nethai river were almost of similar magnitude (Figure 6). But a significant variation in the amount of fish catch was observed among different gears. It is concluded that a significant reduction in the abundance fish was noted in the Nethi River every year.

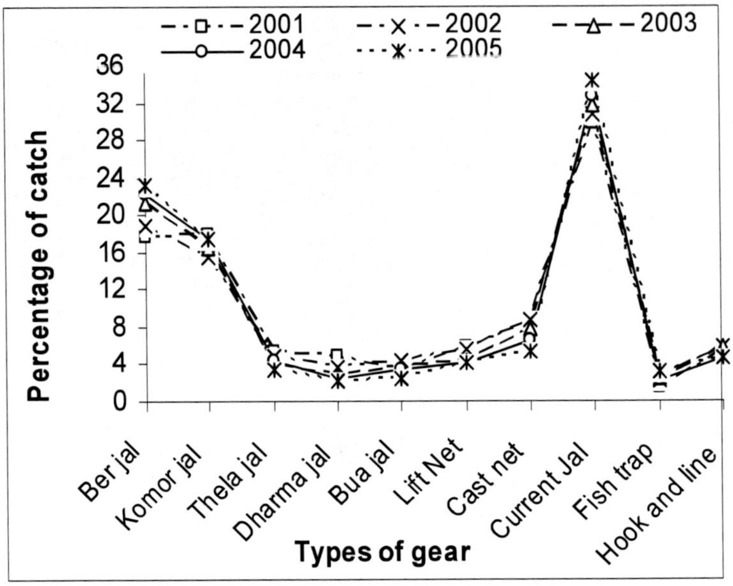

Figure 5. Percent composition of catches by different types of fishing gear, 2001- 2005. (Someswari).

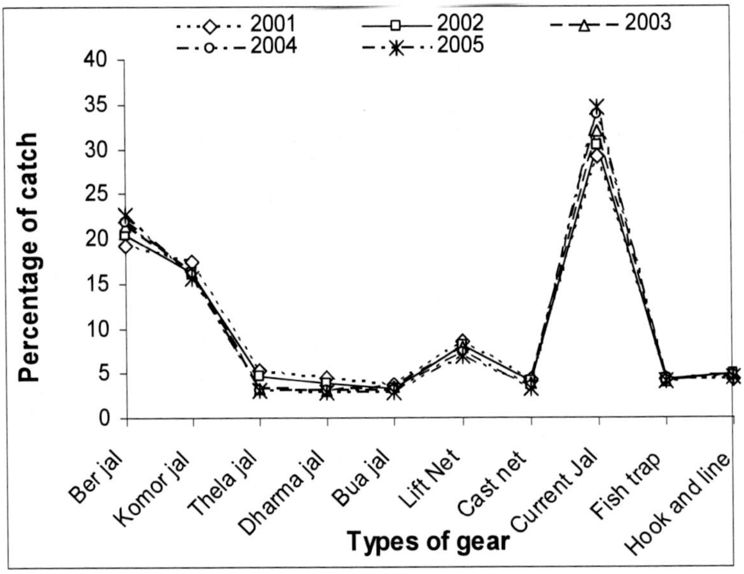

Figure 6. Percent composition of catches by different types of fishing gear, 2001-2005. (Nethai).

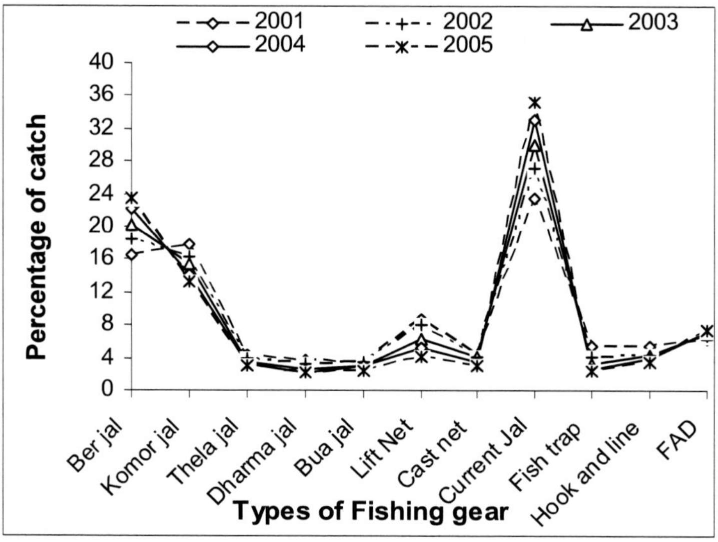

Figure 7. Percent composition of catches by different types of fishing methods, 2001- 2005. (Gharia).

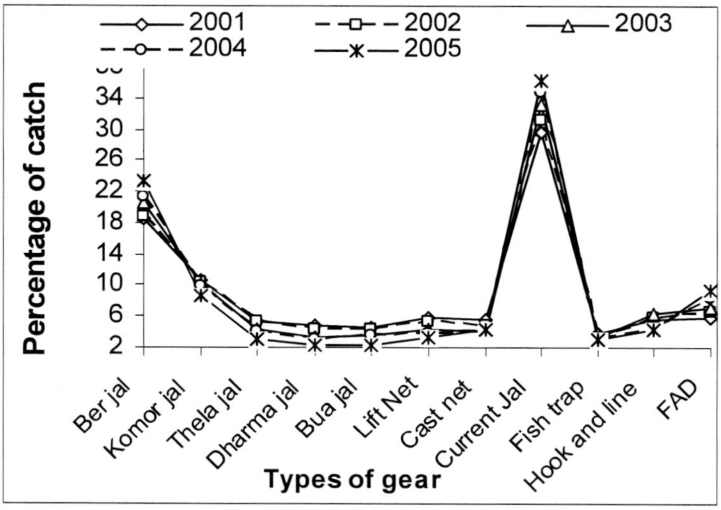

Figure. 8. Percent composition of catches by different types of fishing methods, 2001- 2005. (Nidaya).

Craft And Gears Used Gharia And Nidaya Beel

Like Someswari and Nethai river, fishing efforts were found to be escalating in Gharia and Nidaya beel. Uses of illegal fishing gear were very much common here. The percentage of catch statistics by using current jal and ber jal (moshari jal) were 23.50%, 27.10%, 30.10%, 33.10% and 35.30%; and 16.60%, 18.50%, 20.20%, 22.10% and 23.50% within the year 2001, 2002, 2003, 2004 and 2005, The catch statistics by using thela jal, dharma jal, bua jal, lift Net, cast net, fish trap and hook and line was decreasing and also differed significantly ($P<0.05$) in the year 2001, 2002, 2003, 2004 and 2005. Catch by FAD was also increasing from 6.30% to 7.40% during the last five years. As a result, a significant decrease in fish abundance is observed in the Gharia beel every year. There were no significant difference in the type's use of fishing crafts and gears between the Gharia and Nidaya beel. Catch composition by different gears provides very much similar sceneries. The most striking feature is that abundance of aquatic lives is gradually decreases over times.

Table 11. Uses of different fishing gears, trap and hook and line in the different rivers and beels of northern Bangladesh

SL. No	Fishing methods	Picture/Figure	Capture group	Percentage (%) of Fishing methods			
				Somes -wari	Nethai	Gharia	Nidaya
1.	Boat	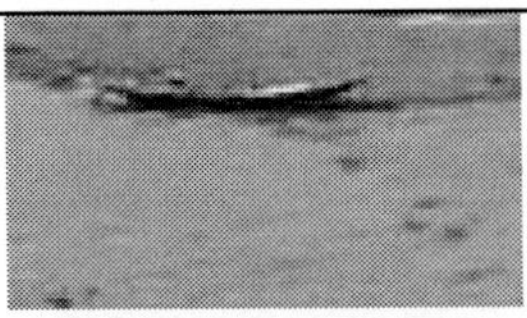	Used boat full time in the river and monsoon in beels for transport of nets, traps and fishes.	-	-	-	-
2.	Ber jal (Seine net)		Major and minor carp, small fish, Cat and small cat fish, Snake headed, Prawn and Reptiles	17.6-23.3	19.6-22.2	16.6-2.4	18.60-23.2
3.	Komor jal (Seine net)		Major and minor carp, small fish, Cat and small cat fish, Snake headed, Prawn and Reptiles	17.8-17.5	17.3-17.1	17.8-14.6	10.50-8.70
4.	Thela jal (Push net)		Snake headed, Minor carp, Small cat fish, Eels and small prawn.	5.4-3.4	5.3-4.8	4.4-3.0	5.4-3.1
5.	Dharma jal	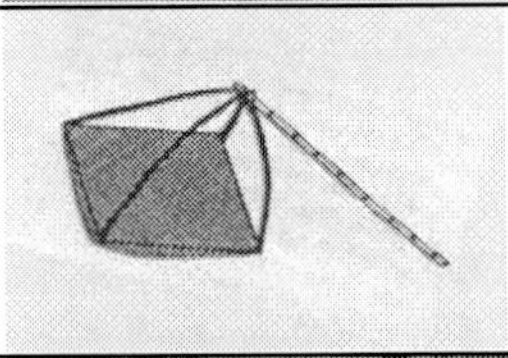	Minor carp, Small fish, prawn, Cat fish, and Crabs	4.8-2.1	4.4-2.4	3.8-2.2	4.8-2.3
6.	Bua jal		Major and minor carp, Small fish, Small cat fish, Prawn and Crabs	3.5-2.3	3.7-2.7	3.5-2.5	4.5-2.2
7.	Lift Net (Vashal Jal)		Major and minor carp, small fish, Cat fish, Small cat fish, Snake headed, Eels, Prawn and Crabs.	4.8-4.0	9.4-5.2	7.8-4.2	5.7-3.3
8.	Cast net (Jaki jal or Kepla Jal)		Major and minor carp, Cat and small cat fish, Snake headed, Eel, Prawn and kakra.	8.4-5.2	4.3-3.1	4.4-3.0	5.6-4.3

Table 11. (Continued)

SL. No	Fishing methods	Picture/Figure	Capture group	Percentage (%) of Fishing methods			
				Somes -wari	Nethai	Gharia	Nidaya
9.	Gill Net (Current Jal)		Major and Minor carp, small fish, Snake headed, Cat and small cat fish, Eel, Prawn and kakra.	2.98-3.44	29.0-34.4	2.35-36.1	29.7-36.1
10	Fish Trap						
a.	Ocha		Major and minor carp, small fish, Snake headed, Cat and small cat fish, Eel, Prawn and kakra.	1.5-3.2	1.8-3.1	5.5-2.4	3.8-3.1
b.	Bitte						
c.	Polo	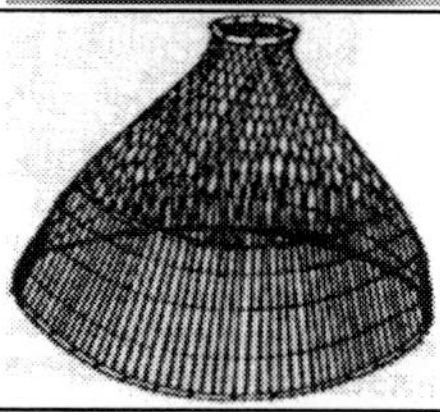					
d.	Dugair	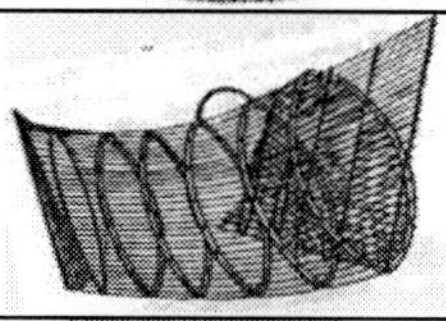					
e.	Bair	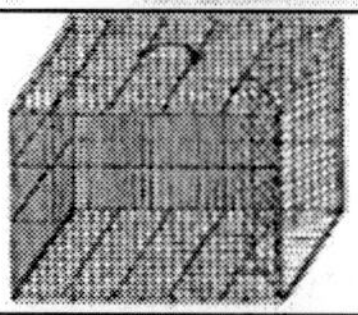					
f.	Guni	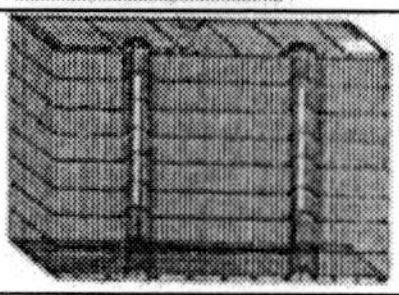					
11.	Hook and line:						
a.	Barshi	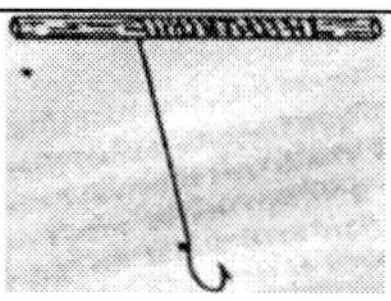	Major and Minor carp, small fish, Snake headed, Cat and small cat fish, Eel and turtles.	5.4-4.6	5.2-5.1	5.4-3.5	5.6-4.3
b.	Chasra	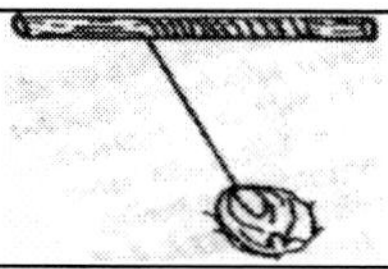					

Table 11. (Continued)

SL. No	Fishing methods	Picture/Figure	Capture group	Percentage (%) of Fishing methods			
				Somes -wari	Nethai	Gharia	Nidaya
c.	Nul barshi	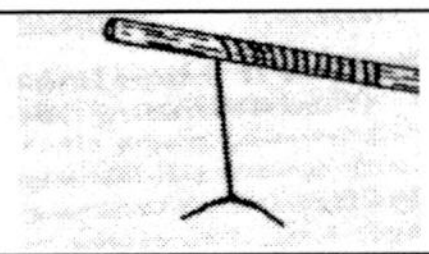					
d.	Spear Jhupi	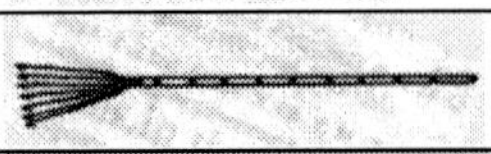					
e.	Fulkuchi	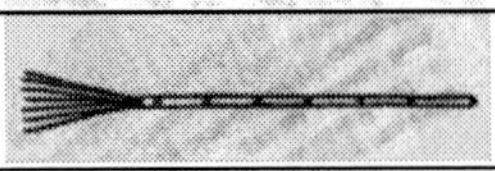					
f.	Aikra	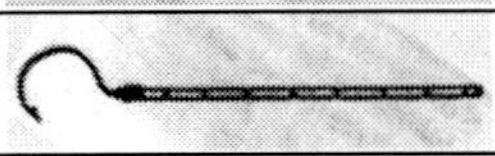					
12.	FAD	-	All groups	-	-	7.3-4.4	5.9-9.4

3.7. Fish Catch and Composition

An organized sampling program was run for a long time to get a true picture of the catch and composition of experimental rivers and beels. The present investigation gave a broad picture of a stock of aquatic wild lives which were obtained through market survey and interaction with fishers' in the river and beel. Present status and position of the investigated rivers and beels is shown in the Table 14.

Someswari River

Fishing activity in the Someswari River indicated the presence of 72 aquatic wild lives (64 species of wild fishes, three species of prawn, one species of crabs, one species of snail and four species of turtles) belonging to 51 genera and 23 families. The annual catch assessment of the experimental river was around 95.77±6.89; 76.99±5.43; 62.33±4.69; 48.13±3.49 and 38.61±2.83 mt in the year 2001, 2002, 20003, 2004 and 2005, respectively consisting of 10 groups viz., Knife fish, major carp and minor carp, small fish, cat fish and small Cat fish, eels, prawn, crabs and reptiles (Table 12).

From the catch statistics it was found that the magnitude of fishing pressure had been increasing every year. As a result, the total abundance of aquatic lives registered a progressive to decrease 19.61%, 34.92%, 49.74% and 59.68% in the year 2001-02, 2002-03, 2003-04 and 2004-05, respectively (Figure 4). Table 14 shows the existing status of the 72 aquatic wild animal of the Someswari River which were identified as E-06 (8%), CR-09 (13%), EN-49 (68%), VU-05 (7%), LR-02 (3%) and NO-01 (1%), respectively (Fig13). Commercial important fishes mohasseer (*Tot tor*), putitor mohasher (*Tor putitora*), nandina, (*Labeo nandina*), local sarpunti, (*Puntius sarana*) and reptiles (*Kachuga kachuga* and *Morenia petersi*) were rarely found in 2001 to 2003. But these species were extinct between 2004 and 2005. It was also found that alonga (*Bengala elanga*), bhangna bata (*Labeo bata*), ghora muikha (*Labeo pangusia*) utti (*Chagunius chagunio*) Tila koksa (*Barilius tileo*) dhela (*Rohtee cotio*) guizza (*Mystus seenghala*) baghair (*Bagarius yarrellii*) and cheka (*Chaca chaca*) were critically endangered, common carp (*Cyprinus carpio*), teri punti (*Puntius terio*) bujuri (*Mystus tengra),* Shotka chingri (*Machrobrachium malcolmsni*) and kakra (*Stylla serrata*) were vulnerable, kakila (*Xenentodon cancila*) and Boal (*Wallago attu*) were lower risk, Gura chingri (*Machrobrachium biramanicus*) was not threatened and other 49 species were endangered in status and position. Cat fish was the most dominant group of fish in decreasing the period of investigate the river and second highest production was recorded in the small cat fish group. The catch of fish in the river had been progressive over the last five years (Table 12).

Table 12. Production of different groups of fishes in the Someswari River during 2001-2005

SL. No.	Groups	Yearly production (mt)				
		2001	2002	2003	2004	2005
A.	Major carps	11.03±2.57^{a}	8.85±1.33^{b}	7.76±1.06^{c}	6.45±1.01^{d}	5.02±1.00^{e}
B.	Minor carp	6.65±1.05^{a}	5.42±0.76^{b}	4.00±0.47^{c}	2.83±0.35^{d}	1.60±022^{e}
C.	Small fish	13.92±1.44^{a}	11.49±1.13^{b}	9.54±1.01^{c}	7.89±0.95^{d}	6.58±0.92^{e}
D.	Knife and Feather back	3.06±0.85^{a}	2.21±0.61^{b}	1.73±0.45^{c}	1.30±0.15^{d}	0.87±0.11^{e}
E.	Snake headed	22.99±4.22^{a}	17.49±3.32^{b}	13.55±2.02^{c}	10.58±1.25^{d}	8.17±1.02^{e}
F.	Cat fish	15.61±3.24^{a}	13.05±2.46^{b}	10.90±1.56^{c}	8.67±1.21^{d}	7.11±0.96^{e}
G.	Small cat fish	2.38±0.82^{a}	1.84±0.74^{b}	1.36±0.52^{c}	1.02±0.24^{d}	0.80±0.01^{e}
H.	Spiny eel	5.32±1.18^{a}	4.20±0.95^{b}	3.62±0.44^{c}	2.80±0.38^{d}	2.24±0.28^{e}
I.	Prawn	6.90±1.11^{a}	5.96±0.88^{b}	5.18±0.52^{c}	4.49±0.41^{d}	3.62±0.33^{e}
J.	Crabs	2.98±0.88^{a}	2.28±0.65^{b}	1.74±0.44^{c}	1.54±0.33^{d}	1.22±0.22^{e}
K.	Reptiles	4.95±1.12^{a}	4.2±0.85^{b}	2.95±0.52^{c}	2.10±0.23^{d}	1.38±0.21^{e}
	Total	95.79±6.89^{a}	76.99±5.43^{b}	62.33±4.39^{c}	49.67±3.49^{d}	38.61±2.83^{e}

Figure in the same row having the same superscript are not significantly different ($P>0.05$). Figures in the parenthesis indicate the range.

In the figure 9, it is found that the percentage of different groups of fishes in the river had sharply decreased in every year and some important native species was extinct from the Someswari River. The capture of fishes, crabs and reptiles in the river was recorded highest status in 2001-02, where as it decreased in 2002-04 and sharply decreased between 2004 and 2005. The production figure differed significantly ($P<0.05$) among the different years (Figure 9). The result of the present study showed that the abundance of different species in the river Someswari were at high risk and total production has been sharply decreasing during last five years.

Nethai River

The Nethai River was represented by 66 aquatic wild lives (58 species of wild fishes, three species of prawn, one species of crabs and four species of reptiles) belonging to 51 genera and 24 families. The total catch statistics of aquatic wild lives in the Nethi River was recorded to be 175.20±11.83, 136.36±9.29, 116.15±8.15, 95.32±6.70 and 76.63±5.59 mt in the year 2001, 2002, 2003, 2004 and 2005, respectively consisting of knife fish, major carp, minor carp, small fish, cat fish, small cat fish, eels, prawn, crabs and reptiles (Table 13). Catch statistics of the river showed that the percentage of fishing pressure was increasing every year. As a result, the total production of the river had decreased to 22.17%, 33.70%, 45.59% and 56.26% in the year 2001-02, 2002-03, 2003-04 and 2004-05, respectively (Figure 12). Table 14 shows the existing status of the 65 aquatic wild animal of the Nethai river which was designated as E-05 (7%), CR-04 (6%), EN-48 (71%), VU-08 (12%), LR-02 (3%) and NO-01 (1%), respectively (Figure 14). Like Someswari river commercial important wild fishes mohasseer (*Tot tor* and *Tor putitora*), nandina, (*Labeo nandina*), local sarpunti, (*Puntius sarana*) and Reptiles (*Kachuga kachuga* and *Morenia petersi*) were rarely found in the Nethai river in the year 2001 to 2003. But these species were extinct between 2004 and 2005. It was also found that the status of dhela (*Rohtee cotio*), guzzia air (*Aorichthys seenghala*), baghair (*Bagarius bagarius*) and Chaga (*Chaca chaca*) was critically endangered; common carp (*Cyprinus carpio*), tit puti (*Puntius ticto*), chapila (*Gadusia chapra*), bailla (*Glossogobus giuris*), bujuri (*Mystus tengra*), baim (*Mastacembalus armatus*), Shotka chingri (*Nematopalaemon tenuipes*) and kakra (*Paratelphusa spinigera*) was vulnerable, Kakila (*Xenentodon cancila*) and boal (*Wallago attu*) were at lower risk, Gura chingri (*Machrobrachium birmanicum*) was not threatened and other 48 species were endangered. In order of numerical abundance cat fish and small cat fish were sule recorded as a dominant group and sub dominant group in the river during five years. The production of fish in the river had decreased in 2002 followed by the year 2003, 2004 and 2005, respectively.

Table 13. Production of different groups of fishes in the Nethai River during 2001 – 2005

SL. No.	Groups	Yearly production (mt)				
		2001	2002	2003	2004	2005
A.	Major carps	25.24±4.66^a	17.73±3.01^b	14.90±2.48^c	14.09±1.52	12.58±1.40^e
B.	Minor carp	6.46±1.80^a	5.25±1.55^b	3.81±0.85^c	2.48±0.68d	1.33±0.28^e
C.	Small fish	27.57±3.38^a	22.64±3.08^b	21.61±2.57^c	17.35±2.07	14.3±2.01^e
D.	Knife and Feather back	6.87±1.05^a	5.22±0.58^b	4.39±0.45^c	3.62±0.33^d	2.63±0.21^e
E.	Clupidae	2.46±0.16^a	2.2±0.14^b	2.0±0.11^c	1.7±0.10^d	1.26±0.08^e
F.	Cat fish	36.34±4.33^a	28.03±3.02^b	22.58±2.48^c	18.00±1.77	14.2±1.14^e
G.	Small cat fish	30.57±4.06^a	24.8±3.88^b	21.39±2.22^c	17.11±2.10	13.59±1.48^e
H.	Spiny eel	10.94±2.11^a	7.46±1.88^b	6.2±1.02^c	4.92±0.35^d	4.17±0.31^e
I.	Prawn	14.72±2.77^a	12.03±1.46^b	10.87±1.39^c	9.36±122^d	7.55±1.01^e
J.	Crabs	5.13±1.35^a	3.90±0.81^b	2.62±0.61^c	2.25±0.38^d	2.10±0.22^e
K.	Reptiles:	8.92±1.59^a	7.10±1.54^b	5.78±1.01^c	4.44±0.82^d	2.92±0.68^e
	Total	175.20±11.83	136.36±9.29^b	116.15±8.15	95.32±6.70	76.63±5.59^e

Figure in the same row having the same superscript are not significantly different (P>0.05). Figures in the parenthesis indicate the range.

Like Someswari river the percentage of different groups of fishes in the river has been sharply decreasing in every year and some important native species became extinct in the Nethai River. Cyprinids and catfishes are dominant groups of fish in the river. However, the production percentage (%) differed significantly (P<0.05) among the different years (Figure 4). Abundance of different species of fish in the river indicated that they are at they are at high risk and total production registered sharp decrease over the five years. However it was found that in the early flood phase of Someswari and Nethai river some important fish species migrated from upstream (Meghaloyan hilly area) into shallow running water (lower stretch) for spawning. At that time, the fishermen easily killed the brood fishes of all species by using with various type of fishing gear and fish trap. As a result, 80-90% of them had to die before they have a chance to spawn for continuation of the species. 15-20% of brood fishes escaped and bred in the available waters but their hatchlings and fingerlings again fall prey to fishermen's gear and fish trap when they are on the down-stream march.

Gharia Beel

Catch statistics indicated that 81 aquatic wild animals are available in Gharia beel (71 species of wild fishes, four species of prawn, one species of crabs and five species of turtles) belonging to 48 genera and 21 families. The annual catch of fish in the beel was around 104.65±5.50, 82.75±4.68, 71.27±4.23, 59.17±3.52 and 48.03±2.97 mt in the year 2001, 2002, 20003, 2004 and 2005, respectively consisting of 11 groups viz., major carp, minor carp, small fish, knife fish and feather back, snake head fish, cat fish, small cat fish, spiny eel, prawn, crabs and reptiles (Table 15). The catches of all the groups of fishes, crabs and reptiles were higher in 2001 but they showed decreasing trends in year 2002, 2003, 2004 and 2005 (Figure 10). However, the total catches in different years were significantly different ($P<0.05$). The catch statistics of Gharia beel showed that fishing pressure was increased every year. Catch statistics of the Gharia beel showed that the abundance of different important species of the beel was facing as extremely high risk of extinction between 2004 and 2005.

Table 14. Status and distribution of different fishes in the Someswari and Nethai river; Gharia and Nudaya beels of northern Bangladesh

Status code: E: Extinct, CR: Critically Endangered, EN- Endangered, VU-Vulnerable, LR- Lower risk, NO- Not threatened (As per IUCN, 2000)

SL No	Group/ Family	Local name	Scientific name	Picture/Figure	Status			
					Somes -wari	Nethai	Gharia	Nidaya
A.	Major carps:							
1	Cyprinidae	Catla	Catla catla	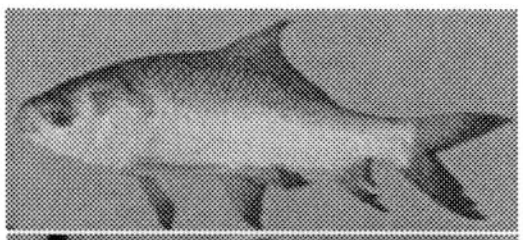	EN	EN	CR	EN
2	Cyprinidae	Rui	Labeo rohita		EN	EN	EN	EN
3	Cyprinidae	Mrigal	Cirrhinus mrigala		EN	EN	EN	EN
4	Cyprinidae	Mahashol	Trt tor		E	E	E	E

Table 14. (Continued)

SL No	Group/ Family	Local name	Scientific name	Picture/Figure	Status			
					Someswari	Nethai	Gharia	Nidaya
	Cyprinidae	Mahashol	Tot putitora	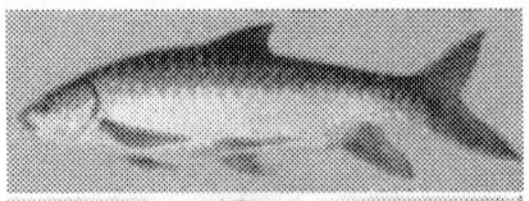	E	E	-	-
5	Cyprinidae	Nandil	Labeo nandina		E	E	E	E
6	Cyprinidae	Kalbaus	Labeo calbasu		EN	EN	EN	EN
7	Cyprinidae	Ghonia	Labeo Gonius		EN	EN	EN	EN
8	Cyprinidae	Comon carp	Cyprinus carpio	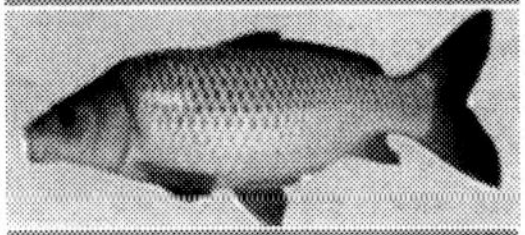	VU	VU	LR	LR
9	Cyprinidae	Silver carp	Hypophthalmicichthys molitrix		-	-	LR	LR
10	Cyprinidae	Grass carp	Ctenopharyngodon idellus		-	-	LR	LR
B.	Minor carp:							
11	Cyprinidae	Alonga	Bengala elanga		CR		CR	EN
12	Cyprinidae	Bata	Cirrhinus reba	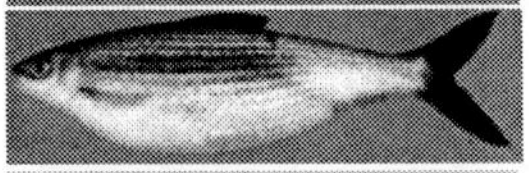	EN	EN	CR	EN
13	Cyprinidae	Bhangna bata	Labeo bata		CR	EN	CR	CR
14	Cyprinidae	Puda /Sharpunti	Puntius sarana	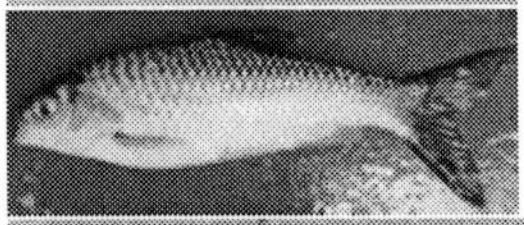	E	E	E	E
15	Cyprinidae	Ghora muikha	Labeo pangusia		CR	EN	-	-
16	Cyprinidae.	Utti	Chagunius chagunio		CR	-	-	-

17	Cyprinidae	Tila koksa	Barilius tileo		CR	EN	-	-
18	Cyprinidae	Thi sarpunti	Puntius gonionotus	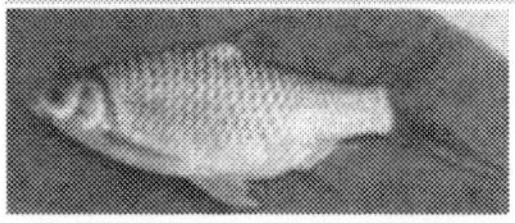	-	-	LR	LR
C.	Small fish:							
19	Cyprinidae	Mola/ Moya	Amblypharyngodon mola		EN	EN	VU	VU
20	Cyprinidae	Laubuca/ kashkhaira	Chela laubuca		EN	EN	EN	CR
21	Cyprinidae	Chap chela	Danio Devario		EN	EN	-	-
22	Cyprinidae	Dhela/ Dhiphali	Rohtee cotio		CR	CR	CR	CR
23	Cyprinidae	Chola punti	Puntius chola		EN	EN	EN	EN
24	Cyprinidae	Taka punti	Puntius conchonius		EN	EN	VU	VU
25	Cyprinidae	Phutani punti	Puntius phutunio	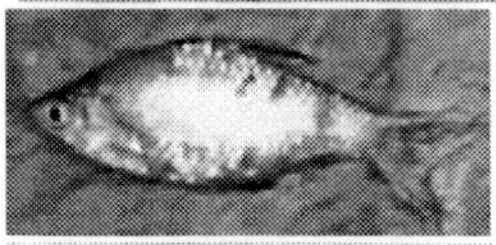	EN	EN	EN	EN
26	Cyprinidae	Jatpunti/ Vali Punti	Puntius sophore		EN	EN	EN	EN
27	Cyprinidae	Teri punti	Puntius terio		VU	EN	EN	VU
28	Cyprinidae	Tit Punti	Puntius ticto	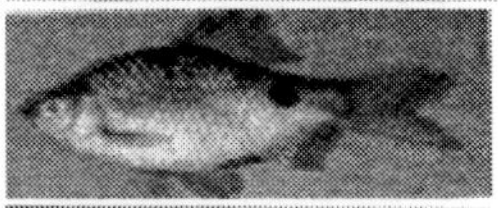	EN	VU	EN	VU
29	Cyprinidae	Fulchela	Salmostoma phulo	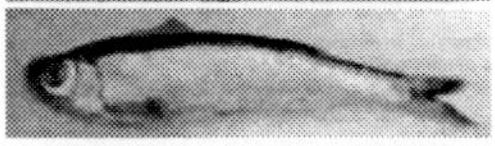	EN	EN	EN	VU
30	Anabantidae	Khailsha	Colisa fasciata	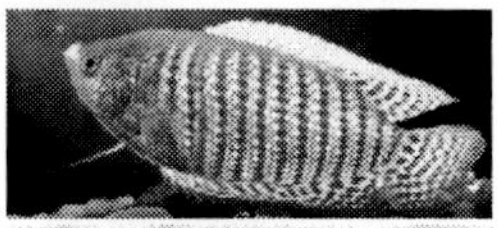	-	-	EN	VU
31	Anabantidae	Lal khailsha	Colisa lalia	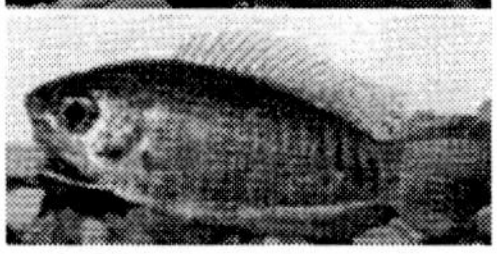	-	-	EN	VU

Table 14. (Continued)

SL No	Group/ Family	Local name	Scientific name	Picture/Figure	Status			
					Somes -wari	Nethai	Gharia	Nidaya
32	Anabantid ae	Chuna Khalisha	Colisa sota	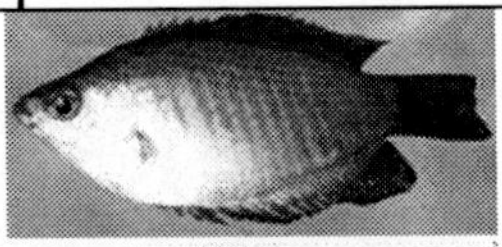	-	-	VU	VU
33	Nandidae	Mini/ Bheda	Nundas nandus		-	-	EN	EN
34	Cyprinidae	Darkina	Esomus danricus		-	-	EN	VU
35	Cyprinidae	Kanpon a	Oryzias melastigm a	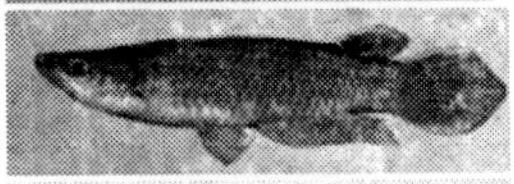	-	-	EN	VU
36	Cyprinidae	Anju	Danio rerio		-	-	EN	EN
37	Mastacem belidae	Gutum	Lepidocep halus gontea	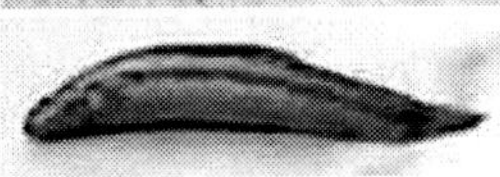	EN	EN	VU	VU
38	Clupeidae	Chapila	Gadusia chapra		EN	VU	EN	VU
39		Kachki	Corica soborna	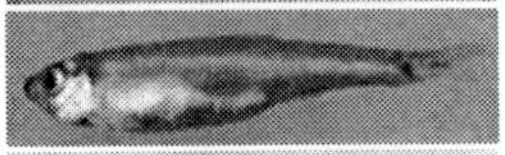	EN	EN	-	-
40	Centropo midae	Chanda	Chanda nama		EN	EN	EN	VU
41	Centropo midae	Chanda	Pseudamb asis baculis		EN	EN	VU	VU
42	Centropo midae	Ranga chanda	Pseudamb asis ranga	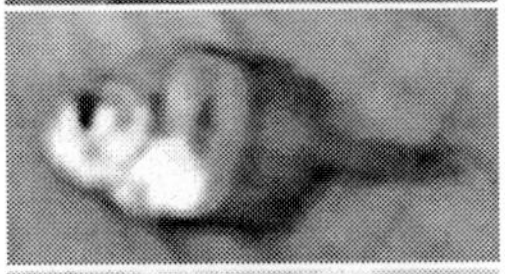	EN	EN	EN	VU
43	Cobitidae	Rani/ Botya	Botia dario		EN	EN	EN	EN
44	Cobitidae	Rani	Botia dayi		EN	EN	EN	EN
45	Tetradonti dae	Potka	Tetrodon cutcutia		EN	EN	EN	EN
46	Gobiidae	Baila/ bele	Glossogob us giuris	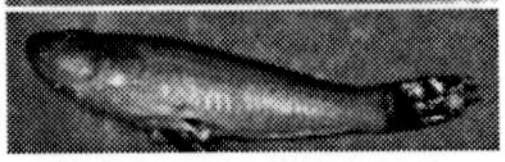	EN	VU	VU	VU

Table 14. (Continued)

SL No	Group/ Family	Local name	Scientific name	Picture/Figure	Status			
					Someswari	Nethai	Gharia	Nidaya
D	Knife fish and Feather back							
47	Notopteridae	Chitol	Notopterus chitala		EN	EN	EN	EN
48	Notopteridae	Foli	Notopterus notopterus		EN	EN	EN	EN
49	Belonidae	Kakila	Xenentodon cancila		LR	LR	VU	VU
E.	Snake heads							
50	Channidae	Gajar	Channa marulius	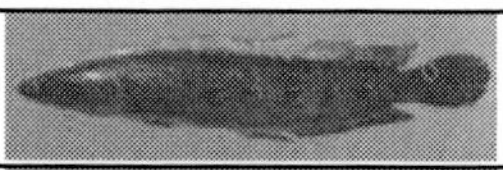	-	-	E	E
51	Channidae	Shol	Channa striatus	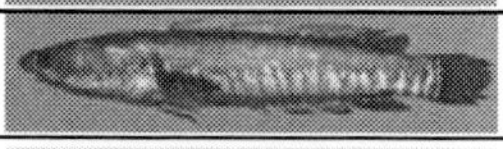	-	-	EN	LR
52	Channidae	Cheng/ Gachuaraga	Channa gachua		-	-	EN	VU
53	Channidae	Taki/ Okol	Channa punctatus	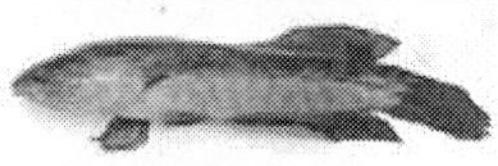	-	-	VU	VU
54	Anabantidae	Koi	Anabas testudineus		-	-	EN	EN
55	Anabantidae	Neftani	Ctenops nobiilis		-	-	EN	EN
56	Anabantidae	Napit koi	Badis badis		-	-	EN	EN
F.	Cat fish:							
57	Schilbeidae	Ayre/ Aor	Mystus aor	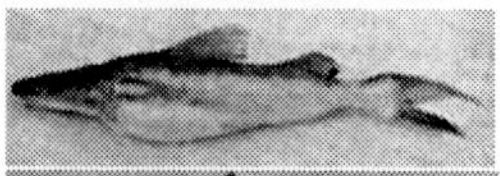	EN	EN	EN	EN
58	Schilbeidae	Guizza/ Guizza	Aorichthys seenghala		CR	CR	CR	EN
59	Bagridae	Baghair	Bagarius yarrellii		CR	CR	CR	CR
60	Bagridae	Rita	Rita rita		EN	EN	EN	EN
61	Chacidae	Chaga	Chaca chaca		CR	CR	-	-

Table 14. (Continued)

SL No	Group/ Family	Local name	Scientific name	Picture/Figure	Status			
					Somes -wari	Nethai	Gharia	Nidaya
62	Siluridae	Boal	Wallago attu	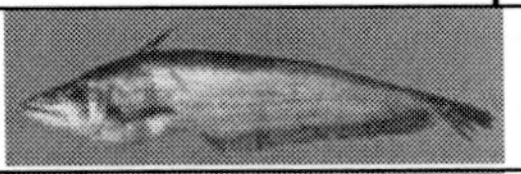	LR	LR	VU	VU
63	Schilbeid ae	Shilong	Silonia silonia		EN	EN	-	-
G.	Small cat fish:							
64	Schilbeid ae	Bujuri	Mystus tengra		VU	VU	LR	LR
65	Schilbeid ae	Tengra	Mystus vitttus		EN	EN	EN	VU
66	Schilbeid ae	Gulsa	Mystus cavasius		EN	EN	EN	EN
67	Siluridae	Kani Pabda	Ompok bimaculatu s		EN	EN	EN	EN
68	Siluridae	Madhu Pabda	Ompok pabda		EN	EN	EN	EN
69	Siluridae	Pabda	Ompok pabo	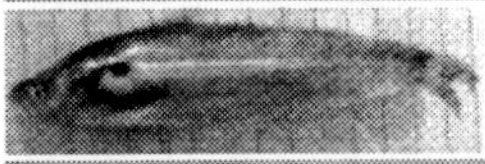	EN	-	EN	EN
70	Schilbeid ae	Kajuli	Ailia coila	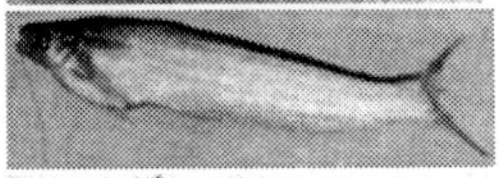	EN	EN	EN	EN
71	Schilbeid ae	Gharua	Clupisoma garua		EN	EN	EN	EN
72	Schilbeid ae	Bacha	Eutropiicht hys vacha		EN	EN	EN	EN
73	Schilbeid ae	Batashi	Pseudontro pius atheronoid es	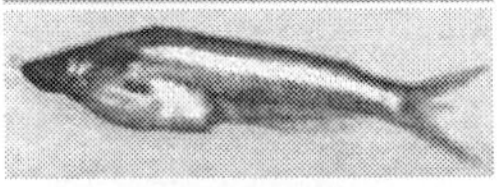	EN	EN	VU	VU
74	Claridae	Magur	Clarius batrachus		-	-	EN	EN
75	Heteropn eustidae	Singi	Heteropne ustes fossilis	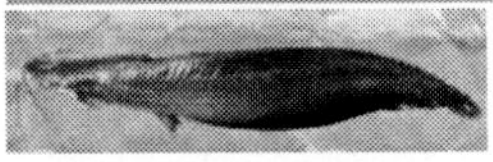	-	-	EN	VU
H.	Spiny eel							
76	Mastace mbelidae	Baim	Mastacem balus armatus	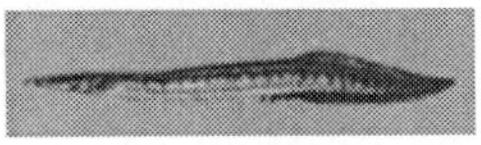	EN	VU	EN	EN

Table 14. (Continued)

SL No	Group/ Family	Local name	Scientific name	Picture/Figure	Status			
					Somes-wari	Nethai	Gharia	Nidaya
77	Synbranchidae	Kuicha	Monopterus cuchia	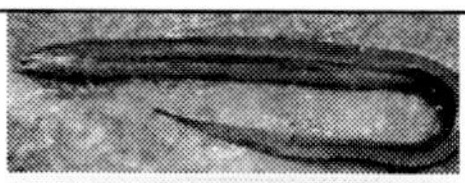	EN	-	EN	EN
78	Mastacembelidae	Tara Baim	Macrognathus aral		EN	-	EN	VU
79	Mastacembelidae	Chikra/ Guchi	Macrognathus pancalus	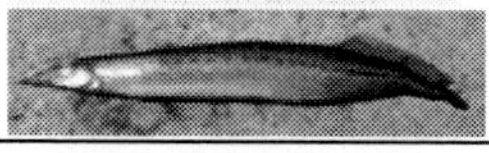	EN	EN	VU	VU
I.	Prawn:							
80	Palaemonidae	Golda Isa	Machrobrachiu rosenbergii		EN	EN	EN	EN
81	Palaemonidae	Gatakia chingri	Machrobrachium villosimanus		-	-	NO	NO
82	Palaemonidae	Gura chingri	Machrobrachium birmanicum		NO	NO	VU	VU
83	Palaemonidae	Shotka chingri	Machrobrachium malcolmsnii		VU	VU	EN	VU
J.	Crabs / Snail							
84	Potamonidae	Kakra	Paratelphusa spinigera	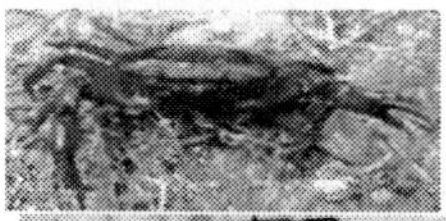	VU	VU	VU	VU
85	Unionidae	Bivalve	Lamellidens marginalis		-	-	-	VU
K.	Reptiles:							
86	Chelonia	Narrow headed Shoft shell	Chiitra indica		-	-	EN	EN
87	Chelonia	Spotted Flapshell	Lissemys punctata	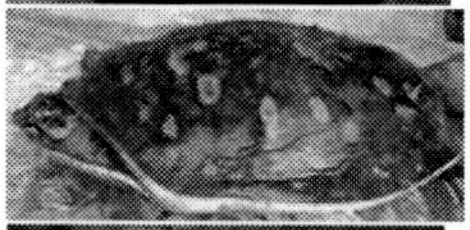	EN	EN	EN	EN
88	Chelonia	Common Roof Turtile	Kachuga tecta		EN	EN	EN	EN
89	Chelonia	Painted Roof Turtile	Kachuga kachuga	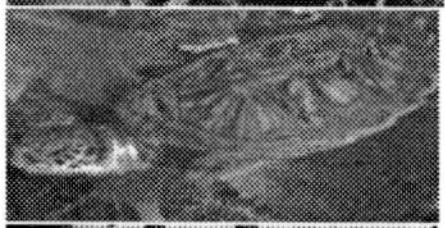	E	E	E	E
90	Chelonia	Bengal Eyed Turtile	Morenia petersi		E	E	E	E

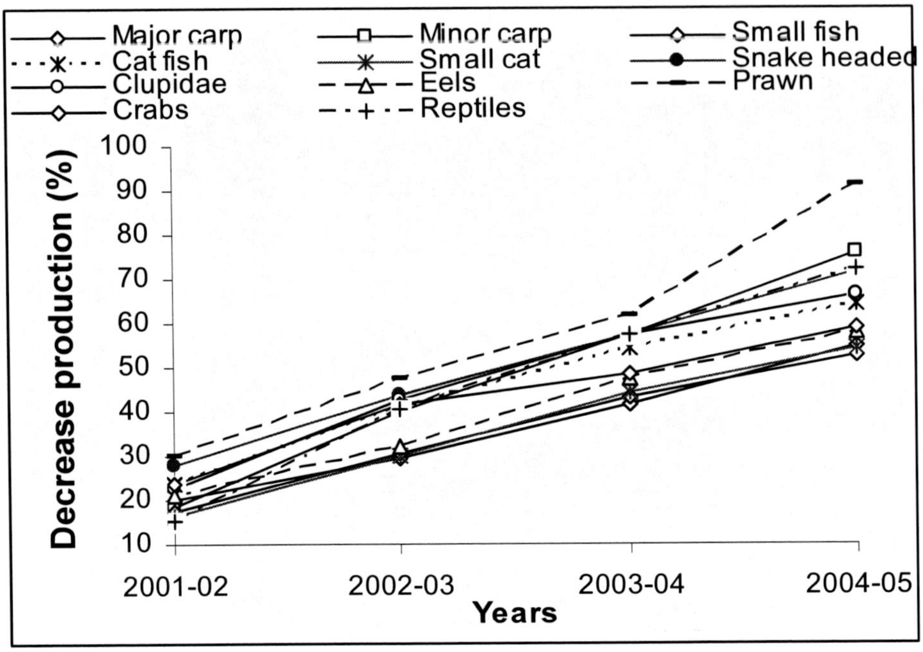

Figure 9. The production percentage of different groups of aquatic lives decreasing in the Someswari River during 2001 to 2005.

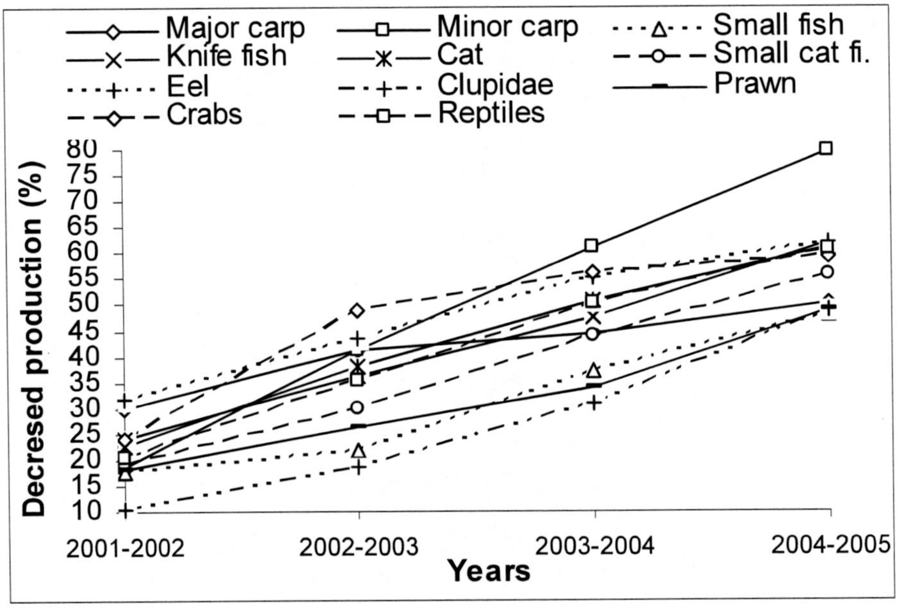

Figure 10. The production percentage of different groups of aquatic lives decreasing in the Nethai River during 2001 to 2005.

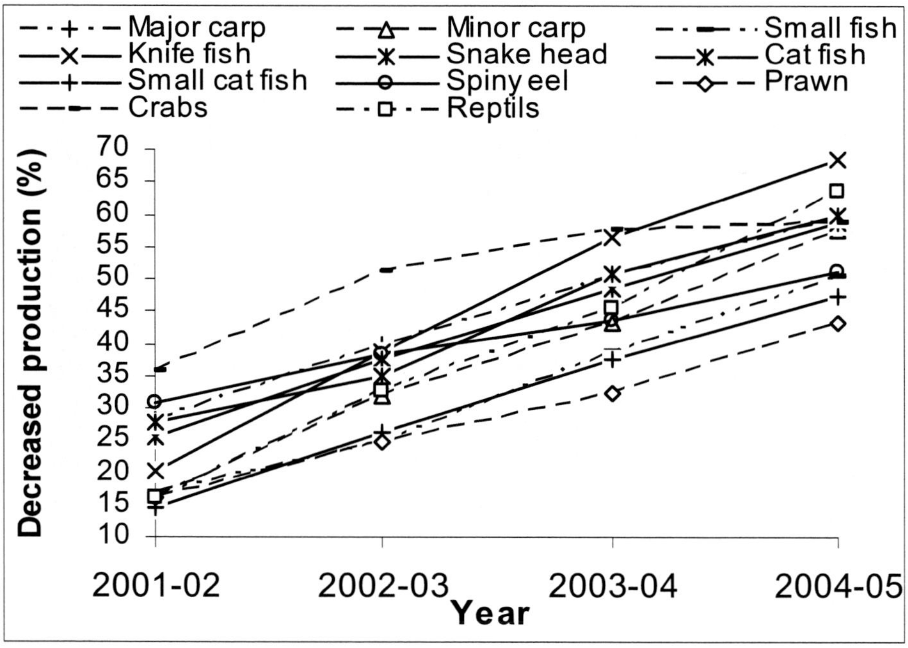

Figure. The production percentage of different groups of aquatic lives decreasing in the Gharia beel during 2001 to 2005.

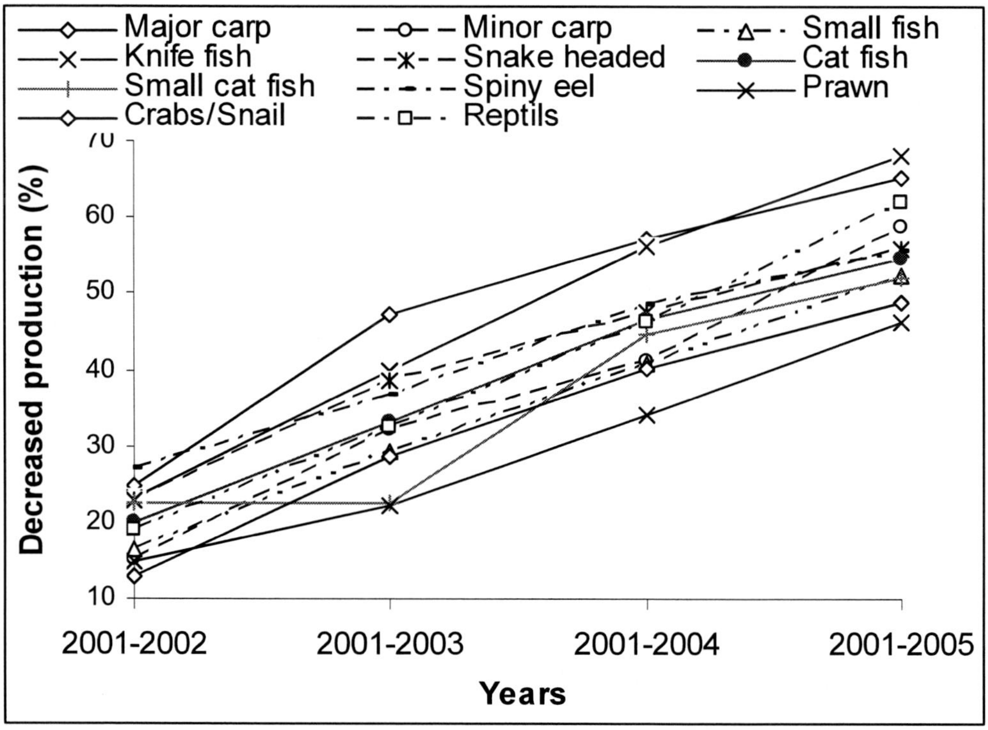

Fgure. The production percentage of different groups of aquatic lives decreasing in the Nidaya beel during 2001 to 2005.

Table 15. Production of different groups of fishes in the Gharia beel during 2001 – 2005

SL. No.	Groups	Yearly production (mt)				
		2001	2002	2003	2004	2005
A.	Major carps	9.26±1.89[a]	6.95±0.65[b]	5.59±0.62[c]	4.61±0.58[d]	3.75±0.58[e]
B.	Minor carp	6.05±1.04[a]	5.07±0.42[b]	4.1±0.40[c]	3.43±0.46[d]	2.57±0.38[e]
C.	Small fish	17.43±3.41[a]	14.51±2.88[b]	13.14±2.5[c]	10.68±182[d]	8.60±1.01[e]
D.	Knife and Feather back	2.76±1.05[a]	2.20±1.01[b]	1.69±0.83[c]	1.20±0.5 8[d]	0.87±0.28[e]
E.	Snake headed	9.62±0.79[a]	7.16±0.60[b]	5.99±0.54[c]	4.94±0.51[d]	3.96±0.41[e]
F.	Cat fish	13.43±2.80[a]	9.73±2.01[b]	8.76±1.80[c]	6.61±1.28[d]	5.38±1.13[e]
G.	Small cat fish	19.87±3.62[a]	16.96±2.54[b]	14.64±2.44[c]	12.44±2.02[d]	10.49±1.22[e]
H.	Spiny eel	8.89±1.46[a]	6.14±1.22[b]	5.49±1.03[c]	5.00±1.01[d]	4.33±0.98[e]
I.	Prawn	8.55±201[a]	7.17±1.47[b]	6.44±1.22[c]	5.78±0.88[d]	4.76±0.74[e]
J.	Crabs	2.62±0.44[a]	1.69±0.32[b]	1.28±0.30[c]	1.11±0.23[d]	1.08±0.16[e]
K.	Reptiles	6.17±0.58[a]	5.17±0.53[b]	4.15±0.47[c]	3.37±0.44[d]	2.24±0.34[e]
	Total	104.65±5.50[a]	82.75±2.68[b]	71.27±4.23[c]	59.17±3.52[d]	48.03±2.97[e]

Figure in the same row having the same superscript are not significantly different (P>0.05). Figures in the parenthesis indicate the range.

The status of the 80 aquatic wild lives of the Gharia beel was ranked as E-06 (7%), CR-08 (10%), EN-48 (60%), VU-13 (16%), LR-05 (6%) and NO-01 (1%), respectively (Figure 15). Commercial important mohasseer (*Tot tor*), nandina, (*Labeo nandina*) were rarely found in the year of 2001. But these species were extinct during 2002 to 2005. Local sarpunti (*Puntius sarana*) Gajar (*Channa marulius*) and Reptiles (*Kachuga kachuga* and *Morenia* petersi) were rarely found in the year of 2001 to 2004, but these species disappeared in the year 2005.

The other status of the aquatic lives of the river was catla (*Catla catla*), along (*Bengala elanga*), Bhegna bata (*Labeo bata*), bata (*Cirrhinus reba*), Dhela (*Rohtee cotio*), Guizza (*Aorichthys seenghala*), Baghair (*Bagarius yarrellii*) and Bengal eyed turtile (*Morenia petersi*) were critically endangered; mola (*Amblypharyngodon mola*), taka punti (*Puntius conchonius*), chuna Khalisha (*Colisa sota*), gutum (*Lepidocephalus gontea*), chanda (*Pseudambasis baculis*), baila (*Glossogobus giuris*), kakila (*Xenentodon cancila*), taki (*Channa punctatus*), boal (*Wallago attu*), batashi (*Pseudontropius atheronoides*), guchi (*Macrognathus pancalus*), gura chingri (*Machrobrachium birmanicum*) and kakra (*Stylla serrata*) were vulnerable; bujuri (*Mystus tengra*), common carp (*Cyprinus carpio*), silver carp (*Hypophthalmicichthys molitrix*), grass carp (*Ctenopharyngodon idellus*) and Thai sarpunti (*Puntius gonionotus*) were lower risk and gatakia chingri *Machrobrachium villosimanus* was not threatened and other 48 species were facing as high risk (endangered) in position.

Nidaya Beel

Results of the present study indicated that fishing pressure was increased every year in the Nidaya beel. As a result, the total production of the beel had decreased to the extent of 19.84%, 32.94%, 45.74% and 56.32% in the year 2001-02, 2002-03, 2003-04 and 2004-05, respectively (Figure) and Table 14, shows the existing status of the aquatic wild animal of the Nidaya beel as E-06 (7%), CR-05 (6%), EN-53 (66%), VU-11 (13%), LR-06 (7%) and NO-

01 (1%), respectively (Figure). The catch composition and abundance of different species of fish and other aquatic lives in case of Nidaya beel were almost identical to that of Gharia beel. Production of fish in Nidaya beel is furnished in table 17.

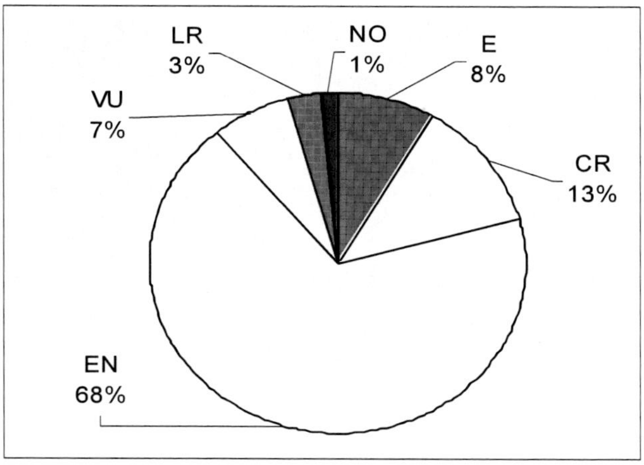

Figure 13. The percentage of status and position of different species of aquatic lives in the Someswari River during 2001-05.

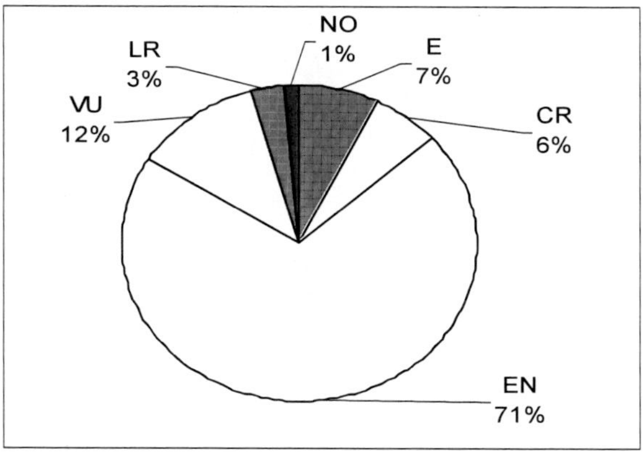

Figure 14. The percentage of status and position of different species of aquatic lives in the Nethai River during 2001-05.

A total of five species of fresh water turtles, viz. *Chiitra indica, Lissemys punctata, Kachuga tecta, kachuga and Morenia petersi* were found in the experimental river and beel. But two important species of turtle was extinct within five years. During investigation period, it was found that fishing effort with smaller mesh size was increased in every year as a result, average size and number of individual fish declined every year.

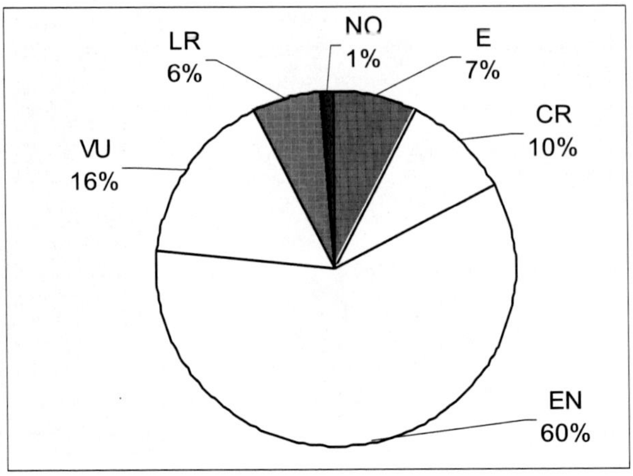

Figure 15. The percentage of status and position of different species of aquatic lives in the Gharia beel during 2001-05.

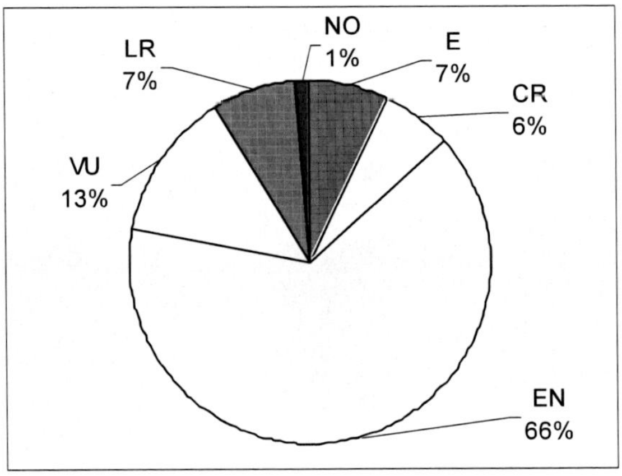

Figure 16. The percentage of status and position of different species of aquatic lives in the Nidaya beel during 2001-05.

During investigation periods, Fresh water pearl bearing mussels were identified in the beel. Shells of bivalve were utilized by rural people for production of lime. Wildlife includes, amphibians (*Buffo melanostictus, Rana tigerina, Rana limnocharis, Rana cyanophyctis* and *Salamandra salamondra*) aves (whistling duck, great crested grebe, great cormorant, red crested pochard, water cock, swamphen, great black headed gull, gray-headed fish eagle, curlew, spotted redshank) and mammals (musk shrew, fishing cat, small Indian mongoose, jackle, flying fox) were identified.

Table 16. Production of different groups of fishes in the Nidaya beel during 2001 - 2005

SL. No.	Groups	Yearly production (mt)				
		2001	2002	2003	2004	2005
A.	Major carps	11.68 ± 2.15^{a}	8.79 ± 1.91^{b}	6.18 ± 1.64^{c}	5.00 ± 1.02^{d}	4.08 ± 1.09^{e}
B.	Minor carp	4.89 ± 0.47^{a}	4.15 ± 0.45^{b}	3.32 ± 0.45^{c}	2.88 ± 0.50^{d}	2.02 ± 0.39^{e}
C.	Small fish	20.16 ± 4.57^{a}	16.83 ± 3.50^{b}	14.26 ± 2.42^{c}	11.94 ± 2.14^{d}	9.60 ± 1.59^{e}
D.	Knife and Feather back	3.13 ± 0.32^{a}	2.41 ± 0.31^{b}	$1.88\pm0.\ 28^{c}$	1.37 ± 0.28^{d}	1.0 ± 0.28^{e}
E.	Snake headed	9.86 ± 1.05^{a}	7.62 ± 0.86^{b}	6.07 ± 0.68^{c}	5.16 ± 0.62^{d}	4.35 ± 0.57^{e}
F.	Cat fish	10.77 ± 2.45^{a}	8.62 ± 1.86^{b}	7.21 ± 1.49^{c}	5.77 ± 1.16^{d}	4.39 ± 0.91^{e}
G.	Small cat fish	22.24 ± 4.85^{a}	17.25 ± 2.63^{b}	14.68 ± 2.50^{c}	12.31 ± 2.42^{d}	10.64 ± 1.36^{e}
H.	Spiny eel	8.27 ± 1.76^{a}	6.04 ± 0.39^{b}	5.23 ± 0.33^{c}	4.27 ± 0.36^{d}	3.68 ± 0.35^{e}
I.	Prawn	8.71 ± 2.41^{a}	7.41 ± 0.49^{b}	6.77 ± 0.49^{c}	5.74 ± 0.44^{d}	4.87 ± 0.37^{e}
J.	Crabs	7.23 ± 1.37^{a}	6.29 ± 0.64^{b}	5.16 ± 0.42^{c}	4.32 ± 0.21^{d}	3.70 ± 0.24^{e}
K.	Reptiles	5.93 ± 1.61^{a}	4.81 ± 0.55^{b}	4.01 ± 0.51^{c}	3.18 ± 0.48^{d}	2.25 ± 0.42^{e}
	Total	112.87 ± 5.98^{a}	90.22 ± 4.77^{b}	74.77 ± 4.10^{c}	61.94 ± 3.47^{d}	50.58 ± 2.98^{e}

Figure in the same row having the same superscript are not significantly different ($P>0.05$). Figures in the parenthesis indicate the range.

4. Discussion

The physico-chemical factors and plankton were found to be more or less normal range. Water temperature of the rivers and beels showed increasing trend in monsoon and post monsoon season and decreasing trend in winter which is supported by Mathew (1975). Transparency was consistently higher in upper region, possibly due to running water of the river and higher in deeper portion of the beel, possibly due to stagnant of water. Rahman (1992) stated that the transparency of productive water bodies should be 40 cm or less. The uniformly average value of oxygen as noted in the rivers and beels agrees well with the findings of Saha *et al.* (1988). pH values agree well with the findings of Rahman and Rahman (2003). Alkalinity levels of the beel were medium to high (Bhuiyan, 1970).

The phytoplankton consisted of 26 genera, which is more or less similar to the observation of Ehshan *et al*, 1997; Hossain *et al*, 1998; Sugunan and Bhattacherjya, 2000. The zooplankton population consisted of 12 genera which are closer to the observation of Ahmed *et al.* (1997) and Sugunan and Bhattacharyya (2000). A total number of 12 species of marginal and submerged vegetation was observed in the floodplain wetlands of rivers and beels, which are comparable with the finding of Sugunan and Bhattacharyya (2000) in case of floodplain of Brahmaputra basin. The swamp forests, mainly represented by hijal tree (*Barringtonia acutangula*) have been reduced to a few small patches in the surveyed area.

The fishing effort with various types of fishing gear such as seine net (especially kaperi jal), gill net (current jal) and FAD had been increasing in every year but use of current jal was increased rapidly during study period. . As a result, an average size and number of aquatic lives declined in the surveyed river, beel and its flood plains wetland. Haroon *et al.* (2002) reported eighteen types of fishing gears from the Sylhet sub-basin and thirteen types from Mymensingh sub-basin which are very similar to this study. Sugunan and Bhattacharyya (2000) found a wide variety of fishing methods employed in the beels of Assam, India which are very similar to the present study. Cast net (Jaki jal) is used whole year in the river. It is a very popular fishing method and used in all over the Bangladesh (Ahmed, 1962).

The catch statistics indicate that fishing pressure of the river and beel was increased in every year. As a result, a decreasing trend in production percentage of Someswari and Nethai rivers and Gharia and Nidaya beels were clearly pronounced within five years which was very much similar report of Moyle and Leidy (1992). They concluded that worldwide 20% of all freshwater species are extinct, endangered or vulnerable. The total catch statistics of aquatic lives in the rivers and beels indicated that percentage of different group of aquatic lives has been sharply decreasing in every year. As a result, commercially important fishes like Mahaseer (*Tot tor* and *Tor putitora*), nandina, (*Labeo nandina*), local sarpunti, (*Puntius sarana*), gazar (*Channa marulius*) and reptiles, *Kachuga kachuga* and *Morenia petersi* were extinct during five years study periods which is more or less similar to the report of IUCN, Bangladesh (1998). According to IUCN, Bangladesh about 56 freshwater fish species is critically or somewhat endangered. Due to over-exploitation and various ecological changes in natural aquatic ecosystem health such as rivers, beels and its flood plains wetland, commercially important aquatic lives are in the verge of extinction which are in agreement with the findings of Sarker (1993).

During winter season, turtles (*Morenia petersi* and *Kachuga tecta*) were caught in the river and its flood plains wetland. Khan (1982) reported that *K. tecta* are mainly distributed between the stretches of Ganges River and the Brahmaputra River. Bengal Eyed turtle, *Morenia petersi* was found in the rivers, beels and its flood plains wetland. Das (1991) mentioned its occurrence in Assam of India. Turtles of the surveyed rivers, beels and its flood plains wetland have been declining because of dewaterization of its habitat for irrigation and destruction of its breeding ground and nesting sites. Over exploitation for local consumption and foreign trade indiscriminately posses a threat to all species of turtles as well. The population of bivalve, *Lamellidens marginalis* as found in the river and its flood plain wetland, has also been decreasing which is consistent with the observation of Ali (1991). Among the available 90 aquatic wild lives of commercially importance seven species had disappeared of the rivers and beels is clearly proved that seven species are disappeared, 52-58 species are facing as extremely high risk of extinction and 7-18 species are more or less at risk of extinction between the year 2001 and 2005 (IUCN, 2000).

The study clearly indicated that the fish stocks of the rivers and beels are subjected to over fishing resulting in gradual decline in aquatic population. In addition, aquatic ecosystem health is changing due to construction of flood control barrage, soil erosion, siltation and drainage structures and agro-chemicals. The downstream of the river system is siltated, which reduces the rate of water flow and causes habitat degradation. Domestic organic wastes (sewage) directly or indirectly passing through canals or drains to the rivers or canals are polluting the aquatic ecosystem health. The genetic stock structure of aquatic populations is reduced due to pollution and destructive fishing practices (Mazid and Hussain, 1995). Indiscriminate killing of fish occurs due to the use of pesticides in improper doses, use of forbidden chemicals, aerial spray of chemicals as used in paddy field which was very much similar to the observation of Mazid (2002).

Indiscriminate destructive fishing practices have caused havoc to the aquatic biodiversity of the rivers and beels. As a result, the ecosystem health and biological diversity of the rivers and beels have been depleting at an unprecedented rate (Hussain and Hossain, 1999). Intervention to control floods, adoption of new agricultural technologies and construction of road networks has altered the ecology of rivers and beels significantly which supported the views of Khan (1993) and Ali (1991). Stock of the wildlife brood fishes in their breeding ground have also suffered significant damages resulting in a reduction of biodiversity as noted by Nishat (1993) and Zaman (1993).

The action plan efforts for saving the stock of aquatic species is usually required which could be accomplished through strick enforcement of fishing rules to prevent fishing with current jal (gill net) and moshary jal (ber jal); prevention of killing brood fish and juveniles from the breeding ground; some specific stretches of the river and beel need to be declared as sanctuaries because sanctuary protects the aquatic lives in all season and may be stocked every year with fingerlings; and develop community based management of open water system.

REFERENCES

Ahmad N. (1962). Fishing gear of East Pakistan. Government of East Pakistan. Directorate of Fisheries. Dhaka. 37 p.

Ahmed, K.K., M. K.I. Haque, G.C. Halder and S.B. Saha. (1997). Ecological studies on the planktonic organisms of Kaptai lake. *Bangladesh J. Agrl., 22*, 63-67.

Alam, A.K.M.K. (2004). The Someswari River (Bangladesh) response to active tectonics, Abstract Volume; Fourth Nepal Geological Congress, Nepal Geological Society, Katmandu, 9-11 April, 2004, 25.

Alam, G.N.(1996). Floodplain Fisheries Development in Bangldesh. Fish. Newsl. Bangladesh *Fish. Res. Inst., 4(1)*, 3-4.

Ali, M.Y. (1991). Towards sustainable development: fisheries resources of Bangladesh Internatioal Union for Conservation of nature and Natural Resources. The World Conservation Union. National Conservation Strategy of Bangladesh and Bangladesh Agricultural Research Council.96p.

Allendroff, F.W. (1988). Conservation Biology of Fishes. Conservation Biol., 2, 145-148.

Bhuiyan, R.R. (1970). Physico-Chemical qualities of some ancient tanks of Sibsagar, Assam. *Environ. Health 12,* 129-134.

Clesceri, L.S., Greenberg, A.E. and Trussell, R.R. (1989). Standard Methods of the Examination of Water and Wastewater (17th ed.). American Public Health Association, American Water Works Association and Water Pollution Control Federation, 1015 Washington D. C., USA 20036, 10-203.

Crains, M.A. and Lackey, R.T. (1992). Biodiversity and management of natural resources: *The issues Fisheries. 17(3)*, 6-10.

Das, I. (1991). Colour Guide to the turtles and Tortoises of the Indian sub continent. R and A publishing Ltd.

Dister E. (1990). Floodplain protection in Central Europe. World Wildlife Found (WWF) Institute of Floodplains Ecology Visiting Card 31/90, Germany.

Ehshan, M.A., Hossain, M.S. Razzaque, A. Mollah M.F.A. and Mazid, M.A. (1997). Limnology of a floodplain; Halti beel. *Univ. J. Zool. RajshahiUniv., 19*, 95-98.

Haroon, A.K.Y., Halder, G.C. Rahman, S.L. Razzaque, M.A. Alam M. and Nurul Amin, S.M. (2002). Sylhet-Mymensingh Basin Fish Stock Assessment. Final Report. Bangladesh Fisheries Research Institute, Reverine Station, Chandpur, Bangladesh 81pp.

Hossain, M.S., Mazid, M.A. Ehshan, M. Rahman, A. S. Islam, A.K.M.S. and Hossain, M.M.M. (1998). Limnological observation on Basukhali-Salimpur-Kola-Barnal (BSKB). *Bangladesh J. Zool., 26(1)*, 79-84.

Hussain, M. G. (1997). Stock improvement and genetic resources conservation of floodplain fishes. *In*. Chu-fo-Tsai and Ali, M. Y. (eds). *Open water fisheries of Bangladesh. Bangldesh Center for Advanced studies.* Dhaka. pp.49-64.

Hussain, M. G. and Hossain, M. A. 1999. Controlled breeding technology and step for conservation of gene pool of certain endangered fish species of Bangladesh. *Fish. News Letter, 7(1-3)*, 2-3.

IUCN, Bangladesh. (1998). List of threatened animals of Bangladesh. Paper presented in the Special Workshop on Bangladesh Red Book of Threatened Animals, 22 February 1998, Dhaka, Bangldesh. 13 pp.

IUCN. (2000). Red book of threatened Fishes of Bangladesh. IUCN – The World Conservation Union xii+116 pp.

Khan, H. R.(1993). Water development activities and their impacts on wetlands. P23-32 .*In*: A. Nisat, Z. Hossain, M. K. Roy, and A. Karim (eds.). *Freshwater wetlands in Bangladesh: Issues and Approaches for management. IUCN,* Gland Switzerland. xii+283p.

Khan, M.A.R. (1982). Chelonians of Bangladesh and their conservation, *J. Bombay Nat. Hist. Soc. 79 (i),* 110-116.

Mathew, P.W. (1975). Limnology and productivity of Govindgarh Lake, Maddya Prodesh, India. *J. Inland Fish. Soc. India, 11*, 16-24.

Mazid M.A. and Hossain, M. S. (1995). Development of fisheries resources in floodplains. FRI publication No. 12. Fisheries Research Institute, Mymensingh, Bangladesh.

Mazid M.A. (2002). Development of Fisheries in Bangladesh: Plans and Strategic for Income Generation and Poverty Alleviation. Fisheries Research Institute, Mymensingh, Bangladesh. 78-79.

Moyle, P.L. and Leidy, R.A. (1992). Loss of biodiversity in aquatic ecosystem: Evidence from Fish Fauna. In P. L. Fielder and H.L. Jani (eds). *Conservation of Biology: the Theory and Practice of Nature Conservation, Preservation and Management.* Chapman and Hall, New York, USA. 562 p.

Moyle, P.B. and Williams, J.E. (1990). Biodiversity loss in the temperature zone; decline of the native fish fauna of California. *Conserv. Biol. 4(3),* 275-284.

Nishat, A. (1993). Freshwater wetlands in Bangladesh: status and issues. P. 9-22 In A. Nishat Z. Hossain, M. K. Roy and A. Karim (eds.). *Freshwater wetlands in Bangladesh: Issues and Approaches for management.* IUCN, Gland Switzerland. xii+283p.

Noss, R.F. (1983). A regional landscape approach to maintain diversity. *Biosci. 33*, 700-706.

Rahman, M.R. and Rahman, M.A. (2003). Studies on the growth, survival and production of calbasu (*Labeo calbasu*, Ham.) at different stocking densities in primary nursing. Bull. Fac. Sci., Unv. Ryuyus, Jpn. 76, 245-255.

Rahman, M.S. (1992). Water quality Management in Aquaculture. BARC Prokashana, 66, Mohakhali, Dhaka 1212, Bangladesh 84 pp.

Saha, S. B., Gupta, M.V. and Sahabuddin, M. (1988). Studies on the comparative efficacy of different fertilizers on the production of silver carp fry. *Bangladesh J. Fish., 11 (2)*, 11-20.

Sarker, S.U. (1993). Faunal diversity and their conservation in freshwater wetlands. In Nisat A. Z. Hussain, Roy M. K. and A. Karim (eds.). *Freshwater wetlands in Bangladesh-issues and approaches for management.* IUCN, The world Conservation Union. pp 105-122.

Stirling, H.P. (1985). Chemical and Biological Methods of Water Analysis for Aquaculturist. Institute of Aquaculture, University of Stirling, Scotland, UK. 119 p.

Sugunan, V.V. and Bhattacharjea, B.K. (2000). Ecology and Fisheries of Beels in Assam. Bull. No. 104, CICFRI, Barrackpore-743101, West Bengal 66 pp.

Zaman, S.M.H. (1993). Agricultural development and sustainablility of wetlands. P 63-178 In A. Nishat, Z. Hossain, M. K. Roy and Ansarul Karim (eds.). *Freshwater wetlands in Bangladesh: Issues and Approaches for management.* IUCN, Gland Switzerland. xii+283p.

Zar, J.H. (1984). *Biostatistics*. Prentice-Hall, Inc. Englewood Cliffs, New Jersey, USA. 718.

In: Aquaculture Research Progress
Editor: Takumi K. Nakamura

ISBN 978-1-60456-247-7

Chapter 9

DEVELOPMENT OF NEW SUSTAINABLE TECHNIQUES FOR SHRIMP CULTURE IN MEXICO USING FLOATING SEAWATER CAGES

Martha Zarain-Herzberg[1]
Centro de Ciencias de Sinaloa. Av. de las Américas 2771 Nte, Culiacán Sinaloa, 80010, México

ABSTRACT

At present aquaculture provides over 40% of the world's seafood supply and according to the FAO this figure will approach 54% by the year 2020.

However aquaculture practices cause environmental problems that include the destruction of natural ecosystems and shrimp disease occurrences. Shrimp culture in floating seawater cages is now being considered as an alternative approach for aquaculture.

The biological viability of culturing the Pacific white shrimp *Litopenaeus vannamei* in floating seawater cages has been validated.

The main purpose of our study was to develop a novel model of a culture system for growing *L. vannamei* with environmental sustainability using floating seawater cages of different dimensions and materials; and to evaluate the effect of different stocking densities, nursery, and growout conditions on efficiency of production. We also evaluated if the shrimp developed the most frequent viral diseases.

For this research we used small (9 m^2) and large (200 m^2) floating cages for growing shrimp. The cages were circular (~200 m^2) made of HDPE polyethylene pipes, and square cage (9 m^2) made of PVC pipes. For both type of cages, PVC-covered polyester nets (Sansuy®) were used to hold the shrimp in nursery and growout stages. The project location was Bahia Navachiste, Guasave, Sinaloa, México.

In the 9 m^2 nursery cages, there was no significant difference (P>0.05) in the survival rate of *L vannamei* culture at densities of 950, 1300 and 1600 shrimp/m^2 for 38 culture days, nevertheless the final weight (g) was significantly lower at 1600 shrimp/m^2, but the number of shrimp per cage was the highest at the density of 1600 shrimp/m^2.

---

[1] Tel.+52 (667) 7599000 ext. 1124. E-mail: martha@computo.ccs.net.mx.

There were no significant differences ($P > 0.05$) in the final weight (g) and survival rate among cultured shrimp in the 9 m^2 growout cages at 200, 250, and 300 shrimp/m^2 densities for 58 days.

The shrimp biomass per square meter obtained at density of 300 shrimp/m^2 was greater (2.6 kg/m^2) when compared with the lowest density assayed.

In 200 m^2 growout cages the growing rate at the lower density (45 shrimp/m^2) is higher (2.1 g/week) and the survival rate is > 80% at both densities assayed (45 and 180 shrimp/m^2) during 62 days. The final weight (g) for each density was 17.7 ± 1.55 g at 45 shrimp/m^2 and 13.3 ± 1.31 g at 180 shrimp/m^2. The FCR as for the 9 m^2 cages was close to 1.

All the monitored shrimp were free of the viral pathogens TSV and WSSV.

1. INTRODUCTION

Each year the world's population continues to increase by about 100 million people, a growth rate of about 2%. According to the Food and Agriculture Organization of the United Nations (FAO), estimates of the world's population by the end of 2050 will exceed 10 billion, which is twice the present amount.

If the present rate of food consumption per capita is maintained, food production must double to satisfy demand. The major concern will be to supply protein according to the three nutritional essentials; protein, carbohydrates, and fat. However, there is a limit for protein production on land and some regions have already reached their production limits, a reason why marine production has to increase its importance as an alternative protein source, and consequently fishing demand will dramatically increase.

Today many marine populations are overexploited and half of the species are fully exploited (48%), thus producing catch close to the maximum sustainable limits. About one quarter is overexploited (17%), 8% are depleted, and only 1% are recovering from depletion [1].

In 2004 the total world production of the fishing sector reached nearly 156 million t, with marine capture of 87 million t and inland capture of nine million t. Mariculture was estimated to be 30 million t, freshwater culture 27 million t, and the remaining three million t came from brackish water culture [2]

In Mexico 87% of the total fish comes from fishery catch. Based on the "2002 Fishing Statistics Annual Data Book" [3], the Pacific Ocean yielded about 80% of the total capture from the total Mexican fishing resources. The shrimp species in the Mexican Pacific ocean represented one of the most important fisheries of this coast and for the country, which species are principally the 5 species of blue shrimp (*Litopenaeus stylirostris*), white shrimp (*Litopenaeus vannamei*), brown shrimp (*Farfantepenaeus californiensis*), red or crystal shrimp (*Farfantepenaeus brevirostris*), and white shrimp (*Litopenaeus occidentalis*).

But in Mexico, 67% of fisheries (79% of species) are either fully (shrimp, shark, octopus, and spiny lobster) or over exploited (anchovy, abalone, sea urchin, and sea cucumber) and cannot grow further.

The fishing sector is the principal source for direct and indirect employment with more than 37,000 employees in the Mexican Pacific, however the fisheries have a series of diverse problems; (1) overfishing by the Mexican fisheries affects shrimp population levels, which are directly related to the quantity and spatial and temporal distribution of the fishery efforts, including artisanal and industrial, (2) the high level of the overfishing of shrimp derives from the superiority of the fish float quantity, which is greater than its biological and economical value for both open sea and artisanal fisheries [4, 5], and (3) the illegal and furtive fishery (probably the greatest problem to be considered for government regulation), and (4) the use of prohibited fishery methods.

During the float capture in the open sea an important cyclical product behavior depending on the environmental and effort variability has been seen. Presently in Mexico, about 60% of the annual shrimp capture comes from open-sea operations and the rest from activities from bay and brackish water. Nationally this kind of fishery ranks at the top for the number of large and small ships used for this activity. Ships for open-sea shrimp capture are, on average, 25-years old and an important number of ships are in need of repair when the fishing season begins.

Shrimp capture in protected water is made with small boats weighing 1/2 to 1 t and with an average length of 4 meters. Propulsion varies from a pair of oars to the use of outboard motors up to 250 hp. The number of small fishing boats is about 56,500 and about 13,000 are from Sinaloa. It is estimated that at least 85% are dedicated to shrimp capture.

During the 2003-2004 season, 670 oceanic ships from Sinaloa, Mexico obtained 11,320 t of shrimp during 196 days of fishing compared to 4,960 t obtained from 140 fishing groups using 4,717 small boats during 206 days, equivalent to 1,160 t per small boat and 50 kg of shrimp per day for each two fishermen. This production does not satisfy the regional population of 38,750 fishermen distributed along 656 km of coast comprised by 154 communities.

During the last 50 years, aquaculture has made important advances according to FAO data. Today aquaculture provides over 40% of the world's seafood supply, and the FAO expects this figure will approach 54% by the year 2020 [5].

Because of the decreasing availability of freshwater, most of this growth will take place in seawater. The aquaculture production for fish, crustaceans, mollusks, and other aquatic species increased from 4% in 1970 to 27% in 2000 and as high as 32% in 2004 [6]. Based on this data, Mexico is number 29, however when considering the annual growth rate, Mexico is number 7. During 2003 aquaculture production was 13% of total volume and 31% of the total generated value for the national fishery [3].

Globally, aquaculture activity grew most quickly by producing crustaceans, which recorded a 19% annual growth for the 2002 – 2004 period with shrimp production the main species [7].

In Latin America shrimp culture in rustic ponds is one of the most important economic activities with an accelerated growth in places with tropical and subtropical conditions. The same is observed in Mexico, with 30% of the product in 2003 coming from aquaculture. Shrimp production was 62,600 t with 85% of the volume from Sonora and Sinaloa [3].

The most important species for aquaculture in Mexico are shrimp, catfish, carp, crawfish, striped bass, tilapia, oysters, and trout, and the most important by production amount are shrimp, tilapia, and oysters [3]. In Mexico the strategic lines defined by the 2001-2006 Aquaculture and Fishery Program established, based on scientific and technical knowledge, a mechanism for the management of fishery and aquaculture resources to push the participation of productive and academic sectors, including the three government levels, to help define and evaluate the opportunities for the development of fisheries and aquaculture.

Nationally, the three most important development zones for shrimp aquaculture in the Mexican Pacific are in the states of Sinaloa, Sonora, and Nayarit. Sinaloa is the state where the aquaculture activity has increased from 4,840 hectares in 1990 to the 37,650 hectares that exist today. Preliminary data stated that in 2006 production was 33,950 t, with a seeding density of 9 postlarvae (PL) / square meter [8].

The culture practice developed in Sinaloa mostly includes semiintensive systems, which require major incomes and technology. We should also add the ongoing concern and controversial debates about the environmental and social and economical impacts [9]. These concerns are considered in the proposal of development patterns considering criteria for sustainable development.

The notion of sustainable development has had a broad political appeal but has proven difficult to define precisely. A definition given by the Board of Sustainable Development of the Nation Research Council says ‘ there is a need for stabilizing future world population while reducing hunger and poverty and maintaining the planet life support system’ [10].

In Mexico even with the existing laws, there is a negative perception because the aquaculture industry is transforming the coastal ecosystems of Sinaloa in a way that affects the sustainability and quality of life of the inhabitants of several rural coastal communities. However when compared to other activities such as agriculture or industry, aquaculture is by far more benign to our ecosystems.

Presently, with aquaculture considered as one of the principal answers to satisfy national and international food demand, this activity is growing with no regard for coastal ecosystems. The possible negative environmental impacts of shrimp culture in traditional ponds are destruction of mangroves, because of the need for open space in mangrove zones, marsh, and coastal lagoons, thus affecting directly the reproduction process of shrimp; impacts on their habitat and the development of commercial species; water pollution caused by nutrients, organic matter, and sediments, which are thrown into the receptor system through effluents; negative effects on saline and ground water and soil caused by the use of antibiotics and other chemical substances; dispersion of diseases; and biodiversity loss in neighbor ecosystems for collecting broodstocks and wild seed [11, 12]. Some experts have estimated that the world's area of mangroves has been reduced by 35% in the past 20 years, with aquaculture implicated as one of the culprits [13].

There are also several different viral diseases known in cultured shrimp, most of them having been discovered because they cause high mortalities in commercial aquaculture systems that represent an important financial loss to producers because of slow growth or early death of the sick organisms. These infectious diseases have caused losses estimated in billions of dollars shown in the shrimp industry in Taiwan in 1987, China in 1992, and Ecuador in 1994 and 1999 [14].

Since 1992 in Mexico there has been an aquaculture-development growth level up to 15%, however disease is a limitation on production, a factor that is a problem that is affecting survival and shrimp growth [14, 15]. Mexico has also faced important economic losses caused by the Infectious Hypodermal and Hematopoietic Necrosis Virus (IHHNV) during the 1980s, the Taura Syndrome Virus (TSV) in 1994 and in late 1999, and the White Spot Syndrome Virus (WSSV). All these diseases are viral and are not susceptible to eradication, but only to control as a best case scenario. They represent the greatest threat to shrimp culture because of their great ability to be dispersed and their high degree of infection [16, 17, 18].

The spread of these viruses can be attributed to water deterioration and the transportation of the pathogens between different latitudes [19].

The discharges coming from shrimp ponds are also considered as one of the most recent and serious pollution sources of coastal water of Sinaloa. The residual waters coming from shrimp farms contain great quantities of nutrients, organic matter (feces and uneaten feed), fertilizers, chemical substances, and antibiotics, all of which produce eutrophication of lagoons and estuary systems. The negative effect produced by liquid effluents depends on several factors; (1) discharge size, (2) chemical composition of effluents (suspended solids, nutrients, organic matter), and (3) receptor water characteristics. [20, 21]. We need to face these problems with better use practices.

2. Mariculture

The development of aquaculture and mariculture needs the support of an integrated program that should include scientific and technology research. This program must have clear goals oriented to aquaculture-technology generation that must agree with the ecologic, economic, political, and social conditions of each region.

The shrimp-activity growth must have a balanced development, compatible with the capacity of the ecosystems. Now is the time to have changes made in several myths and practices, which have been used until now for shrimp culture in Mexico.

Mariculture production is growing rapidly. Mariculture is an activity that involves food production for human consumption. It is an activity in which aquatic organisms, both plants and animals, are cultured in a confined environment in the aquatic medium, which may be completely marine or marine mixed with various amounts with freshwater in the brackish water areas.

Over the last decade, mariculture increased from 13% to 19% of the total production. Fresh water culture increased from 11% to 17% while marine capture decreased from 69% to 56% and brackish water culture increased from 1% to 2%. Inland capture remained stationary in relative importance as only 6% of the total production [2].

From the 186 countries with seacoasts, only 86 countries report mariculture production to FAO. Of these, China reported nearly 22 million t, almost 73% of the total globally. The Philippines and Japan each exceeded one million t and there were 13 other countries where mariculture production was more than 100,000 t. Together these top producers accounted for 97% of the global mariculture production [2].

By production weight in broad groups in 2004, mariculture production was dominated by aquatic plants (46%) and mollusks (43%) whereas diadromous fish (mainly salmonids) accounted for 5%, marine fish for 4%, and crustaceans at 2%. The relative proportions have remained similar over the last decade [22]. The main groups of organisms commercially cultivated in the sea are fish, mollusks, and seaweeds. The FAO statistics data include coastal aquaculture, particularly pond culture, as mariculture.

Bridger et al. [23] recognize four classes of marine aquaculture sites according to the degree of exposure; (1) land-based facility, (2) coastal environments (protected bays and fjords, (3) exposed sites, and (4) offshore sites. Muir [24] contrasts coastal (inshore) aquaculture with offshore aquaculture based on five criteria that include location, hydrography, environment, access, and operation.

The requirements for the cultured organism and structure, shore installation support, and access are the factors that collectively define the potential for marine aquaculture. Geographically, three broad classes of information are essential when planning the development and management of marine aquaculture; (1) suitability of the environment for plants and animals to be cultured, (2) suitability of the environment for the culture structures, and (3) access.

Mexico has a great variety of seashore and marine ecosystems with around 11,600 km of coastline, 358,000 m^2 of continental shelf, 3 million km^2 of an exclusive economical zone, and about 2.9 million ha of territorial water. This offers diverse ecosystems for aquatic species with the need for different ecological and geographic conditions. The coastal and oceanic conditions mentioned contribute to a diverse universe of Mexican sea species that belong to temperate, warm, and cold climates, both deep and shallow waters, shore and open sea, regional or migratory species, and all in-between the extremes mentioned [25].

Though even with all these favorable characteristics to develop fishing activities, mariculture is relatively young and with little diversification, which is why Mexico only contributes globally about 1.5% of fishery production and 0.19% of aquaculture production [3].

EEZs (Exclusive Economic Zones) are the main areas in Mexico into which marine aquaculture can expand from the present day near-shore operations to offshore or the open ocean. An EEZ is the area under national jurisdiction (370 km or up to 200 nautical miles wide) declared in line with the provisions of 1982 United Nations Convention of the Sea Law.

At first glance, opportunities for expansion of marine aquaculture into the EEZ appear to be boundless, though present technology constraints related to depth, sea conditions, and competing uses reduce the EEZ availability. It seems however that there is no relationship between EEZ areas of the top mariculture producers and their production during 2004 [2].

Some events have emphasized applied research on marine aquaculture techniques; candidate species, engineering, environment, and operations. Today, there are four main types of structures popular for marine aquaculture operations: floating cages, net enclosures, long-lines, and rafts.

Cage farming has been practiced at an artisanal level for hundreds of years, originally in freshwater and later in seawater [26]. However, the development of modern cage systems has only occurred over the past 20 to 30 years, primarily in line with the development of the salmon-farming industry [27, 28].

Cages may be installed in bays, lagoons, straits, and open coasts as long as they are protected from strong winds and rough seas. The selection of a suitable site is vital to the success of the culture system for a good site selection solves much of the management problems of cage culture.

In general, cage culture has expanded rapidly, especially in the past two decades, with an increasing awareness of its merits over traditional pond culture, such as (1) its applicability in different types of open-water bodies, (2) its high productivity with minimal input and at lower costs to develop and operate, and (3) the greater socio-economic opportunities it provides to low-income families in the rural areas, particularly those displaced by the reduction of wild catch in over-exploited coastal waters, because cage culture requires comparatively low capital outlay and the use of simple and basic technology.

The choice of species for stocking and rearing in cages is governed by much the same criteria as in the species selection for pond culture, including (1) fast growth in confinement, (2) good consumer acceptance, (3) high tolerance to a wide range of environmental conditions, (4) resistance to disease, (5) ready supply of seed for stocking, and (6) ease of culture and management. These factors are to be considered in selecting sites for cages in freshwater, brackish water, or in marine environments.

As, aquaculture, mariculture uses and transforms natural resources to generate a final product. In this process wastes are generated that may have an impact in the environment.

An approach to improving the sustainability of aquaculture has been the development of Better Management Practices (BMPs). To date, BMPs have been most widely used in coastal aquaculture, particularly shrimp culture. More recently, BMPs have been developed particularly for tropical-marine-fish aquaculture [29].

Better management practices include (1) the proper site selection that should ensure adequate water circulation to satisfy both the needs of the mariculture and the flushing of nutrients and wastes, (2) optimal management, including proper feeding to decrease conversion ratios, considering the use of natural productivity, and (3) culturing different species together (polyculture), the wastes of one that can be converted to protein by the other [30, 31, 32]. Integrated aquaculture is broadly defined as the culture of a range of trophic-level organisms whereby outputs from one species or group can be used as inputs by another species group.

The impacts of sea-cage aquaculture in coastal waters may be relatively insignificant compared to the terrestrial inputs. One study of nutrient impacts of marine aquaculture in tropical systems [33] found that although fish cages theoretically contributed 26% to 32% of nitrogen and 83% to 99% of phosphorus to the coastal water studied, the effects of the cages were largely swamped by large inputs of organic matter from mangrove forests, fishing villages, and other industries.

Mariculture can also be considered as an activity with positive effects, because under certain circumstances it provides seeds for sea ranching and recovery of wild stocks and endangered species, or even improves productivity and biodiversity [34].

Shrimp cage-culture has had limited studies by some investigators in Asia [35, 36]. In Latin America the technology of raising marine shrimp in floating cages started about 20 years ago in Brazil [37], first by determining the growth of *Farfantapenaeus paulensis* in floating cages of small dimensions and low seeding densities [38]. For the species *L. vannamei* different rearing conditions and the experimental feasibility of co-culturing with the seaweed *Kappaphycus alvarezii* have been tested [39, 32].

Growth experiments on juvenile blue shrimp (*L. stylirostris)* were done in floating cages in an estuary in Sonora, Mexico [40]. In Asia suspended (hapa) nets as nursery enclosures for shrimp fry of *P. monodon* have been used as an alternative for shrimp-fry nursery systems [41]. A simple floating hatchery system with partially submerged tanks was used to rear the larvae of *Fenneropenaeus indicus* [42]. In 2006 the biological viability of culturing the Pacific white shrimp *Litopenaeus vannamei* in floating cages at different stocking densities in Bahia Santa María, Mexico was assessed [43, 44].

3. Objective

The main purpose of our study was to develop a novel-model culture system for growing Pacific white shrimp *Litopenaeus vannamei* with environmental sustainability using floating seawater cages of different dimensions and materials; and to evaluate the effect of different stocking densities, nursery, and growout conditions on efficiency of production. We also evaluated if the shrimp cultures developed the most frequent viral diseases.

4. Methodology

4.1. Site Location

The zone's cartography was inspected to locate the best sites. The places were then inspected for both maritime and terrestrial access. We measured variables as bathymetry, currents, tides, and winds.

To determine the bathymetric slope we used a Furuno video-sound; model FCV-667, and a Global Positioning System (GPS, Magellan, Model color track). To determine the current at the study sites, a current meter was used (Sensordata; model SD-30). The tides were estimated by using the registry of tidal prediction that is published for the Port of Topolobampo (CICESE) [45].

The culture was done at Bahia Navachiste, Guasave, Sinaloa, Mexico, which has a surface area of about 220 km^2 and two direct access roads to the sea, one located southeast, limited by mountains, and the other to the west near Isla San Ignacio (25°35'12" and 25°21'19" North latitude and 109°02'41" and 108°38'40" West longitude) about 11 km from the town of Huitussi in Guasave. The culture period was from April to August 2006 (Figure 1).

4.2. Design and Construction of the Floating Cage System

The cage size, flotation system, and construction materials were defined according to the hydrographic conditions of the site and the biological species to be grown.

4.3. Nursery Stage

The work procedure consisted in seeding postlarvae (PL) 20 and 28 (12 mg and 40 mg) *L. vannamei or* Pacific white shrimp after acclimatization and quantification in eight 9 m^2 cages. The assayed densities were 950, 1300, 1600, and 2700 PL/m^2, which were evaluated in duplicate. The postlarvae were fed with commercial food, Malta Cleyton 40% protein, at a rate of 5% of their fresh weight, through feeding plates at sunrise and sunset.

4.3. Growout Stage

The juveniles were transferred to growout cages when they reached 0.5 to 1 g average weight. The growth process was done in four ~200 m^2 cages. The proved seeding densities in these cages were 45 and 180 juveniles/m^2. The densities were run in triplicate in 200, 250, and 300 juveniles/m^2 in 9 m^2 cages. The feeding process consisted in using commercial food, Malta Cleyton 40% protein, at a rate of 5% of the fresh weight during the growth stage, fed at sunrise and sunset.

4.4. Water Quality Variables and Health Status

A daily recording of salinity, temperature, and dissolved oxygen at sunrise and sunset was made. Temperature and dissolved oxygen were recorded using an oximeter (YSI model 55), salinity was measured by means of a refractometer (Sper Scientific 3000011), and pH was measured with a pH meter (Hanna Hi98128).

Weekly, a random sample of 50 to 100 shrimp was taken to record the average weight and the health condition.

The viral-disease detection (TSV and WSSV) was done using histological techniques [46] and by genetic amplification with the PCR technique (Polymerase Chain Reaction) using the commercial Kit IQ2000® (Farming IntelliGene Tech Corp, Taiwan, ROC).

Figure 1. Shrimp floating cage culture location in Bahia Navachiste, Guasave, Sinaloa, Mexico.

4.6. Statistical Analysis

Results were analyzed by one-way analysis of variance (ANOVA) using Tukey's test, $P<0.05$.

5. Results

5.1. Site Location

The culture site was a section of a channel with the maximum depth 7 m and the minimum 1 m, both at high tide. This data were from predicted tidal data at Topolobampo, Sinaloa [45], and the closest site to the study site. From this data, the greatest tidal range during 2005 was 1.7 m. The maximum tidal range during the dates of the study was 1.54 m. The recorded current was dominated by the tidal current, which at the ebb-tide had a maximum speed of 24 cm/ s, and the tidal flow was 6.1 cm/ s, which is not too high for a floating cage design but was a convenient speed for solid waste removal. The waves are insignificant because the site is surrounded and protected by two islands and mountains, one island to the north, the other to the south, and with mountains to the west. In the southwest is 16 km of open sea, and because of this low waves are generated with an approximate height of 0.45 cm and with a separation of between 3.5 and 7.5 seconds.

5.2.Floating Cages

The eight square-cage nursery system was built with PVC tubes of 110-mm diameter working as a floating and support structure and polyester-impregnated PVC nets (Sannet QR Sansuy ®) of 1.5- x 2.5–mm mesh surrounding the structure. The cage dimensions are 3 x 3 x 1 m (9 m^2) (Figure 2). A six square-cage growout system was also made with the same dimensions and characteristics as above, with the only difference the mesh opening of 5 x 5 mm (Sannet FR Sansuy ®). Each cage in the system is joined to the next by use of ropes. The cages are anchored to the sea bottom with an anchor.

We also built four circular growout cages, 16-m diameter and 1.5-high (~200 m^2) constructed with high-density polyethylene tubing (HDPE), galvanized steel joints, and polyester nets with 5- x 5-mm mesh (Sannet FR Sansuy ®) (Figure 3). We used an anchor system of sacks filled with sand.

5.3. Seeding

The postlarvae (PL) were acquired from a certified healthy-broodstock regional-production laboratory. PLs were acclimatized in a site near the floating cage location. PLs were acclimatized slowly so they could adjust physiologically to the site's water variables; temperature 25 C, salinity 35‰, and pH 8, to avoid stress or mortality during the seeding process.

Figure 2. Nursery and growout 9 m^2 cage model.

Figure 3. Panoramic view and detail of 200 m^2 growout cages model .

5.4. Nursery stage

During the first 20 days of culture there were no weight differences among densities of 950, 1300, and 1600 PL / m^2 (Figure 4, Table 1).

The weekly growth rates (g/week) for each density were 0.22 (950 shrimp/ m^2), 0.22 (1300 shrimp/ m^2), and 0.10 (1600 PL / m^2).

There was no significant difference ($P>0.05$) in the survival rate between the assayed densities. However, there was a significant difference ($P<0.05$) at the end of the 38 culture days in the final weight (g) in the density of 1600 shrimp/m^2 and the number of shrimp per cage was 8% higher than for the density of 1300 shrimp/ m^2.

A trial of 28 days of culture was also researched with the density of 2700 PL /m^2. The number of shrimp per cage after this time is 60% higher than the density of the 1600 PL /m^2 and this density recorded a weekly growth rate of 0.14 g/week. For this density the number of shrimp was approximately 60% higher than 1600 PL /m^2.The obtained FCR (Feed Conversion Rate) in every case was low (Table 1).

5.5. Growout Stage

There were no significant differences (P>0.05) in the final weight (g) and survival rate of the cultured shrimp in 9 m^2 cages for 58 days at densities of 200, 250, and 300 shrimp/ m^2 (Figure 5, Table 2). The weekly growth rates (g/week) for each density were 1.5 (200 shrimp/ m^2), 1.5 (250 shrimp/ m^2), and 1.4 (300 shrimp/ m^2).

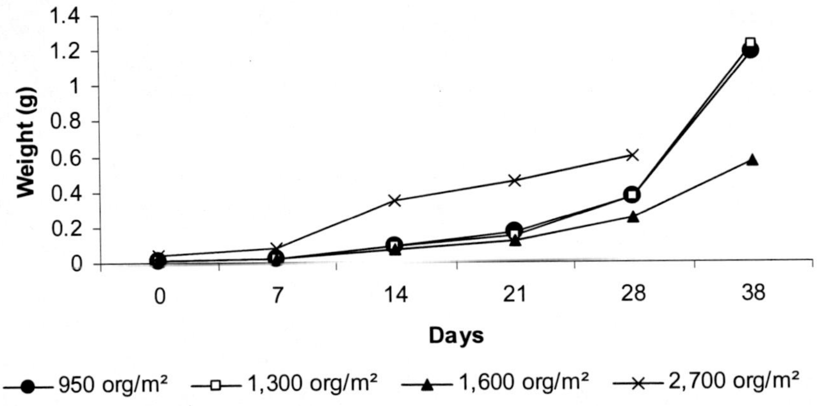

Figure 4. *L. vannamei* nursery curve in 9 m^2 cages at different stocking densities (950, 1300, 1600, and 2700 PL/m^2) for 38 and 28 days.

Table 1. *L vannamei* variables in 9 m^2 nursery cages at different densities (950, 1300, 1600, 2700 shrimp/m^2)

	Treatment (densities) shrimp/m^2			
	950	1300	1600	2700
Initial weight g	0.012	0.012	0.012	0.04
Final weight g	1.18 ± 0.36 [a]	1.22 ± 0.15 [a]	0.57 ± 0.1 [b]	0.59 ± 0.56
Days	38	38	38	28
Survival %	76.65±2.8 [a]	74.5±0.8 [a]	65.8±2.9 [a]	63.6±2.1
FCR	0.15	0.25	0.22	0.25
Growth rate g/week	0.22	0.22	0.1	0.14
Biomass kg/m^2	0.86	1.18	0.60	1.01
Shrimp number/cage	6,580	8,780	9,500	15,550

Data with different super scripts letters are significantly different (P<0.05).

The biomass per square meter obtained in the density of 300 shrimp/m^2 is higher compared to the lowest density assayed. The FCR was < 1 (Table 2).

In 200 m^2 cages the growth rate at the lower density (45 shrimp/m^2) is higher (2.1 g/week) and the survival rate is above 80% at both densities assayed (45 and 180 shrimp/m^2) during 62 days. The final weights (g) at harvesting for each density were 17.7 ± 1.55 g at 45 shrimp/m^2 and 13.3 ± 1.31 g at 180 shrimp/m^2. The FCR as for the 9 m^2 cages is close to 1 (Figure 6 and Table 3).

5.6. Water Quality Variables and Shrimp Health Status

The values of the water quality variables were found acceptable for the growth *L. vannamei* [47]. These values do not vary significantly between the different cages or the square and circular models used during the culture (Table 4).

Weekly shrimp health analyses by histopathology and PCR, showed no presence of viral pathogens during the culture period.

Conclusion

As mentioned in this chapter, there are only a few studies of shrimp culture in floating cages. Our work represents a first attempt to culture shrimp in cages of larger dimensions, similar to those used to culture other fish species, but these had not been used until now to culture shrimp in protected sea waters. Presently, there are a large number of cage systems that might be used in mariculture depending on the ambient conditions and the species culture-characteristics [48, 22, 26].

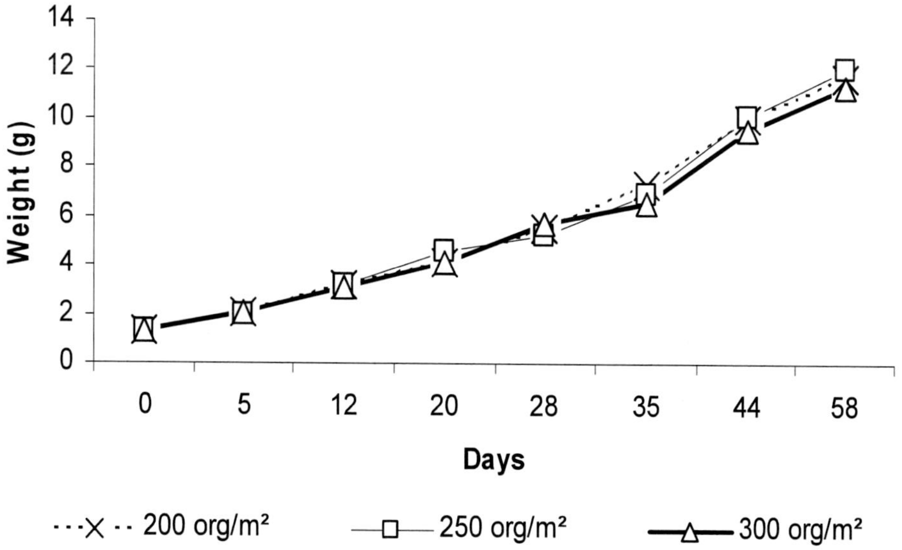

Figure 5. *L. vannamei* growth curve in 9 m^2 cages at different stocking densities (200, 250, and 300 shrimp/m^2) for 58 days.

The models used in this investigation were established according to hydrographic conditions of the site and the biological characteristics of shrimp. During the culture cycle it was proved that the structure and flotation system of both models is stable.

The 9 m^2 cages constructed with PVC tubes and the ~200 m^2 ones constructed with polyethylene tubes (HDPE) were able to resist tides, waves, currents, and winds with a maximum average speed of 40 km/h. The cages used in the study resisted the forces of a tropical storm (June 8, 2006). The quality and mesh opening used in the nets was designed according to the growth stage of the shrimp so that the shrimp could not escape through the mesh opening of the net.

Table 2. *L vannamei* variables in 9 m^2 growout cages at different densities (200, 250, and 300 shrimp/m^2)

	Treatments (densities) shrimp/m^2		
	200	250	300
Initial weight g	1	1	1
Final weight g	11.7±0.2^a	12±0.36^a	11.3±0.35^a
Days	58	58	58
Survival %	81.2±3.4^a	80.4±1.2^a	77.4±4^a
FCR	0.93	0.81	0.87
Growth rate g/week	1.5	1.5	1.4
Biomass kg/m^2	1.9±0.03	2.4±0.05	2.6±0.02

Data with different super scripts letters are significantly different ($P<0.05$).

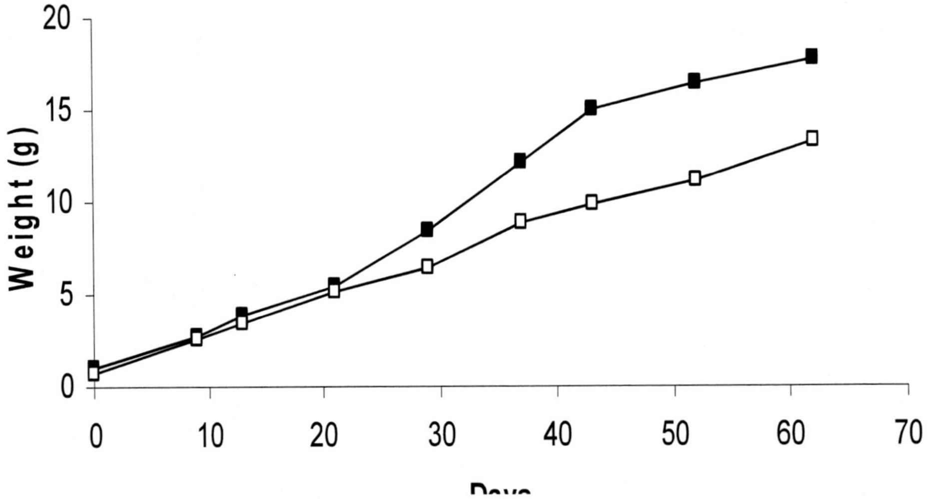

Figure 6. *L. vannamei* growth curve in 200 m^2 cages at different stocking densities (45 and 180 shrimp/m^2) for 62 days.

In previous studies done by our group we proved the biological viability of the white shrimp *L. vannamei* in floating cages [44]. In that past study the nursery stocking density was 700 PL/m^2, 30% greater than that used by Lombardi et al. [32], and at the growout stage we worked with densities of 100, 150, and 200 shrimp/m^2 with excellent growth rate and survival.

Table 3. *L vannamei* variables in 200 m^2 growout cages at different densities (45 and 180 shrimp/m^2)

	Treatments (densities) shrimp/m^2	
	45	180
Initial weight g	1	0.7
Final weight g	17.7 ± 0.84	13.3 ± 0.5
Days	62	62
Survival %	80	80
FCR	1	1
Growth rate g/week	2.1	1.53
Biomass kg/m^2	0.6	1.9

Table 4. Water quality variables, minimum, maximum and mean, measured at sunshine (am) and sunset (pm) in the 9 and 200 m^2 cages during the culture

Variable	Minimum	Maximum	Mean
DO (mg/L) am (200 m^2)	3.6	4.9	4.2 ± 0.34
DO (mg/L) pm (200 m^2)	5.0	5.3	5.4 ± 0.23
DO (mg/L) am (9 m^2)	3.9	5.0	4.5 ± 0.26
DO (mg/L) pm (9 m^2)	5.3	5.8	5.6 ± 0.19
Temperature (C) am	22.7	31.6	28.6 ± 3.33
Temperature (C) pm	24.4	32.4	29.5 ± 2.26
pH am	7.9	8.2	8.01 ± 0.11
pH pm	7.9	8.2	8.02 ± 0.08
Salinity ($^0/_{00}$)	35	39	37 ± 1.03

In our new research we showed the positive effect of high nursery stocking densities (950, 1300, 1600, and 2700 PL/m^2) on the number of shrimp per cage, obtained before they reach 1-g average weight. There were no significant differences (P>0.05) in the survival rate at densities of 950, 1300, and 1600 shrimp/ m^2. In the density of 1600 shrimp/ m^2 the shrimp number per cage was higher, however there was a significantly negative effect (P<0.05) over the shrimp weight.

The postlarvae seeded at a density of 2700 (~385% greater than used by Zarain et al. [43]) had an initial weight of around 40 mg, greater to the weight of the postlarvae seeded at (950-1300 PL/m^2) which had an initial weight of 12 mg. The number of shrimp per cage that weighted > 0.5-g average weight at 2700 PL/m^2 duplicated the amount obtained at a lower density (950 PL/m^2).After 28 days of culture we recorded a weekly growth rate of 0.14 g/ week. However to avoid low survival we recommend transfering the juveniles to the larger cages as soon as they reach 0.5-g average weight.

These results reveal the advantage of using high stocking densities at the nursery stage with the ambient conditions of this study until an average weight limit is reached.

The number of juvenile shrimp obtained in this stage will be used to adjust densities at the growout stage beginning at a minimum average weight of 0.5 g, which is necessary so that the shrimp are not able to escape through the mesh opening (0.5 mm) used in the growout cages. With this procedure the number of nursery cages is optimized.

We found that the use of 9 m^2 cages is adequate at the nursery stage because they are easy to handle for a correct supervision of shrimp with a weight < 1 g, Also the use of 9 m^2 cages helps to choose, at the transfer stage, the appropriate number of cages to adjust seeding densities for the growout stage. We concluded that the high densities used for the nursery stage in this research were favorable to obtain a high number of juveniles for the growout stage.

During this research we used two types of cages for the studies for the growout stage, 9 and ~200 m^2 cages, to quantify the effect of size on the evaluated variables in shrimp culture.

At the growout stage there were no significant differences ($P>0.05$) in final weights and survival rates among shrimp cultured in 9 m^2 cages at densities of 200, 250, and 300 shrimp/m^2 after 58 days of culture. However, the biomass was greater (2.6 kg/m^2) at a density of 300 shrimp/m^2 in comparison to a density of 200 shrimp/m^2. When comparing these results to the ones obtained in a previous research done by our group [43], were we worked with the densities of 100, 150 and 200 shrimp/m^2; we showed a negative effect of the use of higher densities over the weight. In that research we also probed the positive role of submerged structures over the shrimp weight which was increased by the natural productivity.

We believe that water quality conditions at the present study management, natural productivity and the hydrographic characteristics of the site were essential to determine optimum growing conditions

The natural productivity at the site is an important source of food for the growing shrimp. Shrimp can obtain benefits from feeding on microorganisms (phytoplankton and zooplankton) suspended in the water and from those sticking to the net walls, which is demonstrated by the low FCR (Feed Conversion Rates), obtained during the experiment [50, 51] and the rapid growth of *L vannamei* at the assayed densities [52, 53]. The use of feeding trays helps to supply commercial food in adequate amounts and to avoid food loss, which helps to maintain the FCR and avoids an additional contribution of uneaten feed to the ecosystem. Environmental impacts are associated mainly with nutrient input from uneaten food and waste [26], which is why it is sometimes possible to mitigate or eliminate the variety of adverse effects caused by mariculture by using Better Management Practices.

In one of the few existing studies of nutrient impacts of marine aquaculture in tropical systems [30], although raising fish in cages theoretically contributed some percentage of nitrogen and phosphorus, there was no evidence of large scale eutrophication caused by this practice.

In the ~200 m^2 cages of we used densities of 45 and 180 shrimp/m^2 obtaining excellent growth rates (2.1 g/week) at the lowest density and final weights of 17.7 ± 1.55 and 13.3 ± 1.31 g with 80% survival in two months of growout. Sandifer et al. [49] obtained similar results for intensive shrimp culture in earth ponds. And the FCR as for the 9 m^2 cages was close to 1.

This experiment represents a first successful assay of shrimp culture in large floating cages 16-m diameter (~200m^2).

The values of the variables obtained in this experiment particularly final yield is higher compared to the average results obtained from experiments done in general in semiintensive cultures in traditional Mexican pond culture (0.9 kg/m^2). However, comparisons are difficult to make with different published assay results, from either cage or pond culture, because the culture conditions, species, and variables involved are different.

All the analyzed shrimp were free of viral disease. It is possible that the habitat in which they were developed favors a resistance to pathogens. It has been shown that a good nutrient balance in the diet can improve the health of aquatic organisms in culture [54] along with a good water quality [47]

Research done on the shrimp immune system has demonstrated that is possible to increase resistance to disease by ingestion of immunostimulants as beta-glucans, peptidoglycans, lipopolysaccharides, algal derivates, vitamins, carotenoid pigments [55, 56, 57, 58]. An adequate natural nutrition provided by microorganisms present in natural media with the addition of commercial food helped to maintain an adequate state of health during the culture cycle, making the shrimp immune system stronger (data not shown). Viral diseases in traditional shrimp culture in Mexico are a problem that has not been able to be eradicated completely, even with the government's efforts and policies.

Mexico has 1,600,000 hectares of estuary surfaces of which 12,500 km^2 are in 130 lagoons considered as potential areas where mariculture can be easily done, specifically shrimp culture in floating cages. We conclude that the favorable ambient conditions observed during this study at Bahia Navachiste in Sinaloa, Mexico and an adequate management of culture and well-designed cage structures were the factors that made possible the achievement of success for *L vannamei* shrimp culture in floating cages. Additionally this procedure will provide jobs for fishermen displaced from traditional fisheries, will increase mariculture shrimp production, and will reduce pressure on traditional fisheries as a new sustainable technique for large-scale shrimp production.

ACKNOWLEDGEMENT

Funding for this research was providing by a grant from CONAPESCA-SAGARPA of the Mexican Federal Government and FOMIX-CONACyT (Sin-2005-CO1-04). We thank Alfredo E Vazquez Olivares and Rodolfo Rodriguez for their expert technical assistance. Special thanks to Centro de Ciencias de Sinaloa and Fishermen group 'Lucio Ramirez Machado' for providing experimental facilities. Thanks to Dr. Ellis Glazier for editing this English-language text.

REFERENCES

[1] FAO (Organización de las Naciones Unidas para la Agricultura y la Alimentación). (2005). Examen de la situación de los recursos pesqueros marinos mundiales: (Documento técnico de pesca No. 457). Rome, Italy: FAO.

[2] Kapetsky, J.M., and Aguilar-Manjarrez, J. (2007). Geographic information systems remote sensing and mapping for the development and management of marine aquaculture (Fisheries Technical Paper No.458). Rome, Italy: FAO.

[3] SAGARPA (Secretaria de Agricultura, Ganadería, Desarrollo Rural, Pesca y alimentación). (2003). Sistema Integral de Información Agroalimentaria y Pesquera (Anuario Estadístico de Acuacultura y Pesca). México: CONAPESCA-SAGARPA.

[4] Lluch, B.D. (1974). La pesquería de camarón en alta mar en el noroeste: un análisis biológico pesquero (Informe Técnico del Instituto Nacional de la Pesca). México: INP.

[5] Rodríguez de la Cruz, C., and Chávez, E. A. (1996). La pesquería de camarón en alta mar. Pacífico de México: Pesquerías Relevantes de México XXX Aniversario del INP (Dir. de Anal. de Pesq. Instituto Nal. de la Pesca). Mexico: SEMARNAP.

[6] FAO (Food and agriculture of the United Nations) (2004).The state of world fisheries and aquaculture (Fisheries Department). Rome, Italy: FAO.

[7] FAO (Organización de las Naciones Unidas para la Agricultura y la Alimentación). (2007). El estado mundial de la pesca y la acuicultura 2006 (Departamento de Pesca y Acuicultura). Rome, Italy: FAO.

[8] SAGARPA. (2006). Proyecto de evaluación alianza para el campo 2005. Analisis prospectivo de política para la acuacultura y la pesca. Mexico:SAGARPA.

[9] Páez-Osuna, F. (2001) Camaronicultura y Medio Ambiente (El Colegio de Sinaloa). México, DF: UNAM.

[10] National Research Council (1999). Our Common Journey (Natl. Acad. Press). Washington DC:NRC.

[11] Primavera, J.H. (1998). Tropical Shrimp Farming and its Sustainability. In S. de Silva (Ed), *Tropical Mariculture* (pp.257-289) London, UK: Academic Press.

[12] Páez-Osuna, F. (2005). Retos y perspectivas de la camaronicultura en la zona costera. Revista Latinoamericana de Recursos Naturales, 1, 21-31.

[13] FAO World Wetlands Day - Fish for Tomorrow? 2007 [2007 August 9] Available from: http://www.ramsar.org/wwd/7/wwd2007_leaflet_e.pdf

[14] Aguirre, G., and Ascencio, F. (2000). Infectious diseases in shrimp species with aquaculture potential. *Recent Res Deve. Microbiology*, 4, 333-348.

[15] Lightner, D.V., and Redman, R.M. (1998). Shrimp disease and current diagnostic methods. Aquaculture, 164, 201-220.

[16] Lightner, D.V. (2001).The penaeid shrimp virus TSV, IHHNV, WSSV, and YHV: Current status in the Americas, available diagnostic methods, and management strategies. In: C. Lim and C. D. Webster (Eds), *Nutrition and Fish Health* (pp. 79-102) USA: Food Products Press.

[17] Zarain-Herzberg M., and Ascencio-Valle, F. (2001). Taura syndrome in México: follow up study in shrimp farms of Sinaloa. *Aquaculture,* 193, 1-9.

[18] Zarain-Herzberg, M., Hernández N., and Ascencio-Valle F. (2003). Biological Characterization of a less virulent Taura syndrome in Pacific White Shrimp *Litopenaeus vannamei* (Crustacea:Decapoda) cultured in Sinaloa, Mexico: gross signs histopathological lesions, and mortalities. *Journal of the World Aquaculture Society*, 34, 100-105.

[19] Coman, F.E., Connolly, R.M., and Preston, N.P. (2006). Effect of water exchange and abiotic factors on zooplankton and epibenthic fauna in shrimp ponds. *Aquaculture* Research, 37, 1387-1399.

[20] Páez-Osuna, F., Guerrero-Galván, S.R., and Ruiz-Fernández, A.C. (1999). Discharge of nutrients from shrimp farming to coastal waters of the Gulf of California. *Marine Pollution Bulletin*, 38: 585-592.

[21] Boyd, C. E. (2003). Guidelines for aquaculture effluent management at the farm-level. Aquaculture, 226, 101-112.

[22] FAO FOSHSTAT PLUS Universal software for fisheries statistical time series [Version 2.3]. 2006 [2006 August 10]. Available from: http://www.fao.org/ fi/statst7FISOFT /FISHPLUS.asp).

[23] Bridger, C.J., Costa-Pierce, B.A., Goudey, C.A., Stickney, R.R., and Allen, J.D. (2003). Offshore aquaculture development in the Gulf of Mexico: site selection, candidate species, and logistic alleviation. In C.J. Bridger, and B.A. Costa-Pierce (Eds), Open Ocean Aquaculture: From Research to Commercial Reality (273-283) Baton Rouge, Louisiana, USA: *The World Aquaculture Society.*

[24] Muir, J. Offshore Mariculture Options. Lecture delivered at the CIHEAM Advanced Course on Offshore Mariculture, Zaragoza (Spain), 17-22. May 2004. 2007 August 20. Available from http//www.easonline.org/agenda/en/description. Asp?id=264.

[25] Contreras-Espinosa, F. and Castañeda O.L. (1993). Ecosistemas Costeros Mexicanos. México: CONABIO-UNAM-I.Basurco, B and Saralogia, M. (2002). Sea farming today and tomorrow. *European Aquaculture Society.* Special Publication No. 32. 562 pp.

[26] Beveridge, M. 2004. Cage Aquaculture (Third Edition) London: Blackwell Publishing.

[27] Willoughby, S. (1999). Manual of Salmon Farming. Fishing News Books, Oxford, UK.

[28] Phillips, M.J. (1998). Chapter 2- Tropical Mariculture and Coastal Environmental Integrity. In S.S. De Silva (Ed), Tropical Mariculture (pp. 17-69). London, UK: Academic Press,

[29] Sim, S.Y., Rimmer, M.A., Toledo, J.D., Sugama, K., Rumengan, I., Williams,K., and Phillips, M.J. (2005). A guide to Small-Scale Mariner Finfish Hatchery Technology. Bangkok, Thailand: Network of Aquaculture Centres in Asia –Pacific.

[30] Neori, A., Chopin, T., Troell, M., Buschmann, A.H., Kraemer, G.P., Halling, C., Shpigel, M., and Yarish, C. (2004). Integrated aquaculture: rationale, evolution and state of the art emphasizing sea weed biofiltration in modern mariculture. *Aquaculture,* 231, 361-391.

[31] Troell, M., Rönnbäck, P., Halling, C., Kautsky, N. and Buschmann, A.H. (1999). Ecological engineering in aquaculture: use of seaweeds for removing nutrients from intensive mariculture. *Journal of Applied Phycology,* 11, 89-97.

[32] Lombardi, J.V., Marquez, H.L.A., Toledo, R., Pereira, L., Barreto, O.J.S., Paula, E.J., 2006. Cage polyculture of the Pacific white shrimp *Litopenaeus vannamei* and the Philippines seaweed *Kappapphycus alvarezii. Aquaculture*, 258, 412- 415.

[33] Alongi, D.M., Chong, V.C., Dixon, P., Sasekumar, A., and Tirendi, F. (2003). The influence of fish cage aquaculture on pelagic carbon flow and water chemistry in tidally dominated mangroves estuaries of peninsular Malaysia. *Marine Environmental Research,* 55, 313-333.

[34] Bartley, D.M.1(999). Marine ranching: a global perspective. In B.R. Howell, E. Moksness.,and T. Svåsand (Eds). Stock enhancement and sea ranching (pp. 79-90). Oxford, UK: Blackwell Science Ltd./Fishing News Books.

[35] Srikrishnadhas, B. and Sundararaj, V. (1993). Studies on the growth of marine shrimp in floating cages and pen. In: Proceedings of the National seminar on Aquaculture Development in India - Problems and Prospects, 27-29 Nov 1990 (pp 53-58) India: Thiruvananthapuram,

[36] Shanmugam, A., Rajamanickam, S., Kannupandi, T., 1995. Cage culture of India white prawn *Penaeus indicus* in Vellar estuary. *J. Mar. Biol. Assoc. India*, 37, 166-170.
[37] Wasielesky, W.Jr., Poersch, L.H., and Bianchini, A. (1999). Comparaçao entre a sobrevivência e o crescimento do camarão-rosa *Farfantepenaeus paulensis* cultivado em gaiolas e cercados. Nauplius, Rio Grande, 7, 173-177.
[38] Lemus, E. (2003). Raising Marine Shrimp cages a pioneering innovative option. World Aquaculture Society, Book of abstracts (vol I mayo 19-23) Salvador Brazil: *World Aquaculture* 2003.
[39] Paquotte, P., Chim, L., Martin, J.M., Lemos, E., Stern, M., and Tosta, G., (1998). Intensive culture of shrimp *Penaeus vannamei* in floating cages: zootechnical, economic and environmental aspects. *Aquaculture,* 164, 151-166.
[40] Martínez-Córdova, LR., 1988. Culture of blue shrimp *Penaeus stylirostris* in floating cages. The Progressive Fish-Culturist, 50, 36-38.
[41] Rodriguez, E.M., Bombeo-Turburan, I., Fukumoto, S., and Ticar, R.B. (1993). Nursery rearing of *Penaeus monodon* (Fabricius) using suspended (hapa) net enclosures installed in a pond. *Aquaculture,* 112, 107-111.
[42] Walford, J., and Lam, T.J. (1987). Floating hatchery and net cage culture of *Penaeus indicus* in the Straits of Johore, Singapore. *Aquaculture,* 62:11-32.
[43] Zarain-Herzberg M., Campa-Códova, A., and Cavalli, R.O. (2006). Biological viability of producing white shrimp *Litopenaeus vannamei* in seawater floating cages. *Aquaculture,* 259, 283-289.
[44] Zarain-Herzberg M. (2006).Jaulas flotantes: una alternativa de cultivo. Industria acuícola, 2, 17-16.
[45] Bell T.A., and Lightner D.V. (1998). A Hand Book of normal Penaeid histology. Baton Rouge Louisiana: *World Aquaculture Society*.
[46] CICESE. Predicción de mareas en México 2007 [2007september 2]. Available from: http://oceanografia.cicese.mx/predmar/mensual/tpl/tpl0604.pdf.
[47] Boyd, C.E., and Tucker, C.S. (1998) Pond aquaculture Water Quality Management. Boston, Massachusetts, USA: Kluwer Academic Publishers
[48] Huguenin, J. 1997. The design, operations and economics of cage culture systems. *Aquacultural Engineering,* 16,167-203.
[49] Sandifer, P.A., Hopkins, J.S., and Stokes, A.D. (1988). Intensification of Shrimp culture in earthen ponds in South Carolina: Progress and prospects. *Journal of the World Aquaculture Society,* 19(4), 218-226
[50] Otoshi, A.C., Montgomery, A.D., Look, A.M., and Moss, S.M. (2001). Effects of diet and water source on the nursery production of Pacific white shrimp *Litopenaeus vannamei*. *J. World Aquaculture Society,* 32, 243-249.
[51] D'Abramo, L.R. (1997). Triacylglycerols and fatty acid. In: D'Abramo, L.R., Conklin, D.E, and Akiyama, D.M. (Eds), Crustacean Nutrition (pp 71-84) Baton Rouge, USA: World Aquaculture Society.
[52] Moss, S.M., and Prider, G.D. (1995). Characterization of organic particles associated with rapid growth in juvenile white shrimp *Litopenaeus vannamei* Boone, reared under intensive culture conditions, *J. Exp. Mar. Biol. Ecol,* 187, 175-191.
[53] Tidwell, J.H., Shulmeister, G., Mahl, C., and Coyle, S. (2002). Importance of biofilm for water quality and nourishment in intensive shrimp culture, *Aquaculture* 203, 263-278.

[54] Raa, J., 1996. The use of immunostimulatory substances in Fish and Sellfish Farming. *Reviews in Fisheries Science*, 4, 229-288.

[55] Sritunyalucksana, K and Söderhall, K., 2000. The proPo and clotting system in crustaceans. *Aquaculture,* 191, 53-69.

[56] Song, Y.L., and Huang, C.C. (2000). Application of immunostimulants to prevent shrimp disease. In M. Fingerman and R. Nagabhushanam (Eds), Recent Advances in Marine Biotechnology: Immunobiology and Pathology (Vol 5, 173-187), USA: Sciences Publisher, Inc.

[57] Merchie, G., Kontara, E., Laves, P., Robles, R., Kurmaly, K., and Soorgeloos, P. (1998). Effect of Vitamin C and astaxanthin on stress and disease resistance of post larvae tiger shrimp *Penaeus monodon* (Fabricius). *Aquaculture Research,* 29, 579-585.

[58] Philip R., Sahoo, P.K., Screekuman, P.K., Anas, A., Mukherjee, S.C., and Ayyapan, S. (2003). Immunostimulants – source, diversity and mode of application: Immunoestimulants. In I.S.B. Singh, S.S. Pai, R. Philip, and A. Mohandas (Eds), Aquaculture Medicine (1st Edition,135-147) Kochi, India: Centre for Fish Disease and Management, CUSAT.

INDEX

A

B

C

D

E

F

G

H

I

J

K

L

M

N

O

P

Q

R

S

U

V

W

X

Y

Z